Fabrication and Advanced Applications of Nanomaterial-Based Electrochemical Sensors

Fabrication and Advanced Applications of Nanomaterial-Based Electrochemical Sensors will help students understand the concept of nanomaterial-based electrochemical sensors easily by giving simple examples and illustrations. Electrochemical sensors can determine various bioactive compounds and organic molecules, but the further addition of nanomaterials into the electrode can increase the detection limit due to their excellent electrical and chemical properties and their huge surface area. Nanomaterial-based electrochemical sensors can also detect toxic waste and thereby reduce the risk of waterborne diseases to both humans and aquatic animals. This book seeks to enhance environmental awareness and explain how electrochemical sensors contribute to a more sustainable and conscious way of living. The book will be useful for researchers who are fabricating various nanomaterial-based electrodes to determine neurotransmitters, organics, toxic dyes, surfactants, and various bioactive compounds as well as engineering, chemistry, electrochemistry, and nanomaterial students at the undergraduate and postgraduate level.

Key Features:

- The first book to cover novel applications of nanomaterial-based electrochemical sensors.
- Discusses various nanomaterials and composite materials as modifiers for the electrochemical determination of different dyes, pesticides, toxic chemicals, neurotransmitters, food additives, and heavy metals.
- Describes the facilitation of nanomaterial-based electrochemical sensors as compared with other conventional modifiers.

Series in Sensors
Series Editors: Barry Jones and Haiying Huang

Other recent books in the series:

Electrical Impedance: Principles, Measurement, and Applications
Luca Callegaro

Semiconductor X-Ray Detectors
B. G. Lowe, R. A. Sareen

Resistive, Capacitive, Inductive, and Magnetic Sensor Technologies
Winncy Y. Du

Sensors for Safety and Process Control in Hydrogen Technologies
Thomas Hübert, Lois Boon-Brett, William Buttner

Biosensors and Molecular Technologies for Cancer Diagnostics
Keith E. Herold, Avraham Rasooly

Compound Semiconductor Radiation Detectors
Alan Owens

A Hands-On Course in Sensors Using the Arduino and Raspberry Pi
Volker Ziemann

Radiation Sensors with 3D Electrodes
Cinzia Da Vià, Gian-Franco Dalla Betta, Sherwood Parker

Semiconductor Radiation Detectors
Alan Owens

Advanced Chromatic Monitoring
Gordon R. Jones, Joseph W. Spencer

Nanosensors: Physical, Chemical, and Biological, Second Edition
Vinod Kumar Khana

Multi-Component Torque Sensing Systems
Qiaokang Liang

Ultra Fast Silicon Detectors: Design, Tests, and Performances
Marco Ferrero, Roberta Arcidiacono, Marco Mandurrino, Valentina Sola, Nicolò Cartiglia

A Hands-On Course in Sensors Using the Arduino and Raspberry Pi, Second Edition
Volker Ziemann

Multifunctional Sensors: Design, Construction, Methodology and Uses
Bansari Deb Majumder, Joyanta Kumar Roy

IoT Sensors: An Exploration of Sensors for Internet of Things
Vinod Kumar Khanna

Handbook of Carbon Sensors: Understanding and Applications
J. G. Manjunatha

Fabrication and Advanced Applications of Nanomaterial-Based Electrochemical Sensors
Shashanka Rajendrachari, Vinayak Adimule

Fabrication and Advanced Applications of Nanomaterial-Based Electrochemical Sensors

Edited by
Shashanka Rajendrachari and Vinayak Adimule

CRC Press
Taylor & Francis Group
Boca Raton London New York

CRC Press is an imprint of the
Taylor & Francis Group, an **informa** business

Designed cover image: Shutterstock

First edition published 2026
by CRC Press
2385 NW Executive Center Drive, Suite 320, Boca Raton FL 33431

and by CRC Press
4 Park Square, Milton Park, Abingdon, Oxon, OX14 4RN

CRC Press is an imprint of Taylor & Francis Group, LLC

ISBN: 978-1-032-84256-1 (hbk)
ISBN: 978-1-032-84424-4 (pbk)
ISBN: 978-1-003-51263-9 (ebk)

DOI: 10.1201/9781003512639

Typeset in Minion
by Deanta Global Publishing Services, Chennai, India

Contents

About the Editors, viii

List of Contributors, ix

Chapter 1 ▪ Advancement in Electrochemical Sensing Technology 1

Duygu Zabitler, Esra Ülker, Niran Öykü Erdoğan, and Gözde Aydoğdu Tiğ

Chapter 2 ▪ How Nanomaterials Facilitated Electrochemical Sensor Research 33

Gangadhar Bagihalli and Bojappa Ginni

Chapter 3 ▪ How Nanomaterial-Based Electrochemical Sensors Are Used for Biomedical Applications 56

M M Sreejaya, Vaishnavi M, Lina Anil Kumar, Sreelakshmi M, Rejithamol R, Santhy A, and Moumita Gangopadhyay

Chapter 4 ▪ Fabrication of Electrochemical Sensors for Determination of Toxic Dyes 79

Vandna Sharma, Pankaj Kumar, and Vinayak Adimule

Chapter 5 ▪ Advanced Electrochemical Sensing Devices For Analysis of Pesticides, Insecticides, and Other Toxins in Food Products 111

Nikhila N Pai, Shashanka Rajendrachari, and Akshatha R Salian

Chapter 6 ▪ Recent Advances in Nanomaterial-Based Electrochemical Sensors for Pharmaceutical and Biomedical Diagnosis 148

Sudenur Ozbey, Yusuf Yerli, Cem Erkmen, and Sevinc Kurbanoglu

CHAPTER 7 ▪ Applications of Nano Electrochemical Sensors to Determine Vitamins and Hormones 192

SIMI M, LAKSHMI KRISHNAA SURESH, LEKSHMI B R, MOUMITA GANGOPADHYAY, REJITHAMOL R, AND SANTHY A

CHAPTER 8 ▪ Determination of Medical Wastes by Nanomaterial-Based Electrodes 218

RAGHAVENDRA P. BAKALE, SHEELA A. SANGAM, S.N. MATHAD, VINAYAK ADIMULE, GANGADHAR BAGIHALLI, RAJEEV S. JOSHI, AND SANTOSH NANDI

CHAPTER 9 ▪ Advances in Glucose Electrochemical Sensors Using Nanomaterials 243

DHIYA P, REJITHAMOL R, MOUMITA GANGOPADHYAY, SANTHY A, AND VISHNU VARDHAN PALEM

CHAPTER 10 ▪ How Nanomaterial-Based Sensors Contribute to a More Sustainable Environment 258

VARUN DONNAKATTE NEELALOCHANA, SAHANA KAMANNA METRY, JUBATE MOHANTY, SANDEEP SHADAKSHARI, HAREESHA NAGARAJAPPA, SANJAY BALLUR PRASANNA, SANTHOSH AREHALLI SHIVAMURTHY, MANJUNATHA JAMBALLI GANGADHARAPPA, ELEONORA TOMASINO, PAOLO SCARDI, AND NARGES ATAOLLAHI

CHAPTER 11 ▪ Limitations, Challenges, and Future Aspects of Electrochemical Sensing Devices 288

DANIEL YAEZU TIBA, WILLIAN VENANCIO LEMOS BELINO, JOÃO HENRIQUE ASTOLFI FEREIRA, MARIANA MUNIK DA SILVA, LAURA CHALULEU LIEB, AND THIAGO CRUZ CANEVARI

CHAPTER 12 ▪ Advantages of Nanomaterial-Modified Electrochemical Sensors over Conventional Electrochemical Sensors 303

GANGADHAR M KANAGINAHAL, SATEESHKUMAR KANAKANNAVAR, PRASHANT KAKKAMARI, D. NAGARAJU, AND H C ANANDA MURTHY

CHAPTER 13 ■ Carbon-Based Nanomaterials: Synthesis, Characterization, and Applications of Electrochemical Sensors and Biosensors 327

SANJAY BALLUR PRASANNA, HARI PRASAD BADUBANAHALLI RAMESH, SANTHOSH AREHALLI SHIVAMURTHY, KARTHIK CHIMATAHALLI SHANTHAKUMAR, HAREESHA NAGARAJAPPA, VARUN DONNAKATTE NEELALOCHANA, SANDEEP SHADAKSHARI, AND REN-JEI CHUNG

INDEX, 383

About the Editors

Shashanka Rajendrachari is currently working as a Professor in the Department of Chemistry, Global Academy of Technology, Bengaluru, Karnataka, India. He obtained a PhD in powder metallurgy from the Department of Metallurgical and Materials Engineering at the National Institute of Technology Rourkela, India. He was the first recipient of the prestigious Powder Metallurgy Student of the Year award from the Powder Metallurgy Association of India for the year 2015 at the Indian Institute of Technology Bombay. He was awarded the prestigious "Young Scientist-2020" at the 4th International Scientist Awards ceremony for Engineering, Science, and Medicine held in Chennai, India. His name is listed in the top 2% of best scientists in the world for the years 2020, 2021, and 2022 as reported by Stanford University, USA. He has published more than 80 research articles, 18 books, and 17 book chapters.

Vinayak Adimule is presently working as a Professor and Dean in R&D at the Angadi Institute of Technology and Management, Savagaon Road, Belagavi-591108, Karnataka, India. He has 15 years of research experience as a Senior Research Scientist and Associate Research Scientist in R&D organizations. He has published more than 80 research articles and book chapters in Q rated high impact factor journals. He has been granted two Indian patents and two international patents. He has attended, chaired, and presented papers at national and international conferences.

List of Contributors

Dr Santhy A
Department of Chemistry, Amrita Vishwa Vidyapeetham
Amritapuri, India

Vinayak Adimule
Department of Chemistry, Angadi Institute of Technology and Management (AITM)
Belagavi, India

Narges Ataollahi
Department of Civil, Environmental, and Mechanical Engineering, University of Trento
Trento, Italy

Dr. Gangadhar Bagihalli
VTU Recognized Research Centre in Chemistry, KLE Institute of Technology, Hubballi
Karnataka, India

Willian Venancio Lemos Belino
Mackenzie Presbyterian University
São Paulo, Brazil

Ren-Jei Chung
National Taipei University of Technology (Taipei Tech)
Taipei, Taiwan

Thiago da Cruz Canevari
Mackenzie Presbyterian University
São Paulo, Brazil

Cem Erkmen
Istanbul Aydin University
Istanbul, Türkiye

Niran Öykü Erdoğan
Ankara University
Ankara, Turkey

João Henrique Astolfi Fereira
Mackenzie Presbyterian University
São Paulo, Brazil

Manjunatha Jamballi Gangadharappa
Department of Chemistry, FMKMC College, Mangalore University
Madikeri, India

Dr Moumita Gangopadhyay
Department of Chemistry, Amrita Vishwa Vidyapeetham
Amritapuri, India

Bojappa Ginni
Angadi Institute of Technology and Management (AITM)
Karnataka, India

Rajeev S. Joshi
Department of Physics, Central University
Kalburgi, India

Prashant Kakkamari
Department of Mechanical Engineering, KLS Gogte Institute of Technology
Belagavi, India

Gangadhar M Kanaginahal
Department of Mechanical Engineering, Jain College of Engineering and Research
Belagavi, India

Sateeshkumar Kanakannavar
Department of Mechatronics and Automation Engineering, Indian Institute of Information Technology Bhagalpur
Bihar, India

Lina Anil Kumar
Department of Chemistry, Amrita Vishwa Vidyapeetham
Amritapuri, India

Pankaj Kumar
Centre for Liquid Crystal Research, Chitkara University Institute of Engineering and Technology, Chitkara University
Punjab, India

Sevinc Kurbanoglu
Ankara University
Ankara, Türkiye

Laura Chaluleu Lieb,
Mackenzie Presbyterian University
São Paulo, Brazil

Simi M
Department of Chemistry, Amrita Vishwa Vidyapeetham
Amritapuri, India

Sreelakshmi M
Department of Chemistry, Amrita Vishwa Vidyapeetham
Amritapuri, India

Vaishnavi M
Department of Chemistry, Amrita Vishwa Vidyapeetham
Amritapuri, India

S. N. Mathad
Department of Engineering Physics, K.L.E Institute of Technology
Karnataka, India

Sahana Kamanna Metry
Department of Chemistry, SJCE, JSS Science and Technology University
Karnataka, India

Jubate Mohanty
Centre for Nano and Material Sciences, Jain (Deemed-to-be University)
Karnataka, India

H C Ananda Murthy
Department of Applied Sciences, Papua New Guinea University of Technology
Lae, Papua New Guinea

Hareesha Nagarajappa
School of Chemical and Biological Sciences, Department of Chemistry, GM University
Karnataka, India

Santosh Nandi
K.L. E Technological University Dr. M. S. Sheshgiri College of Engineering and Technology
Belagavi, India

Varun Donnakatte Neelalochana
Department of Civil, Environmental, and Mechanical Engineering, University of Trento
Trento, Italy

Sudenur Ozbey
Ankara University
Ankara, Türkiye

Dhiya P
Department of Chemistry, Amrita Vishwa Vidyapeetham
Amritapuri, India

Nikhila N Pai
Department of Chemistry, Moodlakatte Institute of Technology
Kundapura, India

Dr. Vishnu Vardhan Palem
Department of Biomedical Engineering, Sri Ramakrishna Engineering College
Tamil Nadu, India

Sanjay Ballur Prasanna
Department of Chemical Engineering and Biotechnology, National Taipei University of Technology (Taipei Tech)
Taipei, Taiwan.

Lekshmi B R
Department of Chemistry, Amrita Vishwa Vidyapeetham
Amritapuri, India

Dr Rejithamol R
Department of Chemistry, Amrita Vishwa Vidyapeetham
Amritapuri, India

Shashanka Rajendrachari
Global Academy of Technology
Bengaluru, India

P. Bakale Raghavendra
Department of Chemistry, Jain College of Engineering
Belagavi, India

Hariprasad B Ramesh
CSIR-Central Food Technological and Research Institute
Mysuru, India

Akshatha R Salian
Mangalore University
Konaje, India

Sheela A. Sangam
Department of Chemistry, K.L.S Gogte Institute of Technology
Belagavi, India

Paolo Scardi
Department of Civil, Environmental, and Mechanical Engineering, University of Trento
Trento, Italy

Sandeep Shadakshari
Department of Chemistry, SJCE, JSS Science and Technology University
Mysuru, India

Karthik Chimatahalli Shanthakumar
SJCE, JSS Science and Technology University
Mysuru, India

Vandna Sharma
Centre for Liquid Crystal Research, Chitkara University Institute of Engineering and Technology, Chitkara University
Punjab, India

Santhosh Arehalli Shivamurthy
Department of Chemistry (UG), NMKRV College for Women's
Bengaluru, India

Mariana Munik da Silva
Mackenzie Presbyterian University
São Paulo, Brazil

M M Sreejaya
Department of Chemistry, Amrita Vishwa Vidyapeetham
Amritapuri, India

Lakshmi Krishnaa Suresh
Department of Chemistry, Amrita Vishwa Vidyapeetham
Amritapuri, India

Daniel Yaezu Tiba,
Mackenzie Presbyterian University
São Paulo, Brazil

Gözde Aydoğdu Tığ
Ankara University
Ankara, Turkey

Eleonora Tomasino
Department of Civil, Environmental, and Mechanical Engineering, University of Trento
Trento, Italy

Esra Ülker
Ankara University
Ankara, Turkey

Yusuf Yerli
Ankara University
Ankara, Türkiye

Duygu Zabitler
Ankara University
Ankara, Turkey

CHAPTER 1

Advancement in Electrochemical Sensing Technology

Duygu Zabitler, Esra Ülker, Niran Öykü Erdoğan, and Gözde Aydoğdu Tığ

1.1 INTRODUCTION

Biosensors are analytical devices that provide qualitative and quantitative analysis of a specific target molecule. In medical, pharmaceutical, food, environmental, and biotechnological applications, determining biological species in biological or biochemical processes is vital. In addition to using chemical compounds, biorecognition elements such as aptamers, antibodies, and enzymes are also used in these analytical devices when performing analysis. There is an interaction between these recognition elements and the target analyte whose amount needs to be determined, and different types of transducers are used to transform this interaction into a quantitative signal. Signals can be measured optically or electrically. Biosensors utilize three main types of transducers: Optical, electrical, and mechanical (Singh, 2022).

DOI: 10.1201/9781003512639-1

Electrochemical sensor technology stands as a cornerstone of contemporary analytical chemistry, gaining significant traction due to its multifaceted applications and rapid advancements. The growing interest is driven by its high demand and the continuous evolution of the technology. The development of the glass electrode by Cremer in 1906 went down in history as the development that laid the foundation for the history of electrochemical sensors (Power and Morrin 2013). These sensors present numerous benefits compared with traditional analytical techniques, such as enhanced selectivity, increased sensitivity, robustness, ease of use, swift measurement times, portability, and cost-effectiveness. Remarkably, many of these features can be fine-tuned and optimized to meet specific assay requirements, thereby improving performance and expanding their practical applications (Saputra, 2023). One of the advantages of electrochemical sensors that makes them stand out is that, thanks to their theoretical basis, the signals obtained can be easily interpreted (Evtugyn et al, 2022). There are various reasons why these advantages are attributed to electrochemical sensors. The working electrode is modified with a variety of compounds, including nanomaterials, polymers, and bioreceptors such as aptamers or antibodies, which collectively enhance the sensitivity of the detection platform. Thanks to the high conductivity provided by materials such as graphene, carbon nanotubes (CNTs), and metal oxides, electron transport kinetics are also accelerated. Also, it is known that these materials increase the peak currents of the signals that occur when voltammetric procedures are applied (Lochab et al, 2024).

Electrochemical biosensors are versatile tools that can analyse a broad range of samples, including body fluids (blood, urine, saliva), food samples, cell cultures, and environmental samples. Due to their wide applicability, this analytical method is often highlighted in scientific literature (Grieshaber et al, 2008). With its wide application area and adaptability, it has been used in clinical diagnosis (Şimşek et al, 2023), pathogen/virus detection (Ganesh et al, 2024), allergen (Curilli, 2022), (Erdoğan et al, 2023) and toxin analysis (Roushani et al, 2021), cancer diagnosis (Thamilselvan and Kim, 2024) contaminant analysis (Zuo 2024), and clinical biomarker screening (Wehmeyer et al, 2021).

The aim of this chapter is to present the latest innovations in electrochemical sensors and sensing technologies, the materials that are used, how these technologies work, their areas of application, and their advantages. This section highlights the importance of electrochemical sensors in various application areas. In addition, the limitations of current techniques and future research and development directions will be discussed to provide a comprehensive perspective on scientific and technological advances in the field.

1.2 ELECTROCHEMICAL DETECTION TECHNIQUES USED IN SENSORS

Electrochemical sensors convert the sensing events of biological target elements into detectable electrochemical signals (Rezaei and Irannejad, 2019). In the context of biosensing measurements, the intrinsic electrochemical properties of the biological system are exploited to facilitate the acquisition of valuable information. Today, the majority of biosensors are constructed using electrochemical transducers. This is because these transducers are relatively simple to manufacture, easy to use, require small reaction volumes, and are not affected by interferences. In this way, the active component of the bioelectrochemical species acts as the primary transduction element. The principle of operation of an electrochemical transducer is based on the measurement of the accumulation or change in current or charge at the electrode site resulting from a "reaction." In biosensors, a variety of biological recognition components (cells, nucleic acids, antibodies, and microorganisms), as well as enzymes, have been given special consideration due to their distinctive binding ability and biological catalytic activity (Balkourani et al, 2021). In the field of bioelectrochemistry, a variety of electrochemical measurement techniques are employed, depending on the specific property being investigated.

These include:

1. Amperometric techniques.
2. Potentiometric techniques.
3. Electrochemical impedance spectrochemical techniques.
4. Voltammetric techniques.

1.2.1 Amperometric Techniques

The term "amperometry" is used to describe the process of monitoring changes in current (due to electrochemical oxidation or reduction) over time while maintaining a constant potential in the cell or between the working and reference electrodes (Rezaei and Irannejad, 2019). Amperometric measurements are often used as an analytical method because of their high accuracy and sensitivity. The applied voltage acts as a driving force for electrocatalytic redox reactions, resulting in electrical currents proportional to the analyte concentration. A controlled potential system is essential for basic instruments, and the electrochemical cell consists of two electrodes immersed in an electrolyte of suitable composition. A more sophisticated and common design is the use of a three-electrode cell in which one of the electrodes serves as a reference electrode. A working electrode is defined as the electrode where the reaction of interest takes place. In contrast, a reference electrode (such as Ag/AgCl or Hg/Hg_2Cl_2) is defined as the electrode that maintains a constant potential with respect to a working electrode. An inert conductive material (e.g., graphite or platinum) is usually used as an auxiliary electrode. In controlled potential studies, the use of a supporting electrolyte is essential to prevent electromigration effects, reduce solution resistance, and maintain a constant ionic strength. A comprehensive understanding of both practical applications and theoretical principles highlights that in electrochemical analysis, current can be directly measured by adjusting the potential to a specific value or maintaining it constant. The peak current observed over a linear potential range is directly proportional to the bulk concentration of the analyte in solution (Baranwal et al, 2022). In this method, the current is measured directly by increasing the potential to a preferred value or maintaining it at a desired level. The peak current value measured over a linear range of potential is directly proportional to the bulk concentration of the analyte in solution. Amperometric biosensors are highly selective because the potential resulting from the reduction or oxidation reaction used for detection is a characteristic of the analyte species. In cases where the current is measured using a regulated variable potential, the associated methodology is known as voltammetry. In this method, the current response, which may appear as a peak or plateau, is indicative of the analyte concentration. Efficient voltammetric methods with a wide linear dynamic range suitable for small quantities include DC voltammetry and polarography, linear sweep voltammetry (LSV), normal pulse voltammetry (NPV), differential pulse voltammetry (DPV), cyclic voltammetry (CV), square wave voltammetry (SWV), hydrodynamic methods, AC voltammetry, and stripping voltammetry (Rezaei and Irannejad, 2019). Amperometric biosensors represent a novel approach that employs biologically modified electrodes to provide qualitative and quantitative information in a complex matrix. Chronoamperometry (CA) is a type of amperometric technique. It involves applying a potential step and measuring the resulting current from the working electrode as a function of time. CA is a faster technique than other amperometric techniques, with measurements typically made in milliseconds or seconds. This technique requires sophisticated, expensive, specialized equipment and trained professionals to operate it. It is time-consuming and labour-intensive, but the results it produces are extremely accurate (Bansod et al, 2017).

1.2.2 Potentiometric Techniques

Potentiometric sensors are the most widely used practical sensors since the early 1930s. This is due to their simplicity and low cost. Potentiometric devices can be classified into three principal categories: Ion-selective electrodes (ISEs), coated-wire electrodes (CWEs), and field-effect transistors (FETs). The specific composition and structure of the membrane material are crucial for fabricating an electrode that exhibits selective properties towards a single species (Baranwal et al, 2022). In potentiometric devices, the accumulation charge potential is measured at the indicator electrode and compared with that at the reference electrode to obtain data on ion activity in the electrochemical reaction (when there are zero or no significant flow currents across the indicator and reference electrodes). In such circumstances, the measured potential is unequivocally ascribed to the number of electroactive species present in the sample. The Walther Nernst equation is used to establish a definitive relationship between the reduction potential and the concentrations of the analytes.

$$E = E° - \left(\frac{RT}{nF}\right) x\ ln\left(\frac{\alpha Red}{\alpha Ox}\right)$$

In this equation, E° represents the standard state potential, R is the gas constant, T is the constant temperature of the cell in degrees Kelvin, n is the number of electrons exchanged in the redox reaction, F is the Faraday constant, and αRed/αOx is the activity ratio of the reducing and oxidizing species. Direct potentiometry is the direct measurement of analyte concentration through the application of the Nernst equation. The detection limits of potentiometric devices currently range from 10^{-8} to 10^{-11} M. As potentiometric sensors do not chemically affect the sample, they are suitable for measurement at very low concentrations and in very low-volume samples. The potentiometric method represents a suitable alternative to electrical determination of the endpoint in a biochemical titration reaction. The use of titration devices has become the norm in recent times. The availability of suitable indicator electrodes that are both cost-effective and highly reliable makes them ideal for a wide range of titrimetric studies across numerous chemical reactions. The shape and position of potentiometric titration curves provide definitive insight into the location of the endpoint of the titration, while the curve itself offers irrefutable information about the processes occurring concurrently with the reaction (Rezaei and Irannejad, 2019). These techniques are superior to others in the field of electroanalytical techniques because they measure activity within a solution, not concentration. This makes them ideal for qualitative and quantitative analysis. However, these techniques are limited by several factors, including high detection limits, low sensitivity, and the challenge of miniaturizing electrodes (Bansod et al, 2017).

1.2.3 Electrochemical Impedance Spectrochemical Techniques

The electrochemical impedance technique is an outstanding, cost-effective, rapid, and reliable approach for measuring properties in biological systems. In this context, the term "impedance" is used to describe a physical variable that defines the properties of the resistance of an electrical circuit in the presence of an alternating current applied between electrodes. In this system, we quantify the current flow response by applying a small sinusoidal potential and measuring the resultant changes in frequency (f) over a wide frequency range. The mathematical relations obtained by varying the excitation frequency clearly show that the impedance response is a complex number comprising both real and imaginary components as a function of frequency. At high frequencies, we can predict the transfer of electrons. At low frequencies, we can predict the transfer of mass. Electrochemical impedance spectroscopy (EIS) is the gold standard for assessing affinity in label-free electrochemical immunosensors, giving researchers direct control over the antibody–antigen binding process occurring on the electrode surface (Baranwal et al, 2022). EIS is a widely employed technique for the investigation of biosensor fabrication procedures. Furthermore, it plays a significant role in the study of diverse bioanalytical properties in, for example, proteins, microorganisms, nucleic acids, antigens, and antibodies. In EIS, the electrochemical reaction occurring in an electrolytic cell is represented in terms of an electrical equivalent circuit (EEC). By determining this impedance parameter and other resistive capacitive (RC) parameters of the EEC, it is possible to estimate the concentration of metal ions in an electrolytic solution.

The frequency response analyser (FRA) is the most prevalent technique for determining analyte concentration among the various impedance measurements (Bansod et al, 2017). The EIS technique was used by the researchers to study mineral ions. There is no doubt that this agent is a significant contributor to mortality rates in pathological calcification of the cardiovascular system. Consequently, alterations in electrical impedance are monitored through the aggregation of mineral ions in supersaturated serum (Rezaei and Irannejad, 2019). The main disadvantage of this method of detection is the limited regenerative capacity of the working electrode. However, electrochemical impedance detection has several clear advantages, including a low cost, a faster response time, and a higher signal-to-noise ratio (Balkourani et al, 2021).

1.2.4 Voltammetric Techniques

The working electrode is forced to operate within a specified potential range (in opposition to a reference electrode) where an electrochemical oxidation or reduction reaction occurs. The resulting current is then measured (in opposition to a counter electrode) in accordance with the voltammetric detection method. The current response is directly proportional to the analyte concentration when DPV, CV, SWV, and LSV techniques are applied. Aptamer-based biosensors are the future of biosensors, no question. These sensors use voltammetric sensing techniques to take full advantage of their robustness and exceptional low detection limits, even in minimal quantities of the analyte (Balkourani et al, 2021). Voltammetric techniques are the most effective methodology for detecting and quantifying heavy metal ions in diverse complex environmental matrices. These techniques measure current at varying potential points along a current–voltage curve, unlike the fixed potential point used in the amperometric technique. Voltammetry is the best technique for determining trace metals. It is highly accurate and sensitive. These techniques are the optimal choice for suppressing background current and improving detection limits. Reversibility of reactions in an electrochemical cell setup is a reliable method for obtaining qualitative information. This technique is the ideal solution for analysing turbid and coloured solutions, which other techniques simply cannot handle. The utilization of conventional electrodes to conduct these experiments with a linear potential sweep (10 to approximately 1000 mV/s) to ascertain the resulting i–E curve is designated as LSV. These experiments, which involve a double potential step and a reversal of the potential sweep, resulting in a reversal of the current, are often referred to as CV. Voltammetric experiments conducted with a potential step on a dripping mercury electrode (DME) are referred to as polarography or alternatively, as direct current polarography. For concentrations exceeding 10^{-5} M, LSV on a DME or a static mercury drop electrode (SMDE) formed on a glass capillary tip is highly suitable. The utilization of a pulse of a voltage signal with varying shapes and amplitudes in a voltammetric measurement is referred to as pulse voltammetry. This is itself sub-categorized as follows: The main types of pulse voltammetry are NPV or normal pulse polarography (NPP), ladder voltammetry or tast polarography, reverse pulse voltammetry or polarography (RPV), DPV, and SWV. These techniques are suitable for the partial suppression of background current and the improvement of detection limits. Of the various pulse voltammetric methods, differential pulse and square wave voltammetry are the most widely used due to their high sensitivity, which is suitable for trace-level analysis (Bansod et al, 2017).

1.3 RECENT ADVANCES IN ELECTRODE MODIFICATION

Traditional electrochemical sensors consist of three electrodes: The "working electrode," the "counter electrode," and the "reference electrode." Therefore, a reaction is catalysed on the working electrode through electrocatalysis and could be improved by several designs (Singh et al, 2023). An electrode design requires specific properties. These properties can be listed as follows:

1) Relatively moderate-/low-cost electrodes. For example, this factor can be achieved using a low-cost substrate with a more effective surface coating.
2) A long lifespan and corrosion resistance through outstanding mechanical, thermal, and chemical stability.
3) Efficient electrocatalysts to achieve a low overpotential for the intended reaction.
4) To minimize ohmic losses and make sure current distribution is uniform, it is essential to have high electrical conductivity (Walsh et al, 2018).

Many electroanalytical techniques have been developed based on the interaction between the analyte and the modified working electrode. For example, advanced methods like nanomaterials can improve electrochemical sensors' sensitivity and selectivity levels, increasing efficiency in detecting different substances (Mohamad Nor et al, 2023).

Modification of electrode surfaces varies in its effects on their performance depending on the technique used to optimally immobilize the biosensors. This modification is significant in the realm of biosensor research, as it provides a method to enhance biocompatibility, sensitivity, and electron conductivity while reducing surface fouling and noise (Sandhyarani, 2019). Commonly employed methods include functionalizing the electrode surface with self-assembled monolayers (SAMs), electrodeposition of polymers or nanomaterials, conducting polymers, various nanomaterials, and metal–organic frameworks (MOFs).

1.3.1 Self-Assembled Monolayers

SAMs are chemical modifications that organize molecular assemblies of organic molecules deposited on the surface of an electrode by the chemisorption of "polar head groups" from either the vapour or the liquid phase. These heads are placed on the electrode surface to facilitate the placement of "tails" away from the substrate. The process continues until the regions of densely packed molecules expand to cover the entire surface of the substrate with a single, uniform monolayer of molecules (Barhoum et al, 2023). This technique is among the most efficient methods for creating ultrathin organic films with precisely controlled surface thickness, ensuring that the films exhibit the desired properties (Sandhyarani, 2019). Among the most extensively studied examples are silanes, used to modify silica surfaces, and alkanethiols, known for their affinity for coinage metals such as gold, silver, platinum, and copper. Alkanethiols naturally adsorb onto these metals' surfaces. Gold is commonly used because it forms no stable oxide layer under ambient conditions. The thiol groups chemisorb onto the gold surface by forming a gold–thiol bond, resulting in the formation of a densely packed and highly ordered monolayer. SAMs' stability is of outstanding importance in the context of electrochemical sensors. Compared with the standard calomel electrode (SCE), SAMs are stable at potentials ranging from approximately 0.8 to 1.4 V. This stability is crucial in ensuring reliable performance of the sensor, as it allows the SAMs to maintain their structural integrity and functionality under a wide range of electrochemical conditions (Gooding et al, 2003). SAMs offer high sensitivity and selectivity in various fields, from pH measurement to detecting inorganic ions, from biosensors to complex biomolecular systems. The activation of thiol COOH groups with 1-ethyl-3-(3-dimethylaminoprop yl)carbodiimide (EDC) and N-hydroxysuccinimide (NHS) can enhance antibody immobilization on gold electrodes (Barhoum et al, 2023), as shown in the schematic in Figure 1.1a.

1.3.2 Electrochemical Deposition

Electrodeposition is another well-known surface modification technique that is widely used to grow thin films on surfaces. Changing an electrode's electrochemical potential through an external electric field induces either reduction or oxidation reactions on the electrode surface. These reactions can lead to deposition or dissolution on the electrode surface, depending on the specific changes in potential. This method is a valuable tool in various applications, including the development of advanced materials and devices, as it allows precise control over the thickness and composition of the film (Sandhyarani, 2019). Currently, several techniques are in use for electrochemical deposition; these include cyclic voltammetry, potential stepping, and dual pulse deposition. Particle size can be finely adjusted by a change in current density, voltage, and duration of electrolysis. Figure 1.1d shows that a biosensor was constructed by electrochemically depositing a nanoscale-sized porous array of gold nanotubes in the membrane using a gold plating solution (Barhoum et al, 2023).

1.3.3 Coating-Based Techniques

Nanomaterials for coating-based techniques, known as simple, easy, and rapid technique modifications, must be either soluble or colloidal, and they should further show good colloidal stability for proper uniform distribution on the disposable electrode surface (Kaliyaraj Selva Kumar et al, 2020). These nanoscaled materials are applied to the electrode surface through coating methods such as drop casting, dip coating, spin coating, or chemical adsorption (Mohamad Nor et al, 2023). Drop casting generally involves droplets of nanomaterial in solution deposited on the electrode surface, as shown in Figure 1.1c. This technique is known for its simplicity, waste-free nature, and suitability for hybrid nanomaterials. However, film thickness, film homogeneity, and film stability are poor. The dip-coating method could be defined as the deposition of aqueous-based nanomaterial or polymer liquid phase coating solutions on the surface of the electrode. Unlike drop casting, they are well homogenized on the surface of the electrode, but there is a challenge in terms of the amount of waste produced in the process. There is also the concern that this process has the potential

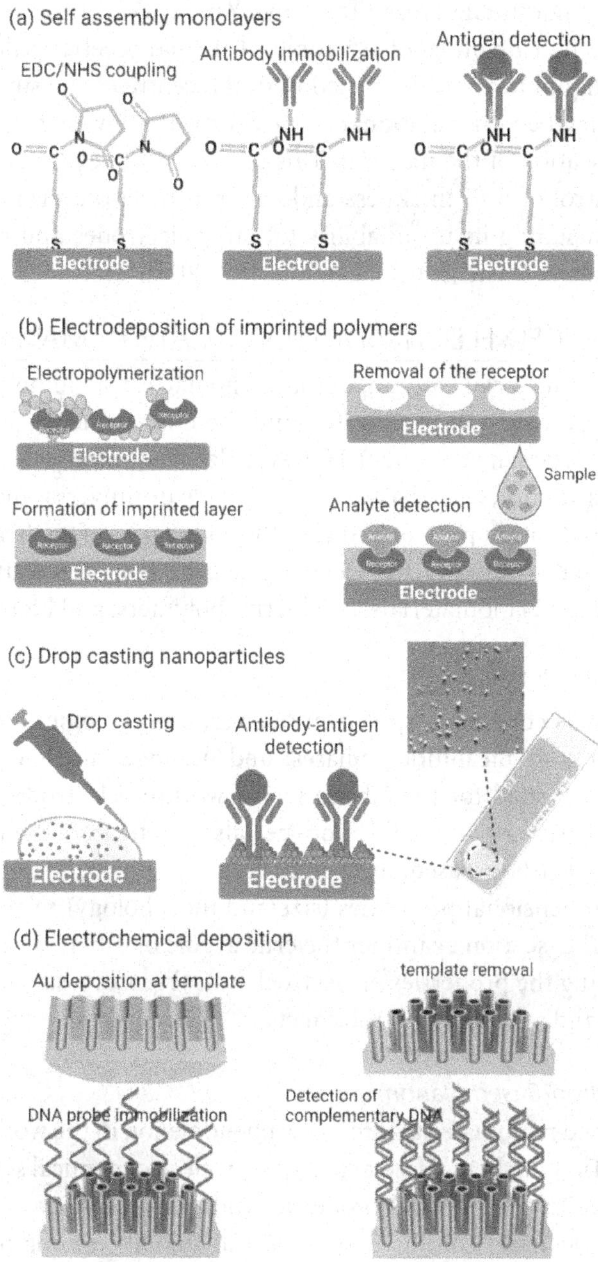

FIGURE 1.1 (a) Self-assembly monolayers; (b) electrodeposition of imprinted polymers; (c) drop-casting technique for disposable electrodes; (d) preparation of a nanotube array electrode by electrochemical deposition of gold followed by template removal and chemisorption of the DNA sample. (Reprinted from *Trends in Environmental Analytical Chemistry*, Barhoum et al. (2023), with permission from Elsevier.

for coating undesirable areas (Tang and Yan, 2017). In the spin-coating process, a small quantity of micrometre- to nanometre-scale material, usually in liquid form, is dispensed onto the centre of the substrate. The substrate may be either stationary or rotating at a low speed during the initial application of the material. This method offers advantages such as precise control of film thickness and uniformity. However, it has disadvantages, including its unsuitability for large electrodes and the generation of extensive waste (Mustafa and Jameel, 2021).

1.4 ADVANCED ELECTRODE MODIFICATION MATERIALS

The electrode material can range from nanomaterials, biomolecules, and conductive polymers to composites, and there is a wide range of design possibilities (Hassan et al, 2022). However, despite the latest developments in materials science, there has yet to be a single or universal material that covers all application principles due to the wide range of applications. The materials used to modify the electrode can be considered under three main headings: Nanomaterials, conductive polymers, and bioreceptors.

1.4.1 Nanomaterials

Nanomaterials covering a spectrum of sizes, morphologies, dimensions, compositions, permeabilities, phases, and homogeneity have been rigorously investigated for the fabrication of working electrodes. The electrochemical performance of nanomaterials depends on their chemical composition (carbon-based, inorganic, organic, or hybrid nanomaterials) and their dimensional properties (size and morphology) relative to bulk materials. This section examines the crucial role of different nanomaterials in defining the properties of electrochemical sensors, such as stability, sensitivity, and selectivity (Barhoum et al, 2023).

1.4.1.1 Carbon-Based Nanomaterials

Carbon-based nanomaterials are a new phenomenon in the world of nanomaterials. They are being used extensively in electrochemical sensors, particularly due to their special properties such as large surface-to-volume ratio, high electrical conductivity, chemical stability, strong mechanical strength, easy functionalization, biocompatibility, and biodegradability, which make these nanomaterials suitable for use in biosensors to enhance signal processing. In addition to diamond and graphite, which are widely recognized as carbon allotropes, the latest additions to the field of synthetic carbon allotropes, including carbon nanotubes, graphene and its derivatives, as well as carbon dots (CDs), have been employed extensively in the development of sensors (Eivazzadeh-Keihan et al, 2022).

1.4.1.1.1 Carbon Nanotubes

The classification of CNTs is typically based on their structural characteristics, namely, their shape and size. This leads to the distinction between single-walled carbon nanotubes (SWCNTs) and multiwalled carbon nanotubes (MWCNTs). The structural configuration of carbon materials allows the fabrication of highly efficacious electrochemical sensors with enhanced durability, as it endows them with a spectrum of surface functionality. The electrical conductivity of the tubes is essential for their participation in electrochemistry due to their electrical properties. MWNTs are regarded as metallic conductors, a highly sought-after attribute for an electrode (Savk et al, 2019). In addition to CNTs, the conjugation of bioreceptors, various metal oxides, and nanomaterials with different structures has been analysed. Due to the instability of enzymes at various stages, including manufacturing, storage, and use, most enzyme-based biosensors have a limited functional lifetime. For this reason, non-enzymatic applications have been investigated. A CNTs-PtNP (carbon nanotubes-platinum nanoparticles) nanocomposite for printable non-enzymatic glucose biosensors was investigated. A fast and template-free method was used to prepare MWCNTs as a conductive nanocomposite. Amperometric and voltammetric methods were applied to study glucose oxidation (Nguyen et al, 2020). By attaching MWCNTs, these could be developed to analyse many substances. Chung and colleagues reported that a needle-shaped biosensor was used to determine amine and glutamate by attaching MWCNTs (Boo et al, 2006). Luo and colleagues have reported the successful simultaneous detection of dopamine, ascorbic acid, and serotonin in rabbit brain tissue using differential pulse voltammetry with an electrode incorporating CNTs (Wang et al, 2003).

Identifying DNA sequences is crucial in the initial recognition of disease and in the detection of viral and genetic material. The history of detecting DNA sequences is characterized by employing a multitude of techniques, the majority of which proved to be both complex and prohibitively expensive (Eivazzadeh-Keihan et al, 2022). In a noteworthy study, Chen et al. developed a biosensor to detect DNA sequences. The biosensor comprised a nanocomposite of copper oxide nanowires and SWCNTs, which exhibited a synergistic effect. The hybrid nanocomposite exhibited highly conductive properties, significantly enhancing the DNA probe's immobilization. In the experimental setup, single-stranded DNA (ssDNA) was immobilized by forming amide bonds between the amino groups of the ssDNA and the carbonyl groups of the immobilization matrix. Double-stranded DNA structures were formed upon the introduction of the complementary target DNA. Subsequently, Adriamycin was employed to facilitate the electrochemical detection of the target DNA. The CuO NWs/SWCNTs-COOH nanocomposite exhibited remarkable selectivity, with a limit of detection (LOD) of 3.5×10^{-15} M and a linear range from 1.0×10^{-12} to 1.0×10^{-6} M (Chen et al, 2016).

1.4.1.1.2 Graphene

Graphene has attracted considerable attention as an emerging carbon nanomaterial. Its physical and electrochemical properties have facilitated its extensive use in electrochemical sensing applications (Singh et al, 2023). According to studies, pristine graphene was used to develop a highly sensitive abscisic acid sensor. The pristine graphene was prepared by organic salt-assisted exfoliation. Compared with chemically converted graphene, this sensor exhibited improved sensing characteristics. Also, this design could determine ascorbic acid, dopamine, and uric acid (Qi et al, 2015).

The global pandemic coronavirus disease 2019 (COVID-19), named Severe Acute Respiratory Syndrome Coronavirus 2 (SARS-CoV-2), has highlighted the urgent need for accurate and affordable diagnostic tests for epidemic management and population monitoring worldwide. In response, a laser-scribed graphene-based electrochemical sensing device for COVID-19 diagnosis incorporating gold nanostructures has been developed. The electrode of this device was modified with the SARS-CoV-2 spike protein antibody and integrated into a hand-held point-of-care (POC) detection system using a smartphone application. This design provides an accessible and affordable diagnostic platform (Aghamirza Moghim Aliabadi et al, 2022).

1.4.1.1.3 Carbon Dots

CDs have attracted considerable interest in biosensor development due to their unique properties. These include optical, electrical, and catalytic properties. CDs' unique absorption properties make them particularly suitable for optical biosensors (Eivazzadeh-Keihan et al, 2022). Numerous studies have investigated the use of CDs and their composites, including fluorine and nitrogen–doped CDs (F, N-CDs) for the detection of catechol (Liu et al, 2019).

1.4.1.2 Metal Nanoparticles

Metal nanomaterials have various applications in various fields, including photoelectric information storage, renewable energy, functional catalysis, and many more. These nanomaterials have expanded their utility by integrating the unique properties of noble metals with the inherent physical and chemical properties of nanomaterials. This combination enhances their effectiveness and expands their potential use in advanced technological applications (Ramya et al, 2022). In this part, gold nanoparticles (AuNPs), platinum nanoparticles (PtNPs), and silver nanoparticles (AgNPs) will be examined.

1.4.1.2.1 Gold Nanoparticles

The most widely used nanoparticles for protein immobilization are undoubtedly AuNPs. As early as the 1990s, researchers began to immobilize various enzymes using gold nanoparticles, creating various enzyme electrodes. These enzyme electrodes effectively preserved the enzymatic activity, thus proving the potential of AuNPs to maintain functional stability in biocatalytic approaches (Luo et al, 2024). A new enzymatic electrochemical biosensor for the detection of cholesterol was investigated, a composite material consisting of cobalt ferrite, molybdenum disulfide, and AuNPs ($CoFe_2O_4$@MoS_2/AuNPs). This composite material was used to determine cholesterol using voltammetric methods quantitatively. The integration of these materials aimed to improve the sensitivity and specificity of the sensor by exploiting the unique electrochemical properties of the composite components (Ayyandurai et al, 2024).

1.4.1.2.2 Platinum Nanoparticles

The use of PtNPs has facilitated a range of catalytic reactions, with these nanoparticles offering a controllable surface colour and exhibiting a wide range of catalytic activities (Ramya et al, 2022). According to the studies, an electrode design incorporating PtNPs was used to determine ascorbic acid, epinephrine, and uric acid. Thirumalraj and collaborators presented a non-enzymatic hydrogen peroxide (H_2O_2) sensor utilizing a graphite/gelatin hydrogel composite (GR/GLN) modified screen-printed carbon electrode (SPCE) decorated with PtNPs. The PtNPs@GR/GLN composite was synthesized by an electrodeposition method using a 0.5 mM K_2PtCl_6 solution containing 5 mM $NaNO_3$ at a potential range of −0.1 to 0.2 V. The PtNPs@GR/GLN modified electrode was effectively employed for the electrochemical detection of H_2O_2.The developed sensor exhibited fast response, higher electrocatalytic activity, lower detection limits, and a wide linear range for H_2O_2 detection. In addition, the developed sensor showed significant potential for determining H_2O_2 in real biological samples. Thus, for the detection of non-enzymatic H_2O_2, the PtNPs@GR/GLN composite is a suitable platform (Thirumalraj et al, 2024).

1.4.1.2.3 Silver Nanoparticles

Materials based on AgNPs show promising properties in a wide range of fields, including catalysis, electronics, chemistry, electrochemistry, magnetic fields, and optical spectral properties. AgNPs are valuable in the development of advanced medical treatments as they have demonstrated impressive properties and potential. These include novel antimicrobial agents based on the antimicrobial properties of silver, drug delivery systems, electrochemical biosensors, coatings for biomaterials and medical devices, materials for tissue repair and regeneration, and strategies for managing complex medical conditions.

In 2024, Sharma and co-workers developed an aptamer-based flexible carbon ink printed electrode modified with AgNPs for the electrochemical detection of the Chikungunya virus. The biosensor was used to assess CHIKV-Ag in a human serum sample and demonstrated a linear detection range from 0.1 ng/mL to 1 μg/mL, with a detection limit as low as 0.1 ng/mL for CHIKV-Ag. Due to its production flexibility and the inherent electrocatalytic properties of the zero-dimensional nanostructure, the proposed approach offers new avenues for cost-effective bioelectronics (Sharma et al, 2024).

Habibi and colleagues presented a precise and highly sensitive electrochemical method for determining clonazepam (CZP). The method employs a glassy carbon electrode modified with a nanocomposite comprising AgNPs and MWCNTs. The studies were conducted using two analytical techniques: Cyclic voltammetry and DPV. The CZP concentration range investigated spanned from 5.0×10^{-8} to 2.5×10^{-6} M, with a detection limit of 6.0×10^{-9} M (Habibi et al, 2014).

1.4.1.3 Hybrid Nanoparticles

Hybrid nanoparticles such as nanoparticle–enzyme (Willner et al, 2007), nanoparticle–polymer (Dakshayini et al, 2019), and biomolecule–nanoparticle (Guo et al, 2009) are also highly effective in electrochemical sensor applications. Rather than using these substances individually, the formation of hybrid structures could provide a more effective application due to their synergistic effect.

Luo et al. used a modified glassy carbon electrode (GCE) coated with AuNPs and poly-o-aminobenzenethiol poly(oATP) films to design and evaluate a novel electrochemical biosensor. The electrochemical low-density lipoprotein (LDL) biosensor has impressive performance characteristics, including wide linear range, low detection limit, high specificity, and satisfactory reproducibility and stability (Luo et al, 2023).

1.4.2 Conductive Polymers

Sensor platforms based on polymeric nanomaterials show several unique properties, such as sensitivity, linearity, selectivity, and ease of manufacture. These polymers play a critical role in the development of advanced sensors. Among the advantages of using polymers, the most prominent is the cost-effectiveness of these materials. In the manufacture of electrodes for sensor applications, conductive polymers such as polyacetylene, polyparaphenylene, polypyrrole, polyaniline, and poly(3,4-ethylenedioxythiophene) (PEDOT) are commonly used. These polymers combine the mechanical flexibility and processability of conventional polymers with the electrical conductivity of metals, giving them exceptional electrical properties. This combination creates highly sensitive and selective sensors that can operate in various conditions (Naveen et al, 2017).

Polyacetylene, for example, was one of the first conductive polymers discovered and paved the way for developing other conductive materials. Polyparaphenylene and polypyrrole are also widely used because of their stability and high conductivity. Polypyrrole is favoured for its ease of synthesis and good environmental stability, making it an excellent candidate for electrode development in electrochemical sensors (Lee and Chang, 2019).

Conducting polymers can be used to analyse many analytes, including heavy metals, by forming hybrid structures with nanoparticles and bioreceptors and by copolymerization. Zhong and his team have successfully developed a sensitive and selective electrochemical sensing platform to detect Pb(II). This platform, shown in Figure 1.2, was constructed using a dopamine polymer doped polypyrrole (PDA-PPy) hydrogel-modified SPCE. The hydrogels were synthesized by hybridizing dopamine hydrochloride and pyrrole in a fast and one-step freeze-drying process, which can be considered green chemistry (Zhong et al, 2021).

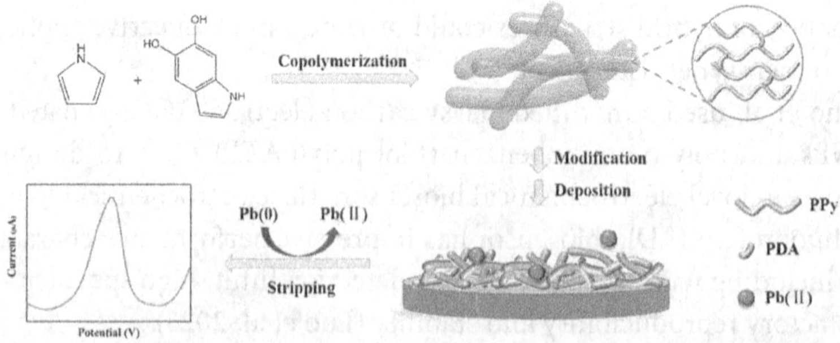

FIGURE 1.2 Electrochemical detection of Pb(II) using the PDA-PPy/SPCE sensor platform. (Reprinted from *Trends in Environmental Analytical Chemistry*, Barhoum et al. (2023), with permission from Elsevier.)

1.4.3 Bioreceptors

The bioreceptor, also known as the recognition element, is an essential part of a biosensor, enabling sensitive and selective analysis by recognizing and interacting with a target analyte (Alhamoud et al, 2019). The key mechanisms of biomolecular recognition are non-covalent interactions, including hydrogen bonding, electrostatic forces, van der Waals forces, and hydrophobic interactions. In the presence of a target molecule, recognition elements either undergo a (bio)chemical reaction that directly gives a readily measurable signal or are labelled with signalling molecules, such as enzymes, to enable signal transduction, as shown in Figure 1.3 (Dincer et al, 2019; Sengupta and Hussain, 2022).

These recognition elements (Figure 1.4) can be either natural or artificial bioreceptors. Naturally occurring bioreceptors are found in living organisms and can be isolated directly from them or synthesized in the laboratory. The most abundant natural bioreceptors used in disposable sensors include nucleic acids, enzymes, antibodies, membranes, bacteriophages, organelles, cells, and tissues. Artificial bioreceptors, on the other hand, can be fully or partially synthetic and are often designed to enhance or replace natural variants. Prominent examples of artificial bioreceptors include aptamers, molecularly imprinted polymers (MIPs), supramolecular receptors, synthetic peptides with receptor properties, macrocycles, and recombinant or genetically engineered natural biomolecules such as antibody fragments, protein domains, or multifunctional molecules such as catalytic antibodies. Compared with their natural counterparts, engineered bioreceptors typically offer improved stability and high affinity at a more cost-effective price, making them ideal for use in disposable sensors.

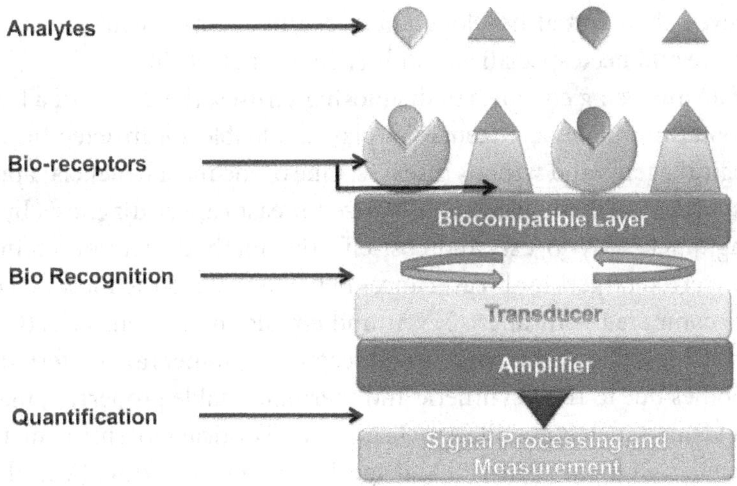

FIGURE 1.3 A biosensor is designed to detect analytes using isolated bioreceptors such as enzymes, immune systems, tissues, organelles, etc. (Reprinted from Sengupta and Hussain (2022), Nanomaterials, 12, 4132, with permission from MDPI under CC BY license.)

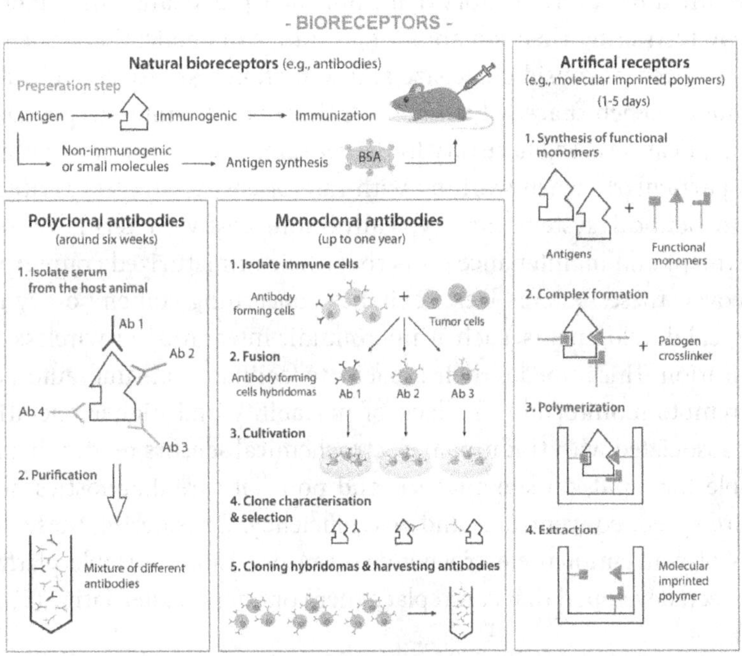

FIGURE 1.4 Antibody- and MIP-based examples illustrate natural and artificial recognition elements. (Reprinted from Dincer et al. (2019), Advanced Materials, 31(30), 1806739, with permission from Wiley.)

However, their initial development and capital investment are resource intensive and need specialized skills (Dincer et al, 2019).

From analysing enzymes to diagnosing viruses, the design of a biosensor containing biological agents also plays a valuable role in detecting many biological agents and various diseases. One of the most beneficial applications, by Mittal et al., has revolutionized breast cancer diagnosis by providing quick, easy-to-use, and cost-effective methods. Glycoproteins are the preferred targets for breast cancer detection due to their low detection limits compared with micro RNAs and circulating tumour cells (CTCs), which are less abundant. Among bioreceptors, aptamers are preferred over antibodies due to their synthetic and thermally stable properties, making them ideal for detection. Electrochemical and optical biosensing methods have achieved high sensitivity and specificity in recent years (Mittal et al, 2017).

1.5 MINIATURIZED ELECTROCHEMICAL SENSORS AND MULTISENSORS

The deployment of conventional electrochemical sensors is frequently constrained by several factors that limit their practicality and efficiency, thereby restricting their suitability for a range of applications. Such sensors are often of considerable size and weight, necessitating a substantial amount of bench space, which renders them poorly suited to portable or field applications. The necessity for larger sample volumes can be impractical, particularly when working with rare or costly samples. Traditional electrochemical systems are typically more costly in terms of initial investment and maintenance costs than their miniaturized counterparts. Moreover, these systems lack flexibility concerning contemporary technological developments, such as microfluidic integration or wireless communication. This impedes their capacity to facilitate real-time, automated, and remote monitoring. The lack of portability and higher operational costs associated with traditional electrochemical sensors render them less suitable for rapid, on-site analyses and point-of-care diagnostics, which require speed, convenience, and cost efficiency. Considering these drawbacks of conventional electrochemical sensors, more portable platforms have been developed that can replace them or offer an alternative (Turner, 2013).

1.5.1 Miniaturized Electrochemical Sensors

Electrochemical sensors are analytical tools with many advantages. Today, it can be seen in the literature that these sensors are constantly evolving.

One of the areas where this development is widespread is miniature electrochemical sensors, which are becoming popular because they can be used easily in applications outside the laboratory environment. These analysers can integrate voltammetric, impedimetric, amperometric, and potentiometric techniques within a single electronic system (Manikandan et al, 2024).

Application areas of miniaturized electrochemical sensors include situations where trace amounts of biomarkers such as organic molecules, metal ions, and biomolecules are present. These biomarkers are analysed using electrochemical signal changes (such as current, potential, impedance, or voltage) that occur as a result of oxidation or reduction reactions. Miniaturized electrochemical sensors can be classified as DNA-based, immobilization, miniature electrochemical sensors that do not require immobilization, paper-based, microfluidic-based, capillary electrophoresis–based, or SPCE (Figure 1.5), according to the material, technology, and biological recognition element properties used (Zuo et al, 2024).

In 2019, Yang et al. modified the carbon paste electrode with a polyglycine/Nafion/MWCNT surface to obtain the p-Gly/Nafion/MWCNTs/CPE platform and conducted an electrochemical sensor study. In this study, Pt wire was used as the counter electrode, and Ag/AgCl wire was used as the reference. Dopamine and tyrosine analyses were carried out using the obtained plastic-based miniature electrochemical platform. Thanks to the electrocatalytic feature provided by the platform, the oxidation of analysed dopamine and tyrosine was carried out quite successfully.

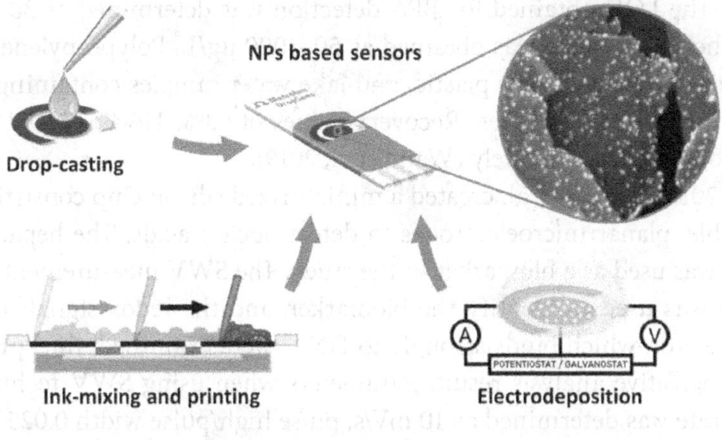

FIGURE 1.5 Diagrammatic view of the three main methods generally used to modify SPEs with metal NPs. (Reprinted from Economou (2018), Sensors, 18(4), 1032, with permission from MDPI under CC BY license.)

Performance comparison of the bare and modified electrodes was done using CV on the redox probe $[Fe(CN)_6]^{3-/4-}$. In addition, LSV was used to detect dopamine and tyrosine, and a potential in the range of +0.1 to +1.1 V (against Ag/AgCl) was applied. As a result of the analyses, the LOD value was calculated as 0.07 μM, and the linear range was found to be 0.2–40 μM. Analyses were performed on human serum samples (Yang et al, 2019).

In their study in 2024, Fernandes-Junior et al. obtained a miniature electrochemical sensor based on polylactic acid (PLA) modified with conductive carbon black commercial filament using a 3D printing prototype and then detected theophylline and paracetamol. This literature study aims to improve the analysis of theophylline and paracetamol, which can cause adverse effects such as tachycardia and renal failure as a result of overdose, with its ease of printing, low cost, and simplicity. Artificial urine and drug tablet samples were used in theophylline analyses with the sensor obtained. The LOD value was determined as 1.2 μmol/L, and the linear range was 5.00–150 μmol/L. For paracetamol, the LOD value was determined as 0.37 μmol/L, and the linear range was 1.00–200 μmol/L. The CV method was used in the characterization of the produced sensor (Fernandes-Junior et al, 2024).

In 2019, Wang et al. produced a disposable miniaturized electrochemical sensor to analyse bisphenol A (BPA). This sensor uses a removable three-electrode system and a coin vibration motor to increase BPA deposition in the voltammetric method. In the three-electrode system, an Au-film electrode, Pt wire, and an Ag/AgCl wire reference electrode were used. The LOD obtained for BPA detection was determined as 30 μg/L, and the linear range was obtained as 50–1000 μg/L. Polypropylene (PP), polyvinylchloride (PVC) plastic, and lake water samples containing BPA were used as real samples. Recovery values of 93%, 116.4%, and 110.9% were obtained, respectively (Wang et al, 2019).

In 2015, Petralia et al. created a miniaturized silicon chip consisting of portable, planar microelectrodes to detect nucleic acids. The hepatitis-B virus was used as a biomarker in the study. The SWV measurement technique was used to identify the biomarker, and the redox signals of the intercalator, which binds strongly to DNA, were examined, thus providing a sensitive analysis result. Parameters when using SWV technique: Scan rate was determined as 10 mV/s, pulse high/pulse width 0.025 V for 0.05 s, stop height 12 mV (Petralia et al, 2015). Another developing sensor type is multisensors, which do not allow many analyses to be performed simultaneously.

1.5.2 Multisensors

Different disciplines have needed electrochemical multisensor systems; many types of environmental and technological monitoring have emerged due to the complex structures of biological and cell cultures. For instance, various sensors or arrays are ideal for concurrently tracking multiple analytes in complex biological environments to obtain more information about a patient's health status or to monitor treatment progress (Pauliukaite et al, 2020). The development of multisensors, which can satisfactorily meet these needs with the advantages of being cheap and miniaturized, speed, and the use of nanomaterials, is becoming more widespread daily (Nikolaev et al, 2018). The literature includes two types of applications involving electrochemical detection of multiple biomarkers. One of these is sensors that are functionalized using separate specific receptors for each target, and the other is sensors that distinguish electrochemical signals transmitted on a single electrode. With the development of artificial intelligence, analysing the electrochemical signals of complex matrices has become more accessible (Sharafeldin et al, 2023). The following summary of literature studies explains the basis and advantages of developing multisensor technology quite well.

Navrátil et al. conducted a predictive study in 2009 using multisensor technology, which is widespread today. This multisensor is designed to combine eight different sensors and is manufactured to enable the analysis of small-scale samples with a volume of 5–200 μL. Five different solid amalgam electrodes, platinum, gold, and glassy carbon electrodes, were used in the eight sensors. The working principle of the resulting multisensor is based on sequential autonomous measurement by connecting to a multi-channel device. Another advantage of this multisensor, which contains eight different sensors, is that it can be operated with eight working electrodes, either different or the same type. Another aspect of this study that enlightens the literature is that each sensor measurement can be made using various methods such as direct current voltammetry (DCV), DPV, coulometry, chronopotentiometry, and square wave voltammetry. A mixture of adenine and guanine was used as an example to demonstrate the capacity and success of the produced device. This sample selection is an excellent choice for this multisensor designed to analyse small-scale samples (e.g., DNA analysis). The device used for the sensor to take measurements was the Multielchem device (Navrátil et al, 2009).

In 2018, Urban et al. produced a microsensor system to analyse hydrogen peroxide, dissolved hydrogen, and oxygen. The sensor principle is based on the ability to measure the three analytes intended to be analysed on the same electrode by changing the measurement protocol and applied potentials. For this, chronoamperometric measurements were made using platinum electrodes. Hydrogen peroxide, dissolved hydrogen, and dissolved oxygen reactants, which it was desired to monitor with this multisensor, were analysed at high concentrations. One of the novelties of this study is integrating an electrochemical cell into a high-pressure stainless steel microreactor environment for the first time in the literature. Oxygen and hydrogen were measured at pressures reaching 70 bar and dissolved concentrations up to 52 mM and 40 mM, respectively. The sensitivities measured were 26 μA/cm²/mM for oxygen and 356 μA/cm²/mM for hydrogen (Urban et al, 2018).

A polymeric microneedle-associated electrochemical sensor array (MNESA) was produced by Zheng et al. in 2023. Then, real-time analyses were performed to detect kidney biomarkers in the skin interstitial fluid (ISF) with this sensor. Detection of these biomarkers is crucial to prevent kidney failure and for early diagnosis of other critical kidney diseases. In this study, four different biomarkers were determined simultaneously, thus preventing false positivity. The biomarkers chosen for analysis were phosphate, uric acid, creatinine, and urea. The microneedles used in the sensor were obtained from a polymeric matrix consisting of cross-linked methacrylated hyaluronic acid. The amperometric and potentiometric responses of MNESA to the analysed biomarkers were "phosphate 0.3–1.8 mM, 3.62 μA/mM; uric acid: 50–550 μM, 4.19 nA/μM; creatine: 50–550 μM, 12.58 nA/μM; urea: 1–16 mM, 44.6 mV/decade." It should be noted that the method used is minimally invasive and self-administered (Zheng et al, 2023).

In 2017, Kongsuphol et al. developed the Miniature Electrophoresis Electrochemical Protein Sensor (MEEPS) platform. Thanks to this platform, analyses of samples with a volume of up to 10 μL can be detected thanks to multiplexed signals. This platform has also been shown to reduce the impact of parasites. The electrophoresis unit that is used increases the multiplexing capacity of the sensor and reduces the background noise effect. The sensor shows multiplexed detections of four cytokines: Interleukin (IL)-6, interferon (IFN)-, tumour necrosis factor (TNF-α)- , and IL-8. The MEEPS plays the roles of both a miniaturized sensor and a multisensor, as it has compact dimensions and can analyse multiple biomarkers simultaneously (Kongsuphol et al, 2017).

In 2024, Gu et al. produced an electrochemical sensor to detect Alzheimer's disease (AD). The features of this sensor are that it is fast, portable, and multiplex. As is known, one of the reliable biomarkers used to detect AD is β-amyloid protein (Aβ). However, it is also known that Fetuin B in the body increases in direct proportion and is associated with the acceleration of AD. Simultaneous determination of two biomarkers associated with this disease on the same platform is crucial for medical diagnosis. The paper-based electrode for this multiplex sensor was prepared via a vacuum filtration method. SWCNTs and AuNPs were modified to complete the electrode fabrication. Real serum and brain tissue samples from mice with AD disease were used to test the sensor. The SWV method was used to prove that this sensor detects analytes in biological samples. The LOD values of the biomarkers were determined as 0.005 ng/mL for Aβo and 0.02 ng/mL for Fetuin B (Gu et al, 2024).

1.6 CONCLUSION

Biochemical sensors are becoming increasingly important in medical, pharmacological, nutritional, and environmental applications, as well as in biotechnology, i.e., the determination of biological samples or biochemical processes. This chapter discusses the modifications that can be made to an electrochemical sensor depending on the working electrode and the conventional and miniaturized design of the chemical sensor, using examples from the literature. As modifications, nanomaterials, polymers, and bioreceptors can be used to analyse many biological and chemical substances. The compatibility of bioreceptors with this design is also a highly innovative approach to diagnosing many diseases, especially cancer. Using hybrid structures, it is also possible to replace substances such as enzymes with nanomaterials or polymers, which are less expensive and require fewer processing steps. Continued work on the design of sensors and electrodes shows potential for the future in several applications.

REFERENCES

Aghamirza Moghim Aliabadi, Hooman, Reza Eivazzadeh-Keihan, Arezoo Beig Parikhani, Sara Fattahi Mehraban, Ali Maleki, Sepideh Fereshteh, Masoume Bazaz, et al. 'COVID-19: A Systematic Review and Update on Prevention, Diagnosis, and Treatment'. *MedComm* 3, no. 1 (2022): 1–42. https://doi.org/10.1002/mco2.115.

Alhamoud, Yasmin, Danting Yang, Samuel Selorm Fiati Kenston, Guozhen Liu, Linyang Liu, Haibo Zhou, Fatma Ahmed, and Jinshun Zhao. 'Advances in Biosensors for the Detection of Ochratoxin A: Bio-Receptors, Nanomaterials, and Their Applications'. *Biosensors and Bioelectronics* 141 (March 2019): 111418. https://doi.org/10.1016/j.bios.2019.111418.

Ayyandurai, Nagarajan, Sethuraman Venkatesan, and Sasikumar Raman. 'A Sensitive Enzymatic Electrochemical Biosensor for Cholesterol Based on Cobalt Ferrite@Molybdenum Disulfide/Gold Nanoparticles'. *ACS Applied Bio Materials* (2024). https://doi.org/10.1021/acsabm.4c00412.

Balkourani, G., A. Brouzgou, M. Archonti, N. Papandrianos, S. Song, and P. Tsiakaras. 'Emerging Materials for the Electrochemical Detection of COVID-19'. *Journal of Electroanalytical Chemistry.* Elsevier B.V., 15 July 2021. https://doi.org/10.1016/j.jelechem.2021.115289.

Bansod, Baban Kumar, Tejinder Kumar, Ritula Thakur, Shakshi Rana, and Inderbir Singh. 'A Review on Various Electrochemical Techniques for Heavy Metal Ions Detection with Different Sensing Platforms'. *Biosensors and Bioelectronics.* Elsevier Ltd, 15 August 2017. https://doi.org/10.1016/j.bios.2017.03.031.

Baranwal, Jaya, Brajesh Barse, Gianluca Gatto, Gabriela Broncova, and Amit Kumar. 'Electrochemical Sensors and Their Applications: A Review'. *Chemosensors.* MDPI, 1 September 2022. https://doi.org/10.3390/chemosensors10090363.

Barhoum, Ahmed, Selma Hamimed, Hamda Slimi, Amina Othmani, Fatehy M. Abdel-Haleem, and Mikhael Bechelany. 'Modern Designs of Electrochemical Sensor Platforms for Environmental Analyses: Principles, Nanofabrication Opportunities, and Challenges'. *Trends in Environmental Analytical Chemistry* 38 (October 2022) (2023): e00199. https://doi.org/10.1016/j.teac.2023.e00199.

Boo, Hankil, Ran A. Jeong, Sejin Park, Keun Soo Kim, Kay Hyeok An, Young Hee Lee, Ji Hyung Han, Hee Chan Kim, and Taek Dong Chung. 'Electrochemical Nanoneedle Biosensor Based on Multiwall Carbon Nanotube'. *Analytical Chemistry* 78, no. 2 (2006): 617–20. https://doi.org/10.1021/ac0508595.

Chen, Mei, Changjun Hou, Danqun Huo, Mei Yang, and Huanbao Fa. 'An Ultrasensitive Electrochemical DNA Biosensor Based on a Copper Oxide Nanowires/Single-Walled Carbon Nanotubes Nanocomposite'. *Applied Surface Science* 364 (2016): 703–9. https://doi.org/10.1016/j.apsusc.2015.12.203.

Curulli, Antonella. 'Recent Advances in Electrochemical Sensing Strategies for Food Allergen Detection'. *Biosensors* 12, no. 7 (2022). https://doi.org/10.3390/bios12070503.

Dakshayini, B. S., Kakarla Raghava Reddy, Amit Mishra, Nagraj P. Shetti, Shweta J. Malode, Soumen Basu, S. Naveen, and Anjanapura V. Raghu. 'Role of Conducting Polymer and Metal Oxide-Based Hybrids for Applications in Ampereometric Sensors and Biosensors'. *Microchemical Journal* 147 (February 2019): 7–24. https://doi.org/10.1016/j.microc.2019.02.061.

Dincer, Can, Richard Bruch, Estefanía Costa-Rama, Maria Teresa Fernández-Abedul, Arben Merkoçi, Andreas Manz, Gerald Anton Urban, and Firat Güder. 'Disposable Sensors in Diagnostics, Food, and Environmental Monitoring'. *Advanced Materials*. Wiley-VCH Verlag, 26 July 2019. https://doi.org/10.1002/adma.201806739.

Economou, Anastasios. 'Screen-Printed Electrodes Modified with "Green" Metals for Electrochemical Stripping Analysis of Toxic Elements'. *Sensors (Switzerland)* 18, no. 4 (2018): 1–23. https://doi.org/10.3390/s18041032.

Eivazzadeh-Keihan, Reza, Ehsan Bahojb Noruzi, Elham Chidar, Mahdokht Jafari, Farahnaz Davoodi, Amir Kashtiaray, Mostafa Ghafori Gorab, et al. 'Applications of Carbon-Based Conductive Nanomaterials in Biosensors'. *Chemical Engineering Journal* 442, no. P1 (2022): 136183. https://doi.org/10.1016/j.cej.2022.136183.

Erdoğan, Niran Öykü, Bengi Uslu, and Gözde Aydoğdu Tığ. 'Development of an Electrochemical Biosensor Utilizing a Combined Aptamer and MIP Strategy for the Detection of the Food Allergen Lysozyme'. *Microchimica Acta* 190, no. 12 (2023). https://doi.org/10.1007/s00604-023-06054-w.

Evtugyn, G. A., A. V. Porfireva, and S. V. Belyakova. 'Electrochemical DNA Sensors for Drug Determination'. *Journal of Pharmaceutical and Biomedical Analysis* 221 (June 2022): 115058. https://doi.org/10.1016/j.jpba.2022.115058.

Fernandes-Junior, Wilson S., Luiz O. Orzari, Cristiane Kalinke, Juliano A. Bonacin, and Bruno C. Janegitz. 'A Miniaturized Additive-Manufactured Carbon Black/PLA Electrochemical Sensor for Pharmaceuticals Detection'. *Talanta* 275 (September 2023) (2024): 126154. https://doi.org/10.1016/j.talanta.2024.126154.

Ganesh, Pattan Siddappa, Saheed Eluwale Elugoke, Seok Han Lee, Sang Youn Kim, and Eno E. Ebenso. 'Smart and Emerging Point of Care Electrochemical Sensors Based on Nanomaterials for SARS-CoV-2 Virus Detection: Towards Designing a Future Rapid Diagnostic Tool'. *Chemosphere* 352 (November 2023) (2024): 141269. https://doi.org/10.1016/j.chemosphere.2024.141269.

Gooding, Justin J., Freya Mearns, Wenrong Yang, and Jingquan Liu. 'Self-Assembled Monolayers into the 21st Century: Recent Advances and Applications'. *Electroanalysis* 15, no. 2 (2003): 81–96. https://doi.org/10.1002/elan.200390017.

Grieshaber, Dorothee, Robert MacKenzie, Janos Vörös, and Erik Reimhult. 'Electrochemical Biosensors - Sensor Principles and Architectures'. *Sensors* 8, no. 3 (2008): 1400–1458. https://doi.org/10.3390/s8031400.

Gu, Jinyu, Liming Wang, Li Zhao, Yingchun Zuo, Shuai Gao, Hui Gu, Yinan Wang, and Yanyan Yu. 'Multiplex Paper-Based Electrochemical Immunosensor for the Simultaneous Monitoring of Blood Biomarkers in Alzheimer's Disease'. *Sensors and Actuators B: Chemical* 406 (February 2024): 135456. https://doi.org/10.1016/j.snb.2024.135456.

Guo, Shaojun, and Shaojun Dong. 'Biomolecule-Nanoparticle Hybrids for Electrochemical Biosensors'. *TrAC - Trends in Analytical Chemistry* 28, no. 1 (2009): 96–109. https://doi.org/10.1016/j.trac.2008.10.014.

Habibi, Biuck, and Mojtaba Jahanbakhshi. 'Silver Nanoparticles/Multi Walled Carbon Nanotubes Nanocomposite Modified Electrode: Voltammetric Determination of Clonazepam'. *Electrochimica Acta* 118 (2014): 10–17. https://doi.org/10.1016/j.electacta.2013.11.169.

Hassan, Mohamed H., Reem Khan, and Silvana Andreescu. 'Advances in Electrochemical Detection Methods for Measuring Contaminants of Emerging Concerns'. *Electrochemical Science Advances* 2, no. 6 (2022): 1–27. https://doi.org/10.1002/elsa.202100184.

Kaliyaraj Selva Kumar, Archana, Yifei Zhang, Danlei Li, and Richard G. Compton. 'A Mini-Review: How Reliable Is the Drop Casting Technique?' *Electrochemistry Communications* 121 (September 2020): 106867. https://doi.org/10.1016/j.elecom.2020.106867.

Kongsuphol, Patthara, Gary C. F. Lee, Sunil K. Arya, Su Yin Chiam, and Mi Kyoung Park. 'Miniaturized Electrophoresis Electrochemical Protein Sensor (MEEPS) for Multiplexed Protein Detections'. *Sensors and Actuators, B: Chemical* 244 (2017): 823–30. https://doi.org/10.1016/j.snb.2017.01.026.

Lee, Sher Ling, and Chi Jung Chang. 'Recent Developments about Conductive Polymer Based Composite Photocatalysts'. *Polymers* 11, no. 2 (2019). https://doi.org/10.3390/polym11020206.

Liu, Li, Sadat Anwar, Haizhen Ding, Mingsheng Xu, Qiang Yin, Yazhong Xiao, Xingyuan Yang, Manqing Yan, and Hong Bi. 'Electrochemical Sensor Based on F,N-Doped Carbon Dots Decorated Laccase for Detection of Catechol'. *Journal of Electroanalytical Chemistry* 840 (March 2019): 84–92. https://doi.org/10.1016/j.jelechem.2019.03.071.

Lochab, Amit, Shalini Baxi, Purnima Tiwari, Shourya Bardiya, and Reena Saxena. 'Electrochemical Sensors for the Determination of Antipyretic and Antibiotic Drugs in Environmental and Biological Samples'. *Microchemical Journal* 199 (September 2023) (2024): 109923. https://doi.org/10.1016/j.microc.2024.109923.

Luo, Ruiyu, Jinghua Li, Guoying Huang, Guiyin Li, Shibin Guo, and Yulin Yuan. 'Electrochemical Biosensor for the Detection of Low Density Lipoprotein Based on Gold Nanoparticles Mediated Bi-Enzymes Catalytic Silver Deposition Reaction'. *Microchemical Journal* 199 (October 2023) (2024): 109927. https://doi.org/10.1016/j.microc.2024.109927.

Manikandan, Ramalingam, Thenmozhi Rajarathinam, Sivaguru Jayaraman, Hyeon Geun Jang, Jang Hee Yoon, Jaewon Lee, Hyun jong Paik, and Seung Cheol Chang. 'Recent Advances in Miniaturized Electrochemical Analyzers for Hazardous Heavy Metal Sensing in Environmental Samples'. *Coordination Chemistry Reviews* 499 (October 2023) (2024): 215487. https://doi.org/10.1016/j.ccr.2023.215487.

Mittal, Sunil, Hardeep Kaur, Nandini Gautam, and Anil K. Mantha. 'Biosensors for Breast Cancer Diagnosis: A Review of Bioreceptors, Biotransducers and Signal Amplification Strategies'. *Biosensors and Bioelectronics* 88 (2017): 217–31. https://doi.org/10.1016/j.bios.2016.08.028.

Mohamad Nor, Noorhashimah, Nurul Hidayah Ramli, Hemalatha Poobalan, Kai Qi Tan, and Khairunisak Abdul Razak. 'Recent Advancement in Disposable Electrode Modified with Nanomaterials for Electrochemical Heavy Metal Sensors'. *Critical Reviews in Analytical Chemistry* 53, no. 2 (2023): 253–88. https://doi.org/10.1080/10408347.2021.1950521.

Mustafa, Haveen Ahmed Mustafa, and Dler Adil Jameel. 'Modeling and the Main Stages of Spin Coating Process: A Review'. *Journal of Applied Science and Technology Trends* 2, no. 02 (2021): 119–23. https://doi.org/10.38094/jastt203109.

Naveen, Malenahalli Halappa, Nanjanagudu Ganesh Gurudatt, and Yoon Bo Shim. 'Applications of Conducting Polymer Composites to Electrochemical Sensors: A Review'. *Applied Materials Today* 9 (2017): 419–33. https://doi.org/10.1016/j.apmt.2017.09.001.

Navrátil, Tomáš, Bogdan Yosypchuk, and Jiří Barek. 'A Multisensor for Electrochemical Sequential Autonomous Automatic Measurements'. *Chemia Analityczna* 54, no. 1 (2009): 3–17.

Nazlı Şimşek, Niran Oyku Erdoğan and Gozde Aydoğdu Tığ. 'Handbook of Nanobioelectrochemistry'. *Handbook of Nanobioelectrochemistry*, 2023, 541–68. https://doi.org/10.1007/978-981-19-9437-1.

Nguyen, Tran N. H., Xin Jin, James K. Nolan, Jian Xu, Khanh Vy H. Le, Stephanie Lam, Yi Wang, Muhammad A. Alam, and Hyowon Lee. 'Printable Nonenzymatic Glucose Biosensors Using Carbon Nanotube-PtNP Nanocomposites Modified with AuRu for Improved Selectivity'. *ACS Biomaterials Science and Engineering* 6, no. 9 (2020): 5315–25. https://doi.org/10.1021/acsbiomaterials.0c00647.

Nikolaev, Konstantin G., Yury E. Ermolenko, Andreas Offenhäusser, Sergey S. Ermakov, Yulia G. Mourzina, Sebastian Urban, Andreas Weltin, et al. 'Multisensor Systems by Electrochemical Nanowire Assembly for the Analysis of Aqueous Solutions'. *Sensors and Actuators, B: Chemical* 39, no. JUN (2018): 1–16. https://doi.org/10.1016/j.snb.2018.07.014.

Pauliukaite, Rasa, and Edita Voitechovič. *Multisensor Systems and Arrays for Medical Applications Employing Naturally-Occurring Compounds and Materials. Sensors (Switzerland).* Vol. 20, 2020. https://doi.org/10.3390/s20123551.

Petralia, Salvatore, Maria Eloisa Castagna, Emanuele Cappello, Fausto Puntoriero, Emanuela Trovato, Antonio Gagliano, and Sabrina Conoci. 'A Miniaturized Silicon Based Device for Nucleic Acids Electrochemical Detection'. *Sensing and Bio-Sensing Research* 6 (2015): 90–94. https://doi.org/10.1016/j.sbsr.2015.09.006.

Power, Aoife C, and Aoife Morrin. 'Electroanalytical Sensor Technology, in Electrochemistry'. *Electrochemistry*, 2013, 141–64.

Qi, Shaopeng, Bo Zhao, Heqing Tang, and Xiaoqing Jiang. 'Determination of Ascorbic Acid, Dopamine, and Uric Acid by a Novel Electrochemical Sensor Based on Pristine Graphene'. *Electrochimica Acta* 161 (2015): 395–402. https://doi.org/10.1016/j.electacta.2015.02.116.

Ramya, M., P. Senthil Kumar, Gayathri Rangasamy, V. Uma shankar, G. Rajesh, K. Nirmala, A. Saravanan, and Alagumalai Krishnapandi. 'A Recent Advancement on the Applications of Nanomaterials in Electrochemical Sensors and Biosensors'. *Chemosphere* 308 (1 December 2022). https://doi.org/10.1016/j.chemosphere.2022.136416.

Rezaei, Behzad, and Neda Irannejad. *Electrochemical Detection Techniques in Biosensor Applications. Electrochemical Biosensors.* Elsevier Inc., 2019. https://doi.org/10.1016/B978-0-12-816491-4.00002-4.

Roushani, Mahmoud, Somayeh Farokhi, and Zeinab Rahmati. 'Development of a Dual-Recognition Strategy for the Aflatoxin B1 Detection Based on a Hybrid of Aptamer-MIP Using a Cu2O NCs/GCE'. *Microchemical Journal* 178 (December 2021) (2022): 107328. https://doi.org/10.1016/j.microc.2022.107328.

Sandhyarani, N. *Surface Modification Methods for Electrochemical Biosensors. Electrochemical Biosensors.* Elsevier Inc., 2019. https://doi.org/10.1016/B978-0-12-816491-4.00003-6.

Saputra, Heru Agung. 'Electrochemical Sensors: Basic Principles, Engineering, and State of the Art'. *Monatshefte Fur Chemie* 154, no. 10 (2023): 1083–1100. https://doi.org/10.1007/s00706-023-03113-z.

Savk, Aysun, Buse Özdil, Buse Demirkan, Mehmet Salih Nas, Mehmet Harbi Calimli, Mehmet Hakkı Alma, Inamuddin, Abdullah M. Asiri, and Fatih Şen. 'Multiwalled Carbon Nanotube-Based Nanosensor for Ultrasensitive Detection of Uric Acid, Dopamine, and Ascorbic Acid'. *Materials Science and Engineering C* 99 (January 2019): 248–54. https://doi.org/10.1016/j.msec.2019.01.113.

Sengupta, J., & Hussain, C. M. Decadal journey of CNT-based analytical biosensing platforms in the detection of human viruses. *Nanomaterials, 12*(23) (2022), 4132. https://doi.org/10.3390/nano12234132

Sharafeldin, Mohamed, and James F. Rusling. 'Multiplexed Electrochemical Assays for Clinical Applications'. *Current Opinion in Electrochemistry* 39 (2023): 101256. https://doi.org/10.1016/j.coelec.2023.101256.

Sharma, Pradakshina, Mohd Rahil Hasan, Ubaid Mushtaq Naikoo, Shaheen Khatoon, Roberto Pilloton, and Jagriti Narang. 'Aptamer Based on Silver Nanoparticle-Modified Flexible Carbon Ink Printed Electrode for the Electrochemical Detection of Chikungunya Virus', 2024.

Singh, Pranveer. *Electrochemical Biosensors: Biomonitoring of Food Adulterants, Allergens, and Pathogens. Electrochemical Biosensors*, 2022. https://doi.org/10.1016/b978-0-323-90632-6.00001-7.

Singh, Rimmy, Ruchi Gupta, Deepak Bansal, Rachna Bhateria, and Mona Sharma. 'A Review on Recent Trends and Future Developments in Electrochemical Sensing'. *ACS Omega*, 2023. https://doi.org/10.1021/acsomega.3c08060.

Tang, Xiaoning, and Xiong Yan. 'Dip-Coating for Fibrous Materials: Mechanism, Methods and Applications'. *Journal of Sol-Gel Science and Technology* 81, no. 2 (2017): 378–404. https://doi.org/10.1007/s10971-016-4197-7.

Thamilselvan, Annadurai, and Moon Il Kim. 'Recent Advances on Nanozyme-Based Electrochemical Biosensors for Cancer Biomarker Detection'. *TrAC - Trends in Analytical Chemistry* 177 (June 2024): 117815. https://doi.org/10.1016/j.trac.2024.117815.

Thirumalraj, Balamurugan, Rajalakshmi Sakthivel, Shen Ming Chen, Chellakannu Rajkumar, Lin kuan Yu, and Subbiramaniyan Kubendhiran. 'A Reliable Electrochemical Sensor for Determination of H2O2 in Biological Samples Using Platinum Nanoparticles Supported Graphite/Gelatin Hydrogel'. *Microchemical Journal* 146 (January 2019): 673–78. https://doi.org/10.1016/j.microc.2019.01.065.

Turner, Anthony P. F. 'Biosensors: Sense and Sensibility'. *Chemical Society Reviews* 42, no. 8 (2013): 3184–96. https://doi.org/10.1039/c3cs35528d.

Urban, Sebastian, Andreas Weltin, Hubert Flamm, Jochen Kieninger, Benedikt J. Deschner, Manfred Kraut, Roland Dittmeyer, and Gerald A. Urban. 'Electrochemical Multisensor System for Monitoring Hydrogen Peroxide, Hydrogen and Oxygen in Direct Synthesis Microreactors'. *Sensors and Actuators, B: Chemical* 273 (June 2018): 973–82. https://doi.org/10.1016/j.snb.2018.07.014.

Walsh, Frank C., Luis F. Arenas, and Carlos Ponce de León. 'Developments in Electrode Design: Structure, Decoration and Applications of Electrodes for Electrochemical Technology'. *Journal of Chemical Technology and Biotechnology.* John Wiley and Sons Ltd, 1 November 2018. https://doi.org/10.1002/jctb.5706.

Wang, Wan, Yanling Fu, Qing Lv, Hua Bai, Haiyu Li, Zhijuan Wang, and Qing Zhang. 'Miniaturized Device with a Detachable Three-Electrode System and Vibration Motor for Electrochemical Analysis Based on Disposable Electrodes'. *Sensors and Actuators B: Chemical* 297 (15 October 2019): 126719. https://doi.org/10.1016/J.SNB.2019.126719.

Wang, Zong Hua, Qiong Lin Liang, Yi Ming Wang, and Guo An Luo. 'Carbon Nanotube-Intercalated Graphite Electrodes for Simultaneous Determination of Dopamine and Serotonin in the Presence of Ascorbic Acid'. *Journal of Electroanalytical Chemistry* 540 (2003): 129–34. https://doi.org/10.1016/S0022-0728(02)01300-1.

Wehmeyer, Kenneth R., Ryan J. White, Peter T. Kissinger, and William R. Heineman. 'Electrochemical Affinity Assays/Sensors: Brief History and Current Status'. *Annual Review of Analytical Chemistry* 14 (2021): 109–31. https://doi.org/10.1146/annurev-anchem-061417-125655.

Willner, Itamar, Bernhard Basnar, and Bilha Willner. 'Nanoparticle-Enzyme Hybrid Systems for Nanobiotechnology'. *FEBS Journal* 274, no. 2 (2007): 302–9. https://doi.org/10.1111/j.1742-4658.2006.05602.x.

Yang, Hui, Zhiqiang Wei, Ting Li, Liuyang Cheng, Yuanhui Li, Xue Zhang, Wenke Li, and Yile Hu. 'Simultaneous Determination of Dopamine and Tyrosine Using Poly-Glycine/Nafion/MWCNTs Functionalized Plastic-Based Miniature Electrochemical Platform'. *International Journal of Electrochemical Science* 14, no. 10 (1 October 2019): 9584–95. https://doi.org/10.20964/2019.10.18.

Zheng, Lewen, Dandan Zhu, Yi Xiao, Xinting Zheng, and Peng Chen. 'Microneedle Coupled Epidermal Sensor for Multiplexed Electrochemical Detection of Kidney Disease Biomarkers'. *Biosensors and Bioelectronics* 237 (July 2023): 115506. https://doi.org/10.1016/j.bios.2023.115506.

Zhong, Jiamiao, Hongli Zhao, Yaxin Cheng, Tong Feng, Minbo Lan, and Shaohua Zuo. 'A High-Performance Electrochemical Sensor for the Determination of Pb(II) Based on Conductive Dopamine Polymer Doped Polypyrrole Hydrogel'. *Journal of Electroanalytical Chemistry* 902 (October 2021): 115815. https://doi.org/10.1016/j.jelechem.2021.115815.

Zuo, Zesen, Hao Zhang, Sanshuang Gao, Chunqiong Wang, Wen-Tong Chen, and Guangzhi Hu. 'Recent Progress in Porous Organic Frameworks for Electrochemical Sensing of Environmental Pollutants'. *Journal of Environmental Chemical Engineering* 12, no. 5 (2024): 113263. https://doi.org/10.1016/j.jece.2024.113263.

CHAPTER 2

How Nanomaterials Facilitated Electrochemical Sensor Research

Gangadhar Bagihalli and Bojappa Ginni

2.1 INTRODUCTION

Electrochemical sensors have become vital instruments in different types of industries, including environmental monitoring, healthcare diagnostics, and food safety. Their use is based on their capacity to transfer chemical information into an electrical signal suitable for measurement and analysis. The work of the sensors, as measured by their selectivity and sensitivity, has greatly improved with the advent of nanoparticles, which have given sensor technology new characteristics and capabilities. The term "nanoparticles" refers to particles with dimensions between 1 and 100 nm on the nanoscale, and also, these are classified based on their dimensions. Figure 2.1 represents the classification of nanomaterials based on their dimensions [1]. These materials exhibit unique chemical and physical characteristics not found in bulk materials. High surface-to-volume ratio, exceptional electrical conductivity, and excellent catalytic activity are some of these characteristics. Because of these characteristics, nanoparticles are particularly applicable for use in electrochemical sensors [2–4],

DOI: 10.1201/9781003512639-2

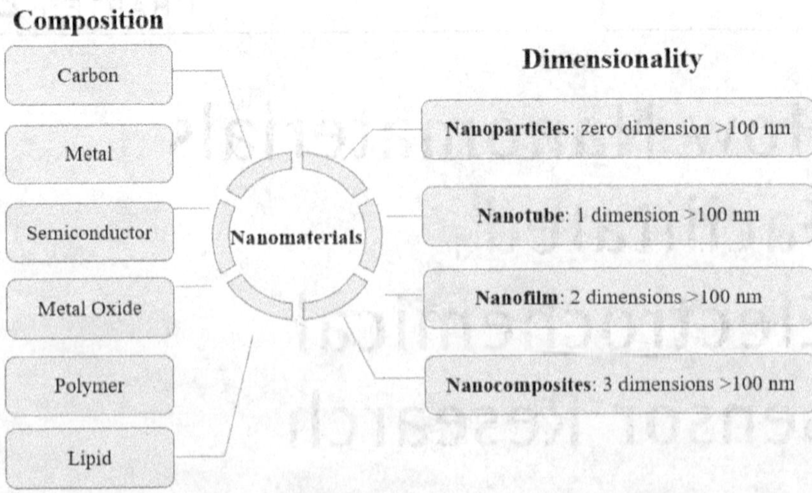

FIGURE 2.1 Classifications of nanomaterials based on dimensionality. (From Ji, Z. et al., *Nanomaterials*, 11, 6, 1599, 2021.)

where they can greatly increase the sensor performance [5]. The increased surface area provided by nanoparticles is one of their most significant contributions to electrochemical sensors. The increased surface area results in more active sites for analyte interaction, which increases sensor sensitivity. For instance, carbon nanotubes (CNTs) and graphene are widely known for their exceptional electrical properties and enormous surface areas, which make them ideal for different types of sensing applications [6, 7]. High-performance electrochemical sensors require efficient electron transport, which is made possible by the conductive network these materials will provide.

In addition to their large surface area and electrical conductivity, nanoparticles have catalytic characteristics that can increase the kinetics of electrochemical reactions. Metal nanoparticles such as gold, platinum, and palladium are widely utilized to improve sensor catalytic activity. These nanoparticles increase reaction speeds, improving the sensitivity and responsiveness of sensors. Platinum nanoparticles, for example, are commonly used in glucose sensors, allowing rapid and precise glucose detection, an important component of diabetes treatment [8]. Because of their versatility, nanoparticles can be utilized in a wide range of electrochemical sensors, including chemical, environmental, and biosensors. A standard application of gold nanoparticles is in the creation of biosensors that can identify biomolecules like DNA and proteins. Moreover, graphene-based sensors are employed in environmental monitoring to identify pollutants and dangerous compounds, giving real-time information on the quality of the air and water [9, 10].

Although the use of nanomaterials has greatly improved electrochemical sensor technology, there are still certain challenges. Before the full potential of nanomaterials can be realized, problems with reproducibility, stability, and potential toxicity must be solved. To get beyond these obstacles and ensure the secure and efficient application of nanomaterial-enhanced sensors, ongoing research into nanoparticle functionalization and manufacture is necessary.

2.2 UNIQUE PROPERTIES OF NANOMATERIALS

While the incorporation of nanomaterials has greatly advanced electrochemical sensor technology, some obstacles remain. Problems with nanomaterials' repeatability, stability, and potential toxicity must be solved before their full promise can be realized. Ongoing research into nanomaterial production and functionalization is critical to overcoming these problems and guaranteeing the safe and effective use of nanomaterial-enhanced sensors. Figure 2.2 illustrates the unique **properties of** nanomaterials.

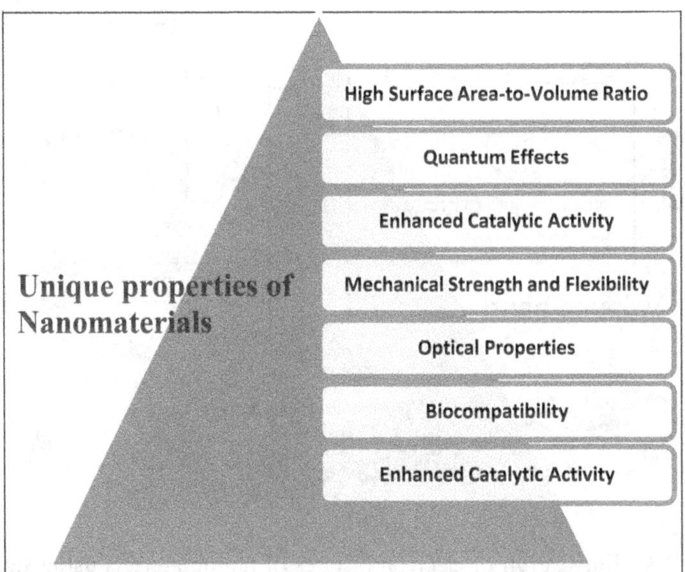

FIGURE 2.2 Unique properties of nanomaterials.

2.2.1 High Surface Area-to-Volume Ratio

One of the most significant characteristics of nanoparticles is their extraordinarily high surface area-to-volume ratio. When a material's dimensions become closer to the nanoscale, its surface area grows dramatically in relation to its volume. This increased surface area increases the number of active sites for chemical reactions, improving the sensitivity and efficacy of sensors.

Implication for Sensors: Increased interactions with target analytes are made possible by the greater surface area, which reduces the detection limits. For example, increasing the electron transfer processes in nanomaterials like graphene and CNTs may improve the efficiency of electrochemical sensors[11, 12].

2.2.2 Quantum Effects

Nanoscale materials frequently show quantum effects that can suddenly change their electrical and optical properties. These effects result in the confinement of charge carriers and the fixed or specific energy levels that come with nanoscale size (Figure 2.3) [13]. Quantum dots (QD) , for

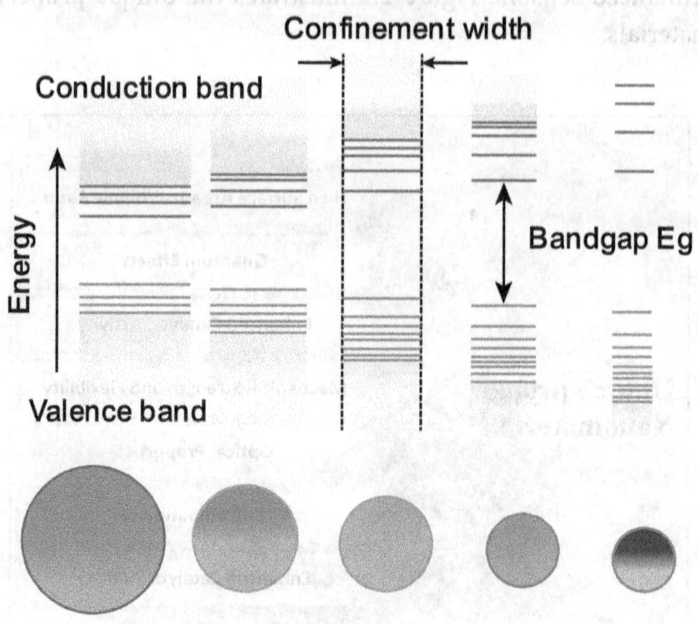

FIGURE 2.3 Illustration of electronic states of nanomaterials using size quantization effect. (From Serena, B. and Donega, C.d.M., *ACS Materials Au*, 2, 3, 237–249, 2022.)

example, exhibit size-tuneable photoluminescence, which can be used for specialized sensing purposes.

Implication for Sensors: QDs' adjustable features enable the development of highly sensitive sensors capable of detecting lower amounts of chemicals via fluorescence or electrochemical signals [14].

2.2.3 Enhanced Catalytic Activity

Nanomaterials typically show enhanced catalytic activity (Figure 2.4) compared with their bulk forms due to their high surface energy and number of active sites. This enhancement can be attributed to several factors, including reduced diffusion distances and increased reactivity at the nanoscale [15].

Implication for Sensors: The adjustable features of QDs enable the fabrication of highly sensitive sensors that can detect low amounts of chemicals by increasing the number of active sites [16].

2.2.4 Mechanical Strength and Flexibility

Nanomaterials regularly exhibit extraordinary mechanical properties, such as great strength and flexibility. For example, carbon nanotubes have much higher tensile strengths than steel while being lightweight.

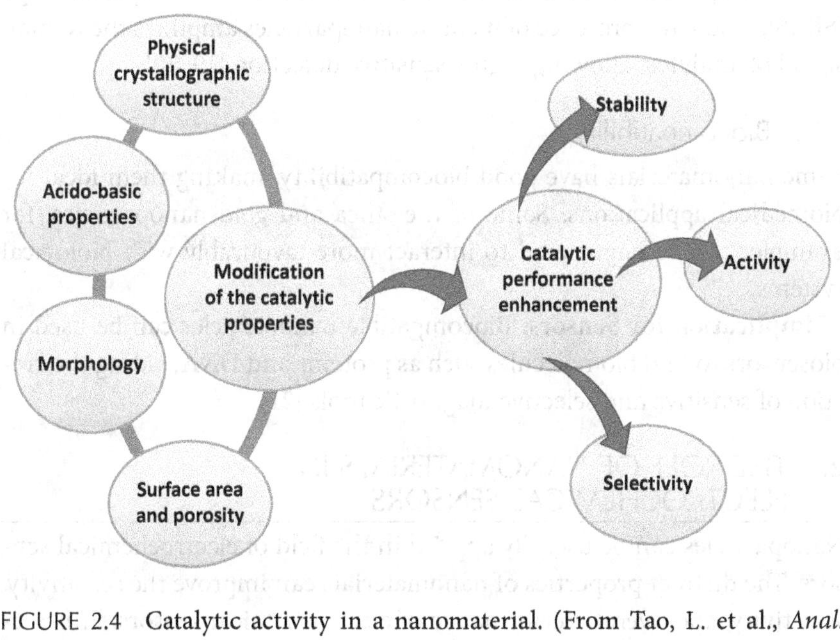

FIGURE 2.4 Catalytic activity in a nanomaterial. (From Tao, L. et al., *Anal. Chim. Acta*, 1207, 339461, 2022.)

Implication for Sensors: Nanomaterials' mechanical strength allows the creation of durable and flexible sensors that may be combined into many substrates, including wearable devices [17].

2.2.5 Conductivity and Electrode Performance

Nanomaterials can have unique electrical properties, such as high conductivity and low resistivity. Graphene and metallic nanowires have excellent conductivity because of their structure and electron mobility.

Implication for Sensors: Nanomaterials' high conductivity allows fast electron transport in electrochemical sensors, which improves sensitivity and response times. For instance, incorporating graphene into electrode materials has considerably improved the performance of electrochemical biosensors [18].

2.2.6 Optical Properties

Many nanomaterials have distinct optical features, including surface plasmon resonance (SPR) in metallic nanoparticles and increased fluorescence in QDs. These qualities can be adjusted depending on particle size, shape, and composition.

Implication for Sensors: Nanomaterials' optical properties can be used in sensor applications, including surface-enhanced Raman spectroscopy (SERS), where the presence of metallic nanoparticles amplifies the Raman signal of analytes, allowing highly sensitive detection [19, 20].

2.2.7 Biocompatibility

Some nanomaterials have good biocompatibility, making them ideal for biomedical applications. Some of the silica and gold nanoparticles, for example, can be engineered to interact more favourably with biological systems.

Implication for Sensors: Biocompatible nanoparticles can be used in biosensors to find biomolecules such as proteins and DNA, aiding the creation of sensitive and selective diagnostic tools [21].

2.3 THE ROLE OF NANOMATERIALS IN ELECTROCHEMICAL SENSORS

Nanoparticles can be usefully applied in the field of electrochemical sensors. The distinct properties of nanomaterials can improve the sensitivity, selectivity, and overall performance of electrochemical sensors. This section goes over the various roles that nanomaterials play in these sensors, as well as how they contribute to sensor technology advancements.

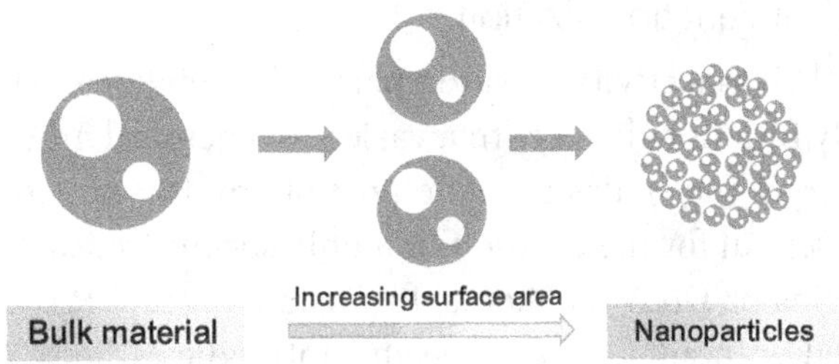

FIGURE 2.5 Enhanced surface area. (From Geim, A. and Novoselov, K., *Nat. Mater.*, 6, 183–191, 2007.)

2.3.1 Enhanced Surface Area

One of the key benefits of using nanoparticles in electrochemical sensors is their large surface-to-volume ratio. Figure 2.5 represents the increase in the surface area of nanoparticles [22]. Nanomaterials like nanoparticles, nanosheets, and nanowires have significantly larger surface area than their bulk counterparts. This expanded surface area provides more active sites for analyte interaction, which improves sensor sensitivity. For example, gold nanoparticles (AuNPs) have been frequently used to modify electrodes, offering a significant increase in sensor performance due to their huge surface area and excellent biocompatibility [23].

2.3.2 Improved Electrical Conductivity

The electrical conductivity of nanomaterials is critical for efficient electron transport between the analyte and the electrode. Carbon-based nanomaterials, like graphene and CNTs, are known for their high electrical conductivity. Graphene's single-layer carbon structure provides good electron mobility, which improves signal responsiveness and lowers sensor detection limits. Similarly, CNTs provide a conductive network that promotes fast electron transfer, making them suitable for many electrochemical sensing applications [24].

2.3.3 Catalytic Properties

Almost all nanomaterials have separate or different catalytic characteristics that enable electrochemical processes. Transition metal nanoparticles, such as platinum and palladium, are good for their catalytic activity, which can greatly increase redox reaction kinetics in sensors. For example, Pt nanoparticles are used in glucose sensors due to their outstanding catalytic characteristics, which allow rapid and sensitive glucose detection. These nanomaterials' catalytic activity not only increases reaction rates but also improves the sensor sensitivity and selectivity [25].

2.3.4 Chemical Functionalization

To increase their selectivity towards particular analytes, nanomaterials can be readily functionalized with a variety of chemical groups. Selective binding of target molecules is made possible by this functionalization, and this is essential for the creation of highly specific sensors. To identify specific proteins or DNA sequences, for instance, graphene oxide can be functionalized with aptamers or specific antibodies. The sensitivity and selectivity of sensors can be greatly increased by controlling the surface chemistry of nanomaterials [26].

2.3.5 Versatility and Multifunctionality

Nanomaterials can be used for a wide range of sensing applications due to their structure, functionalization, and composition. They are suitable for use as conductometric, potentiometric, and amperometric electrochemical sensors in different types of applications. Because of their good characteristics, such as versatility, multifunctional sensors capable of detecting several analytes at once have been developed. Carbon-based sensors have been developed to detect glucose and hydrogen peroxide simultaneously, demonstrating the versatility of nanomaterials in sensor applications [27].

2.3.6 Enhanced Stability and Durability

The long-term performance and dependability of electrochemical sensors depend heavily on their stability and durability. Metal oxides and carbon-based materials are examples of nanomaterials that provide improved stability in a range of environmental circumstances. Zinc oxide and titanium dioxide nanoparticles are well suited for application in environmental sensors because of their good chemical stability and resilience to adverse environments; when these stable nanomaterials are added to sensors, long-term performance consistency is guaranteed [28].

2.4 TYPES OF NANOMATERIALS USED IN ELECTROCHEMICAL SENSORS

Due to the good features of nanomaterials, they has proved indispensable in the creation of compactable electrochemical sensors. Metal nanoparticles, carbon-based nanomaterials, and metal oxide nanoparticles have all been investigated for their ability to increase the performance of sensors. Figure 2.6 illustrates the types of metal nanomaterials used in electrochemical sensors. This section discusses the many types of nanomaterials and their specialized applications in electrochemical sensors.

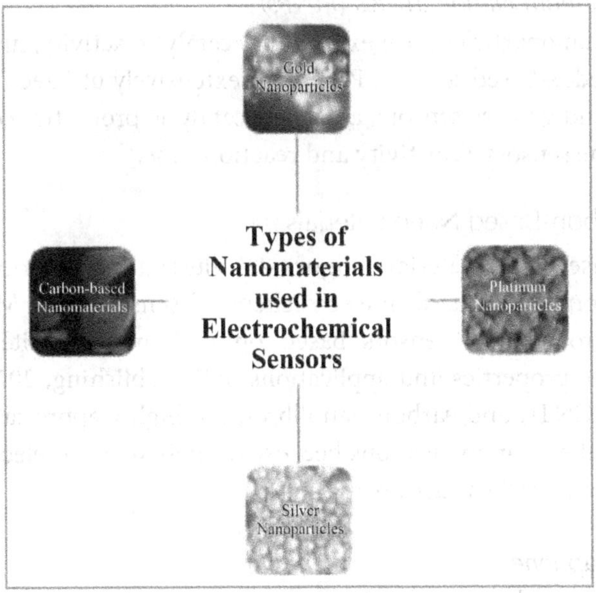

FIGURE 2.6 The types of metal nanomaterials used in electrochemical sensors.

2.4.1 Metal Nanoparticles

Metal nanoparticles, such as gold, silver, and platinum, are commonly utilized in electrochemical sensors due to their large surface area, superior electrical conductivity, and catalytic characteristics [29].

2.4.1.1 Gold Nanoparticles (AuNPs)

Gold nanoparticles are among the most widely utilized nanomaterials in electrochemical sensors. Their favourable surface-to-volume ratio and biocompatibility make them good for increasing sensor sensitivity and specificity. Gold nanoparticles are useful in biosensors to detect biomolecules like DNA, proteins, and enzymes. For example, AuNP-modified electrodes have been used to develop highly sensitive glucose sensors.

2.4.1.2 Silver Nanoparticles (AgNPs)

Silver nanoparticles are well known for their antibacterial capabilities and strong electrical conductivity. AgNPs are employed as sensors to detect infections and other biological organisms. They are also used in environmental sensors for the detection of heavy metals and contaminants because of their strong interaction with diverse analytes [30, 31].

2.4.1.3 *Platinum Nanoparticles (PtNPs)*

Platinum nanoparticles have exceptional catalytic activity, making them ideal for redox-based sensors. PtNPs are extensively utilized in hydrogen peroxide and glucose sensors, and their catalytic properties considerably improve the sensors' sensitivity and reaction time.

2.4.2 Carbon-Based Nanomaterials

Carbon-based nanomaterials (Figure 2.7 illustrates the various carbon-based nanomaterials used in electrochemical sensors from Manjunatha, J.G., Electrochemical sensors based on carbon composite materials: Fabrication, properties and applications, IOP Publishing, 2022), such as graphene, CNTs, and carbon nanofibres, are highly appreciated in electrochemical sensor applications because of their superior electrical characteristics and vast surface areas [32–35].

2.4.2.1 *Graphene*

Graphene, a single layer of carbon atoms organized in a two-dimensional lattice, has received a lot of attention for its high electrical conductivity, mechanical strength, and vast surface area. Graphene-based sensors are utilized for a wide range of applications, including the detection of

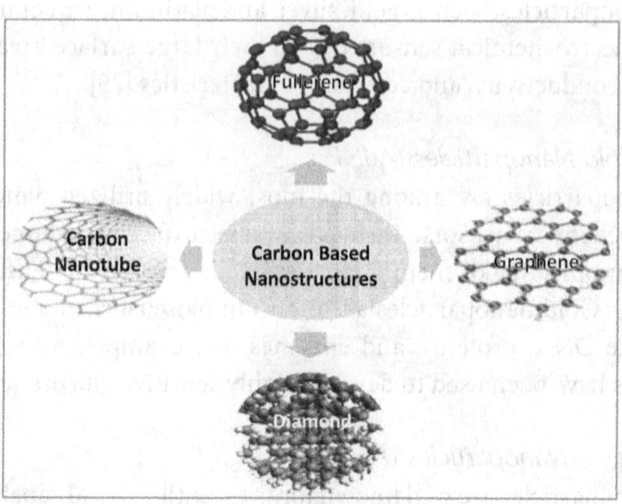

FIGURE 2.7 The various carbon-based nanomaterials used in electrochemical sensors. (From Manjunatha, J.G., *Electrochemical sensors based on carbon composite materials: Fabrication, properties and applications*, United Kingdom: IOP Publishing, 2022.)

glucose, heavy metals, and environmental contaminants. Graphene's high electron mobility increases sensitivity and reduces the detection limits of electrochemical sensors [36–39].

2.4.2.2 Carbon Nanotubes (CNTs)

CNTs, cylindrical nanostructures made by rolling graphene sheets, are one of the common methods for electrochemical sensors. Carbon nanotubes produce a conductive network that enables fast electron transfer, which is required for high-performance sensors. Biosensors, gas sensors, and chemical sensors apply them to detect different types of analytes, such as glucose, gases, and biomolecules [40].

2.4.2.3 Carbon Nanofibres (CNFs)

Carbon nanofibres resemble CNTs but have a stacked cup structure. They have a large surface area, stable electrical conductivity, and mechanical strength. CNFs are applied in electrochemical sensors to detect different types of analytes, such as heavy metals and organic chemicals. Their unusual structure enables efficient electron transport and excellent sensitivity [41, 42].

2.4.3 Metal Oxide Nanoparticles

Metal oxide nanoparticles, such as zinc oxide, titanium dioxide, and iron oxide, are commonly used in electrochemical sensors because of their semiconducting and catalytic capabilities.

2.4.3.1 Zinc Oxide (ZnO)

Zinc oxide nanoparticles are well-known for their superior chemical stability, high isoelectric point, and biocompatibility. ZnO sensors are used to detect gases, glucose, and other biomolecules. Their huge surface area and catalytic characteristics improve the sensitivity and reaction time of sensors.

2.4.3.2 Titanium Dioxide (TiO_2)

Titanium dioxide nanoparticles are valuable due to their photocatalytic capabilities and chemical stability. TiO_2 is utilized in environmental pollution sensors as well as biosensors to detect biomolecules. The capacity to create electron–hole pairs under UV light increases the sensitivity and selectivity of the sensors.

2.4.3.3 *Iron Oxide (Fe_3O_4)*

Iron oxide nanoparticles have magnetic characteristics and strong catalytic activity. Electrochemical sensors use Fe_3O_4 to detect heavy metals, organic chemicals, and biological organisms. Its magnetic qualities enable easy separation and reuse of the sensors, making it suitable for reusable sensor applications [43].

2.5 APPLICATIONS OF NANOMATERIAL-BASED ELECTROCHEMICAL SENSORS

Because of their increased sensitivity, selectivity, and quick response times, nanomaterial-based electrochemical sensors have transformed various fields. These well-characterized sensors use nanoparticles' unique features to achieve good detection performance over a wide variety of analytes. This part of the chapter reveals various uses of nanomaterial-based electrochemical sensors (Figure 2.8), focusing mainly on healthcare, environmental monitoring, food safety, and industrial process control.

2.5.1 Healthcare Diagnostics

The use of nanoparticles in electrochemical sensors has considerably improved healthcare diagnostics by enabling the detection of biomolecules

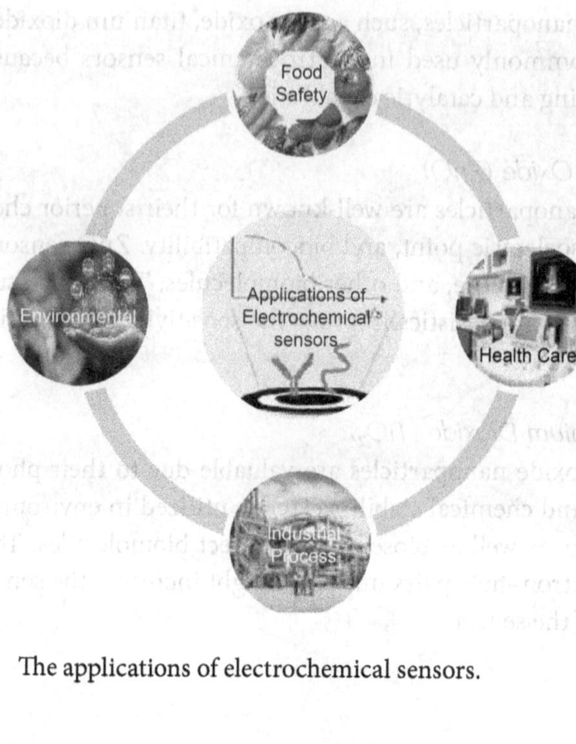

FIGURE 2.8 The applications of electrochemical sensors.

in lower concentrations, resulting in early diagnosis and management of illness.

2.5.1.1 Glucose Monitoring

One of the most useful applications is glucose monitoring for the management of diabetes. Nanomaterials like gold nanoparticles and CNTs increase the glucose sensor's efficiency by increased immobilization of enzyme and promoting efficient electron transfer. Sensors of this type provide fast, accurate, and continuous glucose monitoring, increasing the quality of life for diabetic patients.

2.5.1.2 Cancer Biomarker Detection

Nanomaterial-based electrochemical sensors are used to detect cancer biomarkers. For example, graphene oxide in combination with particular antibodies or aptamers can detect cancer-related proteins and nucleic acids, enabling early detection and treatment. These sensors have the ability to detect biomarkers at extremely low levels with great sensitivity and specificity [44–46].

2.5.2 Environmental Monitoring

Nanomaterial-based electrochemical sensors have been developed for environmental monitoring, since they can detect pollutants and harmful compounds in the air, water, and soil.

2.5.2.1 Heavy Metal Detection

Heavy metals, including lead, mercury, and cadmium, pose serious environmental and health dangers. Nanomaterials with a high affinity for metal ions, such as gold and silver nanoparticles, are utilized in sensors to detect these elements. These sensors provide rapid and accurate detection, which is critical for monitoring and regulating environmental contamination.

2.5.2.2 Gas Sensing

Ensuring food safety is an important application for nanomaterial-based electrochemical sensors. These sensors aid in the detection of pollutants, diseases, and chemical residues in food products [47–49].

2.5.3 Food Safety

Ensuring food safety is an important application for nanomaterial-based electrochemical sensors. These sensors aid in the detection of pollutants, diseases, and chemical residues in food products [50].

2.5.3.1 Pathogen Detection

Silver nanoparticles (AgNPs) and carbon-based nanomaterials are utilized in sensors to detect foodborne pathogens, including *Salmonella* and *Escherichia coli* (*E. coli*). The increased surface area and antibacterial capabilities of these nanomaterials boost the sensitivity and specificity of the sensors, allowing rapid identification and prevention of foodborne diseases [51].

2.5.3.2 Pesticide Residue Detection

Sensors that detect pesticide residues in agricultural goods use nanomaterials such as gold nanoparticles and CNTs. These sensors can detect even trace levels of pesticides, which ensures food safety and conformity with regulatory regulations [52, 53].

2.5.4 Industrial Process Control

Nanomaterial-based electrochemical sensors are used in industrial settings to control and monitor processes, ensuring efficiency, safety, and quality.

2.5.4.1 Process Monitoring

Graphene-based sensors are used in industrial processes to monitor parameters such as pH, temperature, and chemical concentrations. Sensors of this type give the correct data, control, and optimization of manufacturing processes.

2.5.4.2 Corrosion Detection

Metal oxide nanoparticles, including iron oxide, are used in sensors to detect oxidation and reduction reactions on metal surfaces of pipes and infrastructure. These sensors contribute to the early identification and prevention of corrosion-related problems, ensuring the longevity and safety of industrial systems [54].

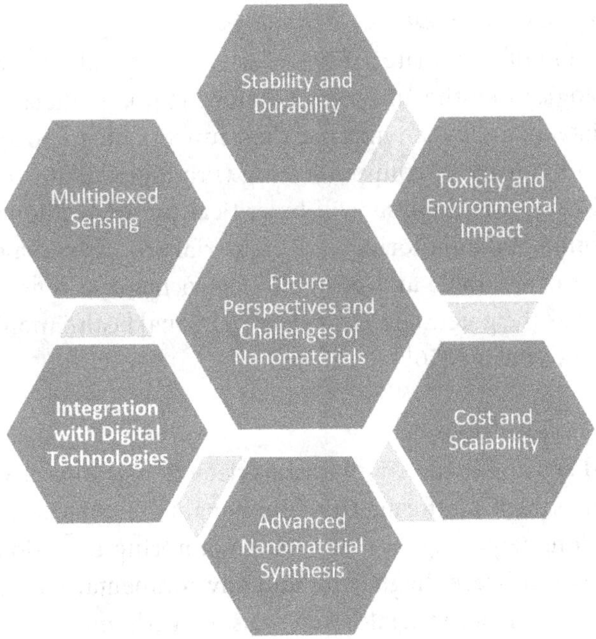

FIGURE 2.9 Future perspectives and challenges.

2.6 FUTURE PERSPECTIVES AND CHALLENGES

The use of nanomaterials in electrochemical sensors has substantially advanced the field, resulting in increased sensitivity, selectivity, and response times. Despite these advances, there remain significant difficulties and prospects for future improvement. This section examines the future prospects and challenges (Figure 2.9) for nanomaterial-based electrochemical sensors.

2.6.1 Future Perspectives

2.6.1.1 Advanced Nanomaterial Synthesis

The production of nanomaterials with exact size, shape, and surface characteristics is critical for building high-performance electrochemical sensors. Future research will most likely focus on creating novel synthesis procedures to create nanomaterials with specialized characteristics for specific sensing applications. For example, the development of hybrid nanomaterials that mix the capabilities of multiple nanomaterials could result in sensors with superior performance characteristics [55].

2.6.1.2 Integration with Digital Technologies

The integration of nanomaterial-based electrochemical sensors with digital technologies like the Internet of Things (IoT), artificial intelligence (AI), and data analytics can improve their functionality and applicability. Smart sensors capable of communicating with digital platforms and providing real-time data analysis will be critical for applications in healthcare, environmental monitoring, and industrial process management. The development of wearable and portable electrochemical sensors with IoT capabilities has the potential to transform personal health monitoring and environmental sensing [56].

2.6.1.3 Multiplexed Sensing

Multiplexed sensing is the simultaneous detection of numerous analytes with a single sensor platform. This functionality is especially significant for applications requiring simultaneous monitoring of various parameters, such as healthcare diagnostics and environmental monitoring. The development of nanomaterial-based sensors with multiplexed sensing capabilities will require breakthroughs in material design and sensor production techniques.

2.6.1.4 Point-of-Care Diagnostics

Nanomaterial-based electrochemical sensors have the potential to transform point-of-care diagnostics by enabling fast, accurate, and low-cost testing. Future research will concentrate on creating portable, user-friendly diagnostic equipment that may be employed in remote and resource-constrained environments. These gadgets will enable early disease identification and monitoring, hence improving healthcare outcomes [57].

2.6.2 Challenges

2.6.2.1 Reproducibility and Standardization

One of the most significant obstacles in the development and marketing of nanomaterial-based electrochemical sensors is ensuring reproducibility and standardization. The synthesis and production of nanomaterials can cause batch-to-batch differences that impair sensor performance. Standardized techniques for nanomaterial synthesis and sensor fabrication are critical to ensuring consistent and reliable sensor performance [58].

2.6.5.2 Stability and Durability

Nanomaterial-based sensors must be stable and durable over time in order to be useful. Nanomaterials can degrade or agglomerate over time, resulting in reduced sensor performance. Research is required to discover ways to enhance the stability of nanomaterials, such as surface functionalization or encapsulation, to ensure long-term sensor dependability [59].

2.6.5.3 Toxicity and Environmental Impact

The possible toxicity and environmental impact of nanoparticles are serious issues that must be addressed. The use of nanomaterials in sensors must be examined for human and environmental safety. Developing biocompatible and ecologically benign nanomaterials will be critical in addressing these challenges. Furthermore, efficient disposal and recycling procedures for nanomaterial-based sensors need to be created [60].

2.6.5.4 Cost and Scalability

The cost and scalability of nanomaterial-based sensor manufacture are critical to the broad adoption of these sensors. High-quality nanomaterial synthesis can be costly and time-consuming, limiting their commercial manufacture. Research should focus on creating cost-effective and scalable synthesis methods to make nanomaterial-based sensors more accessible for commercial applications [61, 62].

2.7 CONCLUSION

The integration of nanomaterials into electrochemical sensor research has significantly transformed the field, offering unprecedented enhancements in sensor performance. This chapter has explored the profound impact of nanomaterials, highlighting their unique properties and the resulting improvements in sensitivity, selectivity, and response times. Each of the diverse types of nanomaterials, including metal nanoparticles, carbon-based nanomaterials, and metal oxide nanoparticles, contributes uniquely to the advancement of electrochemical sensors. The applications of nanomaterial-based electrochemical sensors span various critical fields. In healthcare, these sensors enable early diagnosis and monitoring of diseases, significantly improving patient outcomes. Environmental monitoring has also benefited from these advanced sensors, providing rapid and accurate detection of pollutants and toxic substances, thereby safeguarding public health and the environment. In food safety, nanomaterial-based sensors ensure the detection of pathogens and chemical residues, protecting consumers from foodborne illnesses. Industrial process control has seen improvements in efficiency and safety through real-time monitoring and early detection of potential issues. Despite the remarkable progress, challenges remain. Ensuring reproducibility and standardization, enhancing stability and durability, addressing potential toxicity and environmental impact, and reducing costs for large-scale production are critical areas that require ongoing research and development. Future perspectives point towards advanced nanomaterial synthesis, integration with digital technologies, multiplexed sensing, and the development of point-of-care diagnostics. Overcoming these challenges will unlock the full potential of nanomaterial-based electrochemical sensors.

In conclusion, the field of electrochemical sensor research has been profoundly enhanced by the incorporation of nanomaterials. As research continues to address existing challenges and explore new possibilities, nanomaterial-based electrochemical sensors are poised to make even greater contributions to healthcare, environmental protection, food safety, and industrial processes. The future of these sensors is bright, with the potential to revolutionize various sectors through their advanced capabilities and transformative impact.

AUTHORS' CONTRIBUTIONS

Gangadhar Bagihalli and Bojappa Ginni wrote the draft of the chapter and were involved in taking permissions and corrections.

ACKNOWLEDGEMENTS

Both the authors are thankful to KLE Institute of Technology, Hubballi, Karnataka, India for constant support and encouragement.

CONFLICT OF INTEREST

Both authors declare that they do not have any conflict of interest.

FUNDING

Both authors declare that they have not received any funding from any institutions/organizations.

REFERENCES

1. Ji, Z., Guo, W., Sakkiah, S., Liu, J., Patterson, T. A., & Hong, H. (2021). Nanomaterial databases: Data sources for promoting design and risk assessment of nanomaterials. *Nanomaterials, 11*(6), 1599. https://doi.org/10.3390/nano11061599
2. Cui, Y., Wei, Q., Park, H., & Lieber, C. M. (2001). Nanowire nanosensors for highly sensitive and selective detection of biological and chemical species. *Science, 293*(5533), 1289–1292.
3. Raril, C., & Manjunatha, J. G. (2018). Sensitive and selective analysis of nigrosine dye at polymer modified electrochemical sensor.
4. Amrutha, B. M., Manjunatha, J. G., Aarti, S. B., & Pushpanjali, P. A. (2020). Fabrication of a sensitive and selective electrochemical sensing platform based on poly-l-leucine modified sensor for enhanced voltammetric determination of riboflavin. *Journal of Food Measurement and Characterization, 14*, 3633–3643.
5. Wang, J. (2005). Carbon-nanotube based electrochemical biosensors: A review. *Electroanalysis, 17*(1), 7–14.
6. Geim, A. K., & Novoselov, K. S. (2007). The rise of graphene. *Nature Materials, 6*(3), 183–191.
7. Lim, S. H., & Zhang, Y. (2007). Gold nanoparticle-based biosensors: A review. *Sensors and Actuators B: Chemical, 123*(2), 434–448.
8. Tang, L., et al. (2009). Graphene-based electrochemical sensors for bioanalysis. *Analytical Chemistry, 81*(16), 6641–6649.
9. Gao, Z., et al. (2012). Nanomaterials in electrochemical biosensors. *Nano Today, 7*(4), 390–410.
10. Parappurath, N., Kirubanandam, S., et al. (2018). Nanomaterials history, classification, unique properties, production and market. In *Micro and Nano Technologies, Emerging Applications of Nanoparticles and Architecture Nanostructures* (pp. 341–384). Elsevier. https://doi.org/10.1016/B978-0-323-51254-1.00012-9
11. Manhas, N., Lalita, S. K., & Vanayk, A. (2023). Photoluminescence and supercapacitive properties of carbon dots nanoparticles: A review. *Journal of Metastable and Nanocrystalline Materials, 37*, 1–22.
12. Michael, L. (2023). Physical concepts. In *Food Process Engineering Principles and Data* (pp. 37–46). Woodhead Publishing.
13. Serena, B., & Donega, C. d. M. (2022). Magic-size semiconductor nanostructures: Where does the magic come from? *ACS Materials Au, 2*(3), 237–249. https://doi.org/10.1021/acsmaterialsau.1c00075
14. Xie et al. (2018). Zunhua gabbro. *Acta Geologica Sinica, 92*, 1024–1040.
15. Tao, L., Xu, Y., et al. (2022). Flexible electrochemical sensors integrated with nanomaterials for in situ determination of small molecules in biological samples: A review. *Analytica Chimica Acta, 1207*, 339461. https://doi.org/10.1016/j.aca.2022.339461

16. Papa, F., Vasile, A., & Dobrescu, G. (2022). Effect of the modification of catalysts on the catalytic performance. *Catalysts, 12*(12), 1637. https://doi.org/10.3390/catal12121637
17. Jose, S., González, F., & Centeno, T. A. (2011). The role of the electric conductivity of carbons in the electrochemical capacitor performance. *Journal of Electroanalytical Chemistry, 657*(1–2), 176–180. https://doi.org/10.1016/j.jelechem.2011.03.025
18. Elham, A., Masoumeh, G., & Saeed, S. (2019). Electrochemical sensing based on carbon nanoparticles: A review. *Sensors and Actuators B: Chemical, 293*, 183–209. https://doi.org/10.1016/j.snb.2019.04.075
19. Panda, A. K., Keerthi, M., et al. (2022). Biocompatible electrochemical sensor based on platinum-nickel alloy nanoparticles for in situ monitoring of hydrogen sulfide in breast cancer cells. *Nanomaterials, 12*, 258.
20. Vinayak, A., & Shashanka, R. (Eds.). (2022). *Optical and sensor characteristics of nanocomposites.* Trans Tech Publications Limited.
21. Manyangadze, M., Chikuruwo, N., et al. (2020). Enhancing adsorption capacity of nano-adsorbents via surface modification: A review. *South African Journal of Chemical Engineering, 31*, 25–32. https://doi.org/10.1016/j.sajce.2019.11.003
22. Geim, A., & Novoselov, K. (2007). The rise of graphene. *Nature Materials, 6*, 183–191.
23. Sun, X., Zhang, Y., Chen, G., & Gai, Z. (2017). Application of nanoparticles in enhanced oil recovery: A critical review of recent progress. *Energies, 10*(345). https://doi.org/10.3390/en10030345
24. Białas, K., Moschou, D., Marken, F., et al. (2022). Electrochemical sensors based on metal nanoparticles with biocatalytic activity. *Microchimica Acta, 189*, 172.
25. Białas, K., & Moschou, D. (2022). Electrochemical sensors based on metal nanoparticles with biocatalytic activity. *Microchimica Acta, 189*(4), 172. https://doi.org/10.1007/s00604-022-05252-2
26. Ivanišević, I. (2023). The role of silver nanoparticles in electrochemical sensors for aquatic environmental analysis. *Sensors, 23*, 3692. https://doi.org/10.3390/s23073692
27. Curulli, A. (2020). Nanomaterials in electrochemical sensing area: Applications and challenges in food analysis. *Molecules, 25*(23), 5759. https://doi.org/10.3390/molecules25235759
28. Islam, T., Hasan, M. M., Awal, A., et al. (2020). Metal nanoparticles for electrochemical sensing: Progress and challenges in the clinical transition of point-of-care testing. *Molecules, 25*, 5787.
29. Manjunath, J. G. (2017). A new electrochemical sensor based on modified carbon nanotube-graphite mixture paste electrode for voltammetric determination of resorcinol. *Asian Journal of Pharmaceutical and Clinical Research, 10*(12), 295.
30. Monnappa, B., Manjunatha, J. G., et al. (2019). Electrochemical sensor for the determination of alizarin red-S at non-ionic surfactant modified carbon nanotube paste electrode. *Physical Chemistry Research, 7*(3), 523–533.

31. Vinayak, A., Manhas, N., & Nandi, S. (2024). Additively manufactured electrochemical and biosensors. In S. Rajendrachari (Ed.), *Practical Implementations of Additive Manufacturing Technologies* (Materials Horizons: From Nature to Nanomaterials). Springer, Singapore.
32. Manjunatha, J. G., & Uslu, B. (2022). Carbon-based composite materials for electrodes. *Materials, 15*(14), 4908.
33. Manjunath, J. G. (2023). *Graphene based sensors.* IOP Publishing. ISBN: 978-0-7503-5578-0
34. Farma, R., Deraman, M., Manjunatha, J. G., et al. (2013). Physical and electrochemical properties of supercapacitor electrodes derived from carbon nanotube and biomass carbon. *International Journal of Electrochemical Science, 8*(1), 257–273.
35. Moulya, K. P., Manjunatha, J. G., & Aljuwayid, A. M. (2023). Polymer modified carbon paste electrode for the electrochemical analysis of tartrazine. *Results in Chemistry, 5*, 100809.
36. Charithra, M. M., & Manjunatha, J. G. (2022). Electrochemical detection of amoxicillin as an antibiotic drug by using surface modified carbon-based sensors. In *Electrochemical Sensors Based on Carbon Composite Materials: Fabrication, Properties and Applications* (pp. 10–1). IOP Publishing.
37. Elham, A., Masoumeh, G., & Saeed, S. (2019). Electrochemical sensing based on carbon nanoparticles: A review. *Sensors and Actuators B: Chemical, 293*, 183–209. https://doi.org/10.1016/j.snb.2019.04.075
38. Vinayak, A., Sharma, K., & Lakshminarayana, P. (2024). Biomedical applications of surfactant modified graphene based electrochemical sensing devices. In *Advances in Surfactant Biosensor and Sensor Technologies.* Springer, Cham. https://doi.org/10.1007/978-3-031-60832-2_83
39. Pai, M., Batakurki, S., Vinayak, A., & Yallur, B. C. (2022). Optical graphene for biosensor application: A review. *Applied Mechanics and Materials, 908*, 51–68.
40. Vinayak, A., et al. (2024). Recent advances in metal-organic framework derived nanocomposites in gas and chemical sensors: A review. *Sensing Technology, 2*(1), 2317385.
41. Manjunatha, J. G., & Hussain, C. M. (Eds.). (2022). *Carbon nanomaterials-based sensors: Emerging research trends in devices and applications.*
42. Manjunatha, J. G. (2022). *Electrochemical sensors based on carbon composite materials: Fabrication, properties and applications.* IOP Publishing.
43. Vinícius, D. N., et al. (2019). Carbon nanostructure-based sensors: A brief review on recent advances. *Advances in Materials Science and Engineering*, Article ID 4293073, 21 pages. https://doi.org/10.1155/2019/4293073
44. Vinayak, A., Basappa, C. Y., & Adarsha, H. J. (2022). Advanced sensors based on carbon nanomaterials. In J. G. Manjunatha & C. M. Hussain (Eds.), *Carbon Nanomaterials-Based Sensors* (pp. 259–268). Elsevier. https://doi.org/10.1016/B978-0-323-91174-0.00014-7
45. Kannan, P., & Maduraiveeran, G. (2023). Metal oxides nanomaterials and nanocomposite-based electrochemical sensors for healthcare applications. *Biosensors, 13*, 542. https://doi.org/10.3390/bios13070542

46. Manjunatha, J. G. (2018). A novel voltammetric method for the enhanced detection of the food additive tartrazine using an electrochemical sensor. *Heliyon, 4*(11), e00964. https://doi.org/10.1016/j.heliyon.2018.e00964
47. Yunus, G., Singh, R., Raveendran, S., & Kuddus, M. (2023). Electrochemical biosensors in healthcare services: Bibliometric analysis and recent developments. *PeerJ, 11*, e15566. https://doi.org/10.7717/peerj.15566
48. Manjunatha, J. G. (2019). Electrochemical polymerised graphene paste electrode and application to catechol sensing. *The Open Chemical Engineering Journal, 13*(1), 17–22.
49. Manjunath, J. G. (2017). Electroanalysis of estriol hormone using electrochemical sensor. *Sensing and Bio-Sensing Research, 16*, 79–84. https://doi.org/10.1016/j.sbsr.2017.11.006
50. Prinith, N. S., Manjunatha, J. G., et al. (2022). Electrochemical analysis of food additive vanillin using poly(aspartic acid) modified graphene and graphite composite paste sensor. *ChemistrySelect, 7*(48), e202203572. https://doi.org/10.1002/slct.202203572
51. Vinayak, A., Batakurki, S., & Keri, R. (2023). Preparation of nanomaterials using coumarin and their various applications. In A. Husen (Ed.), *Secondary Metabolites Based Green Synthesis of Nanomaterials and Their Applications* (pp. 185–203). Springer. https://doi.org/10.1007/978-981-99-0927-8_8
52. Razie, H., Rina, D. M., et al. (2019). Electrochemical sensor for environmental monitoring system: A review. *AIP Conference Proceedings, 2169*(1), 030007. https://doi.org/10.1063/1.5135534
53. Zeng, L., Peng, L., Wu, D., et al. (2019). Electrochemical sensors for food safety. In *Nutrition in Health and Disease – Our Challenges Now and Forthcoming Time*. IntechOpen. https://doi.org/10.5772/intechopen.88223
54. Vinayak, A., Sheetal, B., & Basappa, C. Y. (2024). Toxicity, safety, and disposal of surfactant-based electrochemical sensors and biosensors. In *Surfactant Based Electrochemical Sensors and Biosensors* (pp. 473–488). Elsevier. https://doi.org/10.1016/B978-0-443-15493-5.00019-0
55. Baranwal, J., & Barse, B. A. (2022). Electrochemical sensors and their applications: A review. *Chemosensors, 10*(10), 363. https://doi.org/10.3390/chemosensors10100363
56. Saravanan, A., Kumar, S., Hemavathy, P. R. V., et al. (2022). A review on synthesis methods and recent applications of nanomaterials in wastewater treatment: Challenges and future perspectives. *Chemosphere, 307*, 135713. https://doi.org/10.1016/j.chemosphere.2022.135713
57. Hossain, T. M. T., Akter, S., et al. (2020). Reconceptualizing integration quality dynamics for omnichannel marketing. *Industrial Marketing Management, 87*, 225–241. https://doi.org/10.1016/j.indmarman.2019.11.005
58. Amiri, M., Bezaatpour, A., et al. (2018). Electrochemical methodologies for the detection of pathogens. *ACS Sensors, 3*(6), 1069–1086. https://doi.org/10.1021/acssensors.8b00267

59. Sismaet, H. J., Pinto, A. J., & Goluch, E. D. (2017). Electrochemical sensors for identifying pyocyanin production in clinical *Pseudomonas aeruginosa* isolates. *Biosensors and Bioelectronics, 97*, 65–69. https://doi.org/10.1016/j.bios.2017.05.013
60. Singh, A., Sharma, A., et al. (2021). Recent advances in electrochemical biosensors: Applications, challenges, and future scope. *Biosensors, 11*(9), 336. https://doi.org/10.3390/bios11090336
61. Lumaret, J. P., Errouissi, F., et al. (2012). A review on the toxicity and non-target effects of macrocyclic lactones in terrestrial and aquatic environments. *Current Pharmaceutical Biotechnology, 13*(6), 1004–1060. https://doi.org/10.2174/138920112800399311
62. Singh, A., Sharma, A., et al. (2021). Recent advances in electrochemical biosensors: Applications, challenges, and future scope. *Biosensors, 11*(9), 336. https://doi.org/10.3390/bios11090336

CHAPTER 3

How Nanomaterial-Based Electrochemical Sensors Are Used for Biomedical Applications

M M Sreejaya, Vaishnavi M, Lina Anil Kumar, Sreelakshmi M, Rejithamol R, Santhy A, and Moumita Gangopadhyay

3.1 INTRODUCTION

The advancement in the understanding of nanomaterials heralded a new scope in the field of material science where nanomaterials were integrated into different systems for technological applications. This integration of nanomaterials in material sciences has also influenced the field of biosensing. Electrochemical, magnetic, mechanical, optical, and thermal biosensors are some of the techniques primarily used in biosensing applications. The beginning of electrochemical biosensing can be dated back to 1962, when Clark et al. designed an electrode by incorporating the enzyme glucose oxidase to determine the glucose level in the blood (Clark and Lyons 1962). While the electrochemical sensors faced issues of sensitivity, stability, toxicity, miniaturization challenges, and interference during biomedical applications, they were also considered a good analytical

DOI: 10.1201/9781003512639-3

method due to the simplicity of instrument handling, portability, and comparatively low cost. Due to these advantages, significant progress was made in research based on electrochemical sensors (Kimmel et al. 2012; Akiba and Anzai 2016). This also motivated researchers to look for techniques that can overcome the limitations of electrochemical sensors. The incorporation of nanomaterials (Zhu et al. 2015) like metal nanoparticles, metal oxides, carbon nanotubes, graphene, and quantum dots into the electrochemical sensors has significantly changed their properties and enhanced their applicability (Chen and Chatterjee 2013). Compared with conventional electrochemical sensors, nanomaterial-based electrochemical sensors have several advantages, such as increased sensitivity and stability, improvement in response time, and reduced detection limit. Furthermore, the properties of nanomaterials, such as good mechanical stability, high surface area, and better absorptivity, have also been proved to improve the analytical performance of nanomaterial-based electrochemical sensors.

Research on the applicability of nanomaterial-based electrochemical sensors in biomedicine has progressed over the years from understanding the basic interactions between biomolecules and nanoparticles to employing the sensors in the detection of biomolecules like glucose, cholesterol, proteins, DNA, and many other molecules that are biomarkers (X. Liu, Huang, and Qian 2021). Nanomaterial-based electrochemical sensors can detect cancer biomarkers and hence are used in cancer diagnosis. This method surpasses the other existing methods of cancer detection using biomarkers. This technique is also easy to commercialize due to its ease of handling. Today, as well as cancer, various other diseases are also being diagnosed using these sensors. Apart from the detection of biomolecules, nanomaterial-based electrochemical sensors are also used in the real-time monitoring of different drug and hormone levels and their management (Haroon and Stine 2023; Leau, Lete, and Lupu 2023).

The scope of research in nanomaterial-based electrochemical sensors is enormous. The aim of transforming nanomaterial-based electrochemical sensors into a point-of-care diagnostic is achieved by advantages like miniaturization, low cost, and fast and efficient real-time monitoring. With the development of technology, attempts to incorporate data analytics and artificial intelligence into these sensors can further enhance their performance. The progress in the field of nanomaterials can be reflected in electrochemical sensors by including a variety of nanomaterials in the electrochemical system. Also, attempts are being made to reduce the number of hazardous materials used during their synthesis and fabrication. Sustainable options, like the usage of green solvents, and a recycling strategy must be developed for the sensors. Development in the fields of nanotechnology, bioengineering, and material science in the future will also be reflected in nanomaterial-based electrochemical sensors, which could ultimately lead to the formation of highly sensitive, personalized, and advanced biomedical diagnostic tools that can revolutionize the healthcare sector.

3.2 TYPES OF NANOMATERIALS NEEDED FOR ELECTROCHEMICAL SENSING

Electrochemical sensing is a rapidly evolving field benefiting from a diverse array of nanomaterials, each offering unique advantages for detecting specific analytes. Metal, metal oxide, and carbon nanomaterials are widely used for electrochemical sensing for biomedical applications.(Fritea et al. 2021; Fazio et al. 2021)

3.2.1 Metal Nanoparticles

Gold nanoparticles (AuNPs) are well known for their electronic, optical, and catalytic properties, making them highly suitable for electrochemical sensing applications. Transferrin is an acute phase protein, which is known to decrease in conditions such as inflammation, cancer, and various diseases (Yoshiga et al. 1997). An ultrasensitive electrochemical biosensor for transferrin (Tf) detection was developed using a glassy carbon electrode (GCE) modified with gold nanoparticles (AuNPs) through electroless deposition (Rabbani et al. 2024). The biosensor's performance was characterized using cyclic voltammetry (CV) and electrochemical impedance spectroscopy (EIS), revealing its effective sensing capabilities. Differential pulse voltammetry (DPV) was used to quantify Tf-antigen by detecting the increases in the anodic peak current. The GCE-AuNPs biosensor exhibited high sensitivity and selectivity for Tf, with the ability to detect even low concentrations of the antigen. The sensor demonstrated excellent stability and ease of use without the need for sample pretreatment. The biosensor shows promise for early diagnosis of atransferrinemia and could be applied to detect other serum biomarkers. Silver nanoparticles (AgNPs) are well regarded for their strong antibacterial properties and excellent conductivity, which make them highly effective in electrochemical sensing. A core–shell nanocomposite of carbon black and polyaniline, modified with silver nanoparticles, was developed for glucose detection (Piñón-Balderrama et al. 2024). The modified glassy carbon electrode exhibited enhanced electrochemical properties, with silver nanoparticles providing electrocatalytic activity for glucose detection. The composite was synthesized using an amphiphilic monomer to create a dispersible structure. The glucose detection mechanism involved the catalytic oxidation of glucose by silver oxide, producing a significant oxidation peak. The composite's porous structure and high surface area improved the distribution of silver nanoparticles and facilitated electron transfer. The sensor demonstrated high sensitivity and linear response for glucose detection in alkaline solution. Platinum nanoparticles (PtNPs) are distinguished by their superior catalytic activity and electrocatalytic properties, which are crucial for various electrochemical sensing applications. Expanding on the application of nanoparticles, the focus has also extended to the precise sensing of chemotherapeutic agents like methotrexate, which is crucial for monitoring cancer treatment. Methotrexate, a widely used chemotherapeutic agent, requires precise sensing to monitor its levels in cancer patients, as accurate detection is crucial for optimizing its therapeutic efficiency and minimizing toxic side effects. A sensor combining platinum nanoparticles and single-walled carbon nanotubes was developed for the quantification of methotrexate (Figure 3.1) (Tarlekar and Chatterjee 2024). The presence of platinum nanoparticles on the carbon nanotubes enhanced the electrochemical oxidation of methotrexate, resulting in notable negative shift in the peak potential and an increase in peak current. The synergistic effect of platinum nanoparticles and single-walled

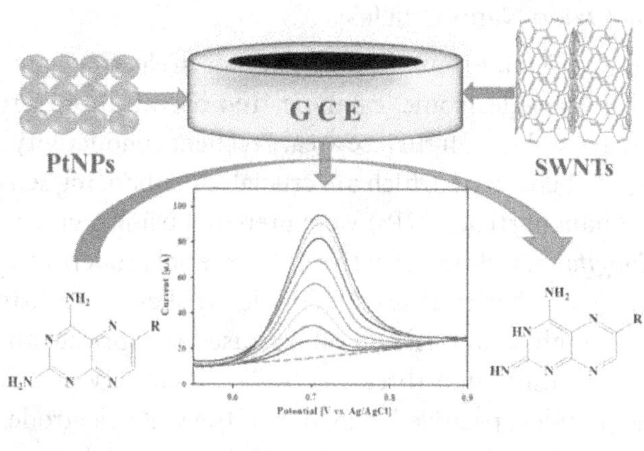

FIGURE 3.1 Mechanism of electrochemical sensing of methotrexate by platinum nanoparticles.

carbon nanotubes improved the electroactive surface area and facilitated more efficient charge transfer. This resulted in a higher oxidation current and lower onset oxidation potential for methotrexate compared with sensors using only platinum nanoparticles or carbon nanotubes. The sensor demonstrated high sensitivity and stability, with excellent reproducibility in detecting methotrexate. This approach also successfully quantified methotrexate in pharmaceutical formulations and biological samples. Palladium nanoparticles (PdNPs) are highly valued in electrochemical sensing due to their exceptional catalytic activity and ability to facilitate various redox reactions. DNA sensing using nanoparticles helps in detecting genetic material, enabling advancements in biomedical diagnostics, environmental monitoring, and forensic investigations. The electrochemical behaviour of palladium nanoparticles (PdNPs) deposited on a GCE was investigated using two deep eutectic solvents (DES), reline and ethaline (Phuong, Trang, and Tam 2024). CV and chronoamperometry revealed that deposition in reline followed a progressive 3D nucleation mechanism. Palladium nanoparticles synthesized in ethaline demonstrated superior sensitivity for DNA sensing compared with those synthesized in reline. The DNA sensors showed that the PdNPs from ethaline exhibited a more pronounced change in current upon hybridization with target DNA, indicating higher sensitivity and better performance. This improvement is attributed to the small size and greater surface area of PdNPs synthesized in ethaline, enhancing the sensor's ability to detect DNA. Both sensors displayed high selectivity, with the PdNPs from ethaline offering superior overall performance.

3.2.2 Metal Oxide Nanoparticles

Metal oxides are an important class of materials in electrochemical sensing due to their diverse electronic, catalytic, and chemical properties. These materials often exhibit high surface area, excellent conductivity, and stable electrochemical behaviour, which are crucial for enhancing sensor performance. CdO nanoparticles (NPs) were prepared using a green route with *Polyalthia longifolia*, yielding effective antibacterial properties against various bacteria (Ghotekar et al. 2024). For electrochemical sensing, a paste of CdO NPs, graphite, and silicon oil was used to fabricate an electrode, which was then tested in a three-electrode system. CV and EIS studies revealed the pseudocapacitive behaviour of the CdO electrode. The CdO electrode demonstrated efficient sensing of glucose and dextrose, showing an increase in redox peak intensity with higher analyte concentrations.

This confirms the potential of CdO electrodes for practical applications in monitoring glucose levels in human blood. Researchers synthesized ultrasmall cerium oxide nanoparticles (USCeOxNPs) using a deep eutectic solvent through a solvothermal method (Gomes-Junior et al. 2024). These nanoparticles, characterized as spherical with cubic crystalline structures, were used to develop an electrochemical sensor for detecting dopamine in biological fluids. The sensor, incorporating carbon black (CB) and USCeOxNPs, demonstrated enhanced current response, good sensitivity, and excellent repeatability. The modified sensor showed improved electrocatalytic activity and stable responses in dopamine redox processes. This study highlights the potential of the USCeOxNPs-DES/CB sensor for practical applications in biological dopamine detection. Researchers developed black titanium oxide nanotube (BTNT) electrodes by electrochemically reducing titanium oxide nanotubes (TNTs) for enhanced electrocatalytic properties when iron(III) oxide (Fe_2O_3) nanoparticles were deposited (S. Sharma, Sidhartha, and Chappanda 2024). The Fe_2O_3 nanoparticles were electrodeposited onto the BTNTs using a chronoamperometric technique, resulting in improved sensitivity and selectivity for dopamine sensing compared with unreduced TNTs. The BTNTs-Fe sensors exhibited higher anodic peak current responses and better electrocatalytic activity due to increased electron transfer rates and reduced band gaps. The sensors demonstrated excellent selectivity towards dopamine over other common interfering substances, indicating their potential for use in biological fluid analysis. The study highlights the effectiveness of using simple reduction and deposition methods to enhance the electrochemical sensing performance of nanotube-based sensors. A novel electrochemical sensor using CuO nanostructures modified on a glassy carbon electrode (CuO/GCE) has been developed for the simultaneous determination of hydroquinone (HQ) and ascorbic acid (AA) (Buledi et al. 2020). The CuO nanomaterial was synthesized via an aqueous chemical growth method and characterized by various techniques. The modified electrode showed excellent electrocatalytic activity and well-resolved oxidation peaks for HQ and AA in borate buffer solution. Differential pulse voltammetry was employed for the simultaneous detection of HQ and AA, with the sensor demonstrating good sensitivity, selectivity, and stability. The sensor was successfully applied to measure AA in fruit juices and both AA and HQ in cosmetic products, confirming its practical applicability. The study highlights the sensor's potential for reliable and efficient detection of these analytes in real samples.

3.2.3 Carbon-Based Nanomaterials

Carbon nanomaterials, including carbon nanotubes (CNTs), graphene, and carbon dots, have revolutionized the field of electrochemical sensing due to their remarkable electronic, structural, and chemical properties. Their high surface area, excellent electrical conductivity, and ability to facilitate efficient electron transfer make them ideal candidates for enhancing the performance of electrochemical sensors. An electrochemical sensor developed for simultaneous determination of amlodipine, atorvastatin, and acetylsalicylic acid utilized a carbon nanotube–ferrocene and nickel chromium oxide nanocomposite modified glassy carbon electrode (El-Gohary, Galal, and Atta 2024). This sensor leverages the enhanced electrical conductivity and electrocatalytic activity of the carbon nanotubes to improve the resolution and sensitivity of the measurements. The ferrocene component contributed to effective redox reactions, while the nickel chromium oxide nanoparticles further enhanced catalytic performance and increased the oxidation peak currents of the target drugs. The sensor achieved significant improvements in both sensitivity and selectivity by combining these materials, demonstrating excellent performance in detecting the drugs in human serum samples. The synergistic effects of the nanocomposite components allow effective simultaneous determination of the compounds with clear separation of their oxidation peaks. The sensor shows a high degree of selectivity, maintaining accurate detection even in the presence of potential interfering substances. Acetylcholine (ACh) is a critical neurotransmitter involved in nerve cell communication and various neurological functions, with imbalances linked to neurodegenerative diseases such as Alzheimer's and Parkinson's disease. In other research, a novel electrochemical sensor for ACh detection was developed using a reduced graphene oxide/zinc oxide (rGO/ZnO) composite (Figure 3.2) (Pitiphattharabun et al. 2024). The composite, synthesized through modified Hummers' and hydrothermal methods, combined the high specific surface area and electron mobility of rGO with the stability and catalytic efficiency of ZnO. Electrochemical studies demonstrate that the rGO/ZnO composite enhances electron transfer efficiency and catalytic activity, resulting in improved sensitivity and selectivity for ACh detection. The sensor's performance includes prominent oxidation and reduction reactions, reflecting the effective electrocatalysis of ACh. This study highlights the sensor's potential in offering a cost-effective, sensitive, and selective method for ACh detection, addressing the need for reliable tools in diagnosing and monitoring neurological disorders. Metronidazole (MTZ) is a

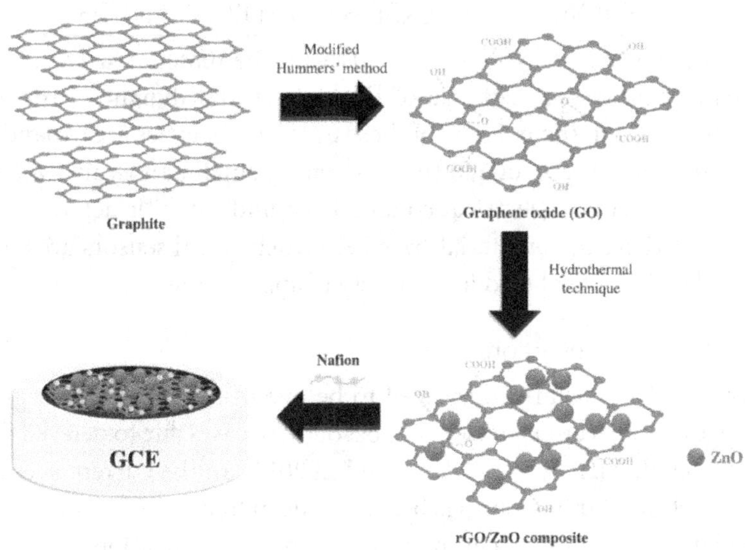

FIGURE 3.2 Schematic representation of the reduced graphene oxide/ZnO composite applied to a glassy carbon electrode.

widely used antibiotic for treating infections in both human and veterinary medicine, but its residues in food products, such as milk, pose significant health risks and contribute to antibiotic resistance. Other research introduced a novel electrochemical sensor based on a carbon dot–silver nanoparticles–copper oxide (CDs-Ag@Cu2O-GA) composite for detecting MTZ traces in food products (Aslam et al. 2024). The sensor, constructed on a glassy carbon electrode, leverages the synergistic properties of carbon dots and metal oxides to enhance electron transfer and signal amplification. The excellent stability, sensitivity, and selectivity of the composite are attributed to its numerous electroactive sites and efficient interaction with MTZ. The sensor's high performance in detecting MTZ in milk demonstrates its potential for ensuring food safety and regulatory compliance. This research addresses the need for reliable, cost-effective, and sensitive methods for monitoring antibiotic residues in food.

The integration of these nanomaterials—carbon, metal, and metal oxide nanoparticles—highlights the versatility and effectiveness of electrochemical sensors in analysing a wide range of analytes. Each type of nanomaterial contributes distinct properties, such as high surface area, catalytic activity, and enhanced electron transfer, which collectively advance the sensitivity, selectivity, and applicability of electrochemical sensing technologies.

3.3 APPLICATIONS IN DIVERSE BIOMEDICAL FIELDS

Nanomaterials are being extensively used in the manufacturing of electrochemical biosensors because of the unique and dynamic properties that they impart to them. Some of these unique properties are expandable surface-to-volume ratio due to their extremely small nanoscale size, high mechanical strength, electrical conductivity, and catalytic activity.(Yoon et al. 2022) These nanomaterial-based electrochemical sensors are widely used in the biomedical field for a variety of applications.

3.3.1 Microbial Detection

Microbial infections are considered to be one of the most urgent global challenges now. Death is caused in most of these cases due to delayed diagnosis (Su et al. 2011; M. Xu, Wang, and Li 2017). In the current scenario, early detection of infections has become crucial to provide effective treatment plans. This, in turn, can prevent the spread of infection and reduce the death rate. Using electrochemical immune sensors, we can detect and identify the presence of such infectious microbes (Burlage and Tillmann 2017).

The detection of microbes happens when the signal is passed on through the biorecognition element on the nanomaterial surface of the working electrode. The final step includes the injection of a secondary enzyme-labelled antibody that follows the addition of a specific enzymatic substrate (Guntupalli et al. 2013).

Along with immunosensors, different methods have been used in order to enhance the selectivity and sensitivity of electrochemical biosensors. A molecularly imprinted polymer (MIP)–based electrochemical biosensor is one such important method for enhancing selectivity and sensitivity. MIPs have been produced using chemical, photochemical, or electrochemical methods in the presence of a target template molecule(s). MIPs have been developed using complete microbial cells as a template; these have been used for many applications, including microbial fuel, specific cell capture, cell sorting, separation, and microbial detection. Rapid capture and identification of *Salmonella typhimurium* in minced beef samples was achieved using fabricated MIP biosensors (Figure 3.3) (Khalid et al. 2022). Here, the biosensor was integrated into electro-polymerized polydopamine matrices created on electrodes screen coated with gold nanoparticles. The assay optimization and validation were achieved using cyclic voltammetry, resulting in a high selectivity and sensitivity against the imprinted cells of *S. typhimurium*.

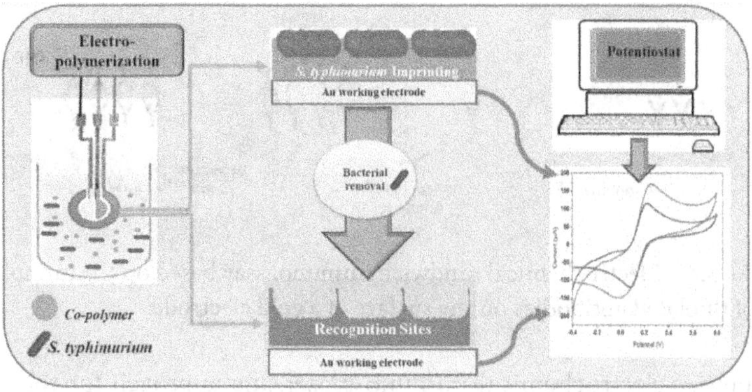

FIGURE 3.3 Schematic representation of the main components and fabrication steps of a *Salmonella*-based biosensor.

Bacteriophages have been used as high-efficiency bio-receptors for the development of electrochemical biosensing platforms. These platforms possess high specificity against their bacterial host (Aliakbar Ahovan et al. 2020; El-Shibiny, El-Sahhar, and Adel 2017). They are preferred over immunosensors due to their important properties, like thermal stability and sensitivity to changes in surrounding pH or solvents (Aliakbar Ahovan et al. 2020).

For example, the selective determination of *Listeria innocua* serovar was achieved by the development of an impedimetric phage-based biosensor (Tolba et al. 2012). The biorecognition element for this biosensor was a gold nanoparticle integrated electrode with bacteriophage endolysin, which was chemically conjugated onto the screen-printed electrode surface via classical 1-ethyl-3-(3-dimethylaminopropyl)carbodiimide/N-hydroxysuccinimide chemistry.

3.3.2 Detection of Toxins

There are several ways in which toxins can either enter or be produced in our body. Apart from the toxins that enter our body through environmental exposure, diet, and medication, microbial extracellular-secreted toxins like hemolysins, leukotoxins, and enterotoxins are the major causative agents of poisoning in the body, as these persist in the body even after disinfection. For this reason, it has become essential to determine their presence and biological activity (Tian et al. 2021; Kumar, Guleria, and Mehta 2017). A method for electrochemical detection of staphylococcal enterotoxin B (SEB) was found by developing a nanostructured aptasensor using a screen-printed electrode modified with reduced graphene oxide–gold nano-urchins (Mousavi Nodoushan et al. 2019).

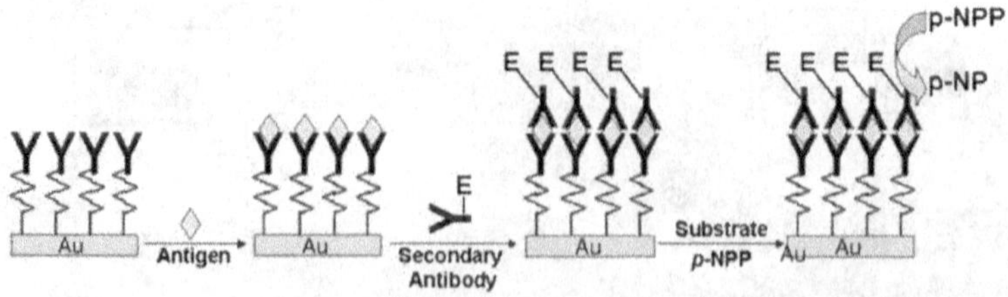

FIGURE 3.4 Electrochemical sandwich immunoassay based on the immobilization of thiolated antibodies on the surface of a gold electrode.

Another electrochemical immuno-biosensing approach for the determination of SEB was developed by chemically immobilizing a selective antibody on the surface of a gold electrode through chemical cross-linking with the ammine or sulfhydryl reactive heterobifunctional cross-linkers. The chemical conjugation of the selected antibody was done on the electrode surface, which was modified by using rGO-chitosan-AuNPs. The biosensor thus formed exhibited high sensing performance because of its flat two-dimensional configuration and large surface area (Figure 3.4) (Chatrathi, Wang, and Collins 2007; A. Sharma et al. 2016).

3.3.3 Cancer Diagnosis

In the field of cancer biology, various biosensor techniques have been exploited to study various new aspects of metastatic cancer. These electrochemical biosensors have been used for inspecting various biological analytes such as extracellular metabolites, nucleotides, expressed protein biomarkers, or whole cancer cells. Various types of these biosensors have now been developed for extensive and efficient usage for the benefit of humanity (Arafa et al. 2022).

Caspase-3 (Cas-3) is a very important intercessor in the extrinsic apoptosis signalling cascade. So, several electrochemical biosensors were developed to ensure its rapid and sensitive detection (Yu et al. 2021; Deng et al. 2019; Frączyk 2021). The cellular microenvironment and cancer cell viability were investigated electrochemically using a 2D and 3D sensing cell culture flask integrated with an electrochemical station for recording the electrochemical signals obtained due to metabolic activity and/or extracellular changes. We can inspect the cell activity during the cultivation and growth cycles electrochemically using standard cell culture flasks with nano-fabricated sensor chips made from metal titanium and platinum affixed to them (Kieninger et al. 2018). These systems have been successfully used for measuring the electrochemical behaviours of breast cancer cells and brain tumour cells (Flamm et al. 2015).

Electrodeposited iridium oxide films were implanted on a potentiometric pH sensor for sensing cellular acidification. The resulting biosensor helped in pursuing 3D-cell biosensor–based cell cultures (Oliveira et al. 2021). An aptasensor was designed that enabled the impedimetric biosensing of Cyt-C using Ag nanocluster conjugation (Shamsipur et al. 2017).

Another biosensor was developed for the direct recognition of Bcl-2 and Bax in cancer cell cultures (Zhou, Wang, and Zhu 2016). Survivin (Sur) is one of the anti-apoptotic biomarkers. A microchannel cytosensor was designed to selectively capture it using an anti-survivin oligonucleotide sensor (Li et al. 2013). A trans-membrane protein known as epithelial cell adhesion molecule (EpCAM) was electrochemically detected using an impedimetric immuno-based biosensor, which is a gold nanoparticle–coated electrode with incorporated anti-EpCAM monoclonal antibody (J. Xu et al. 2019).

3.3.4 Viral Infection Diagnosis

Biosensors have a crucial role in monitoring infectious diseases. Molecularly imprinted polymers (MIPs) have extensive biomedical applications because of their peculiarities, like intrinsic binding affinity to specific proteins and other biological materials. The capability of MIPs to be used for accurate viral diagnosis in real time has been proved by many researchers (Gast, Sobek, and Mizaikoff 2019a). Different epitopes of adeno-associated viruses (AAV) immobilized on glass beads using polymeric nanoparticles of acrylate and methacrylate have been identified and screened using fabricated MIPs (Piletska et al. 2020).

Viral hexon protein is the most accessible and abundant surface protein of the human adenovirus type 5 (hAdV5) icosahedral capsid. Another biosensor was developed for the detection of these proteins using silica-based nanoparticles. The nanoparticles were incubated with respective proteins for the identification of the virus. It was used as the template molecule to create specific binding sites for the selective detection of the entire virus. To prove the sensor's selectivity, two different viruses, including hAdV5 and minute virus of mice (MVM), were exposed to the sensors with no obvious binding affinity (Figure 3.5) (Gast, Sobek, and Mizaikoff 2019b).

A model mammalian virus, PRRSV-1 (Porcine Reproductive and Respiratory Syndrome Virus), was used to impart virus-neutralizing capacity to hydrogel-based nano-MIPs (Graham et al. 2019). Three

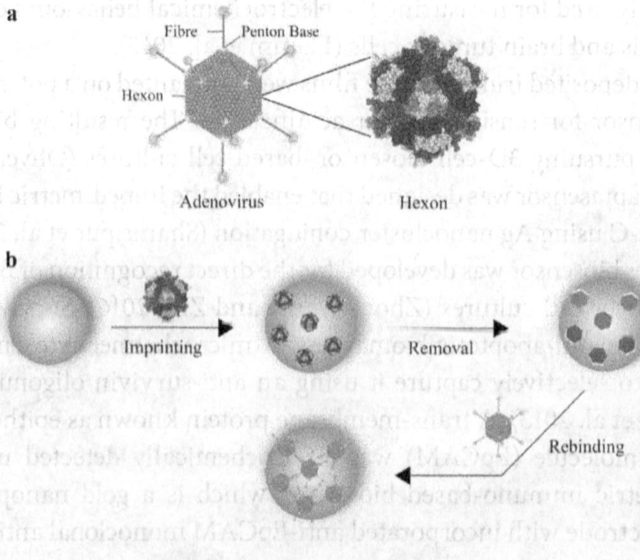

FIGURE 3.5 (a) Structure of overall icosahedral adenovirus and zoom into the structure of one hexon protein; (b) different types of molecular imprinting including imprinting of hexon template, followed by removal and final rebinding of entire virus.

different nano-MIPs were created using polyacrylamide (pAA), poly-N-hydroxymethylacrylamide (pNHMA), and poly-N-isopropylacrylamide (pNIPAM). During the optimization procedure of the biosensor, both the specificity of virus neutralization and the influence of incubation time on sensor performance were studied. These MIPs were shown to carry out effective virus neutralization in just 2.5 minutes, and their effect depended on the concentration (Arshad et al. 2020).

Dengue virus was detected in the early stage of the infection in serum samples by exploiting the self-polymerization of dopamine in the presence of target non-structural protein 1 (NS 1), which was achieved by the construction of an impedimetric MIP-based electrode sensor modified with nanofibers of polysulfone (Arshad et al. 2020).

Hepatitis A virus (HAV) was selectively detected within 20 minutes using a novel pH-responsive molecularly imprinted nanoprobe developed via molecular imprinting. The detection was based on resonance light-scattering (RLS). The sensor showed high selectivity towards HAV over similar viruses such as Japanese encephalitis virus, rabies virus, leprosy virus, due to specific imprinting sites that matched HAV's size and surface properties (Luo et al. 2020).

3.3.5 Heavy Metal Detection

The property of electrochemical sensing of different targets like heavy metals is highly dependent on the type and structure of the material of the working electrodes. Thus, surface modifications of working electrodes have been improvised to ensure highly specific recognition elements and high selectivity towards different heavy metal ions.

Arsenic ions can be directly detected by a method using cyclic voltammetry. This was achieved by the development of an electrochemical setup using electrodeposited Pt nanoparticles on a glassy carbon electrode (PtNPs/GCE) (Dai and Compton 2006).

Direct voltammetric detection of Cu^{2+} ions was achieved by constructing a sensing platform by chemically synthesizing a polymeric nanocomposite consisting of polyurethane doped with Pt nanoparticles (PU/PtNPs) with an average size of 2–5 nm. The applications of this sensor were studied using different biological samples, like serum, urine, and acidified tap water samples (Aragay and Merkoçi 2012).

Gold nanoparticle-based electrochemical sensors have been used widely to determine potentially hazardous levels of hexavalent chromium (VI), which in excess can be highly toxic (Figure 3.6) (Wyantuti, Ishmayana, and Hartati 2015). Successful electrochemical sensing of one of the most toxic elements—mercury—was also achieved using these

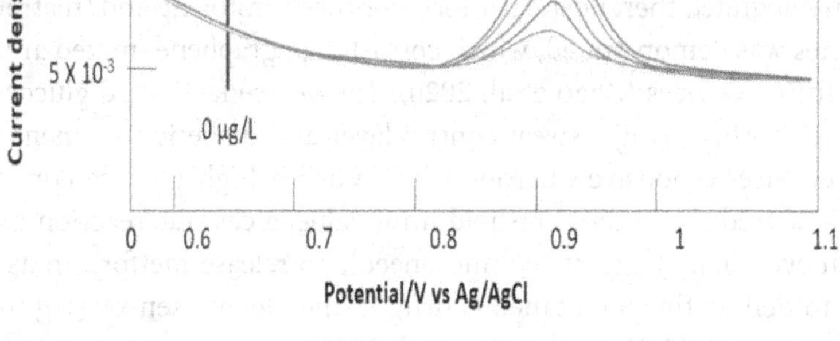

FIGURE 3.6 Voltammogram indicating range of Cr(VI) concentration (0–0.25 μg/L) with deposition potential, deposition time, scan rate, and amplitude modulation at 800 mV, 120 s, 50 mV/s, and 500 mV, respectively.

nanomaterial-based sensors. Here, poly(ester-urethane) urea conjugated with gold nanoparticles (PUU/AuNPs) was used to modify a carbon paste electrode (CPE) in order to achieve the highly sensitive detection of mercury ions in the tissue of fishes (March, Nguyen, and Piro 2015).

Carbon nanofibers (CNFs) were synthesized and characterized for use in the electrochemical detection of Pb^{2+} ions. Nanoelectrode arrays were fabricated from the vertically aligned forms of carbon nanofibers (VACNFs). These nanoelectrodes increased the anodic stripping voltammetric performance for the determination of Pb^{2+} ions (Robinson et al. 2016).

The selective and sensitive impedimetric detection of Cd^{2+} ions was achieved by modifying a disposable screen-printed electrode with a nanocellulose/ligand/CNT/Co_3O_4 nanoparticle (Magar et al. 2022).

3.3.6 Drug Delivery

Targeted and localized drug delivery systems have become quite popular due to the rising demand for efficient functionality and therapy with fewer side effects (Zhao et al. 2020). Electrochemical sensors potentially provide these systems with precise temporal and spatial control (Z. Liu et al. 2019).

For example, an electrochemical method was used for precisely controlling drug release from a sulfonated silica NP–doped composite (Magar et al. 2022). The SiO_2 NP dopants not only provided a potential drug delivery library but also maintained several favourable electrochemical properties. [41]

Controlled release of doxorubicin (DOX) was successfully achieved by the electron injection of DOX@nanographene oxide/platinum micromachines. Here, the reduced nanographene oxide is used as a drug-loading platform (Zhao et al. 2020).

An integrated theranostic platform for the monitoring and treatment of diabetes was demonstrated, which consisted of graphene–mixed arrays of skin-fitting devices (Zhao et al. 2020). The device facilitated glucose and pH monitoring using a sweat control layer and diabetic treatment using temperature-responsive microneedles. When a high glucose concentration is sensed above the threshold limit value, a cascade reaction occurs, which eventually triggers the microneedle to release metformin as feedback to deliver the percutaneous drug to the glucose sensor (Figure 3.7) (Huang et al. 2017; Kusumoputro et al. 2020).

The probable concerns that arose during the development of electrochemical sensor–based health devices were environmental burden,

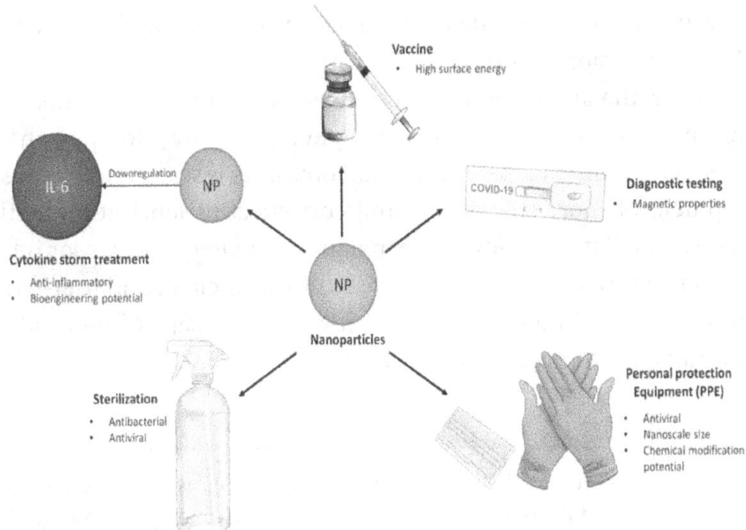

FIGURE 3.7 Wide application of nanoparticles in the healthcare industry during the COVID 19 pandemic.

resource waste, and low human compliance when too many duplicated devices were developed (Huang et al. 2017; Kusumoputro et al. 2020).

Unnecessary waste was avoided by the integration of a controlled drug delivery system into a smart contact lens (Keum et al. 2020). This multifunctional intelligent contact lens was designed with platinum nanostructures, which resulted in the lens possessing properties like low electrical resistance, high sensitivity, and stability for real-time electrochemical detection of glucose.

3.4 CONCLUSION

In conclusion, the integration of nanomaterials into electrochemical sensing technologies has significantly advanced the field of biomedical diagnostics and environmental monitoring. Nanomaterials, including metal nanoparticles, metal oxides, and carbon-based materials, each offer unique properties that enhance the sensitivity, selectivity, and efficiency of electrochemical sensors. Metal nanoparticles such as gold, silver, platinum, and palladium contribute to improved detection capabilities for biomolecules, toxins, and drug levels due to their excellent catalytic activity and conductivity. Metal oxides and carbon-based nanomaterials, such as graphene and carbon nanotubes, further enhance sensor performance through their high surface area and stability. These advancements enable more accurate and reliable detection of a wide range of analytes, including glucose, cancer biomarkers, and heavy metals.

Moreover, the application of these sensors extends to real-time monitoring and point-of-care diagnostics, paving the way for personalized medicine and improved healthcare outcomes. As research progresses, the development of more sustainable and cost-effective fabrication methods will be crucial. Future innovations in nanotechnology and sensor integration hold promise for creating highly sensitive, efficient, and user-friendly diagnostic tools, ultimately transforming the landscape of medical diagnostics and environmental safety.

REFERENCES

Akiba, Uichi, and Jun-ichi Anzai. 2016. "Recent Progress in Electrochemical Biosensors for Glycoproteins." *Sensors* 16 (12): 2045. https://doi.org/10.3390/s16122045.

Aliakbar Ahovan, Zahra, Ali Hashemi, Laura Maria De Plano, Mazaher Gholipourmalekabadi, and Alexander Seifalian. 2020. "Bacteriophage Based Biosensors: Trends, Outcomes and Challenges." *Nanomaterials* 10 (3): 501. https://doi.org/10.3390/nano10030501.

Arafa, Kholoud K., Alaa Ibrahim, Reem Mergawy, Ibrahim M. El-Sherbiny, Ferdinando Febbraio, and Rabeay Y. A. Hassan. 2022. "Advances in Cancer Diagnosis: Bio-Electrochemical and Biophysical Characterizations of Cancer Cells." *Micromachines* 13 (9): 1401. https://doi.org/10.3390/mi13091401.

Aragay, Gemma, and Arben Merkoçi. 2012. "Nanomaterials Application in Electrochemical Detection of Heavy Metals." *Electrochimica Acta* 84 (December): 49–61. https://doi.org/10.1016/j.electacta.2012.04.044.

Arshad, Rabia, Amina Rhouati, Akhtar Hayat, Mian Hasnain Nawaz, Muhammad Arfat Yameen, Adnan Mujahid, and Usman Latif. 2020. "MIP-Based Impedimetric Sensor for Detecting Dengue Fever Biomarker." *Applied Biochemistry and Biotechnology* 191 (4): 1384–94. https://doi.org/10.1007/s12010-020-03285-y.

Aslam, Hafiza Khadija, Sehrish Bilal, Sana Mir, Sobia Tabassum, Mazhar Amjad Gilani, Mustansara Yaqub, and Muhammad Asim. 2024. "A Robust and Simple Non-Enzymatic Electrochemical Sensor Based on Carbon Dots-Metal Oxide Composite for the Detection of Metronidazole Traces in Food Products." *Food Chemistry* 460 (December): 140297. https://doi.org/10.1016/j.foodchem.2024.140297.

Buledi, Jamil A., Sidra Ameen, Nadir H. Khand, Amber R. Solangi, Iqleem H. Taqvi, Muhammad H. Agheem, and Zahid Wajdan. 2020. "CuO Nanostructures Based Electrochemical Sensor for Simultaneous Determination of Hydroquinone and Ascorbic Acid." *Electroanalysis* 32 (7): 1600–1607. https://doi.org/10.1002/elan.202000083.

Burlage, Robert S., and Joshua Tillmann. 2017. "Biosensors of Bacterial Cells." *Journal of Microbiological Methods* 138 (July): 2–11. https://doi.org/10.1016/j.mimet.2016.12.023.

Chatrathi, Madhu Prakash, Joseph Wang, and Greg E. Collins. 2007. "Sandwich Electrochemical Immunoassay for the Detection of Staphylococcal Enterotoxin B Based on Immobilized Thiolated Antibodies." *Biosensors and Bioelectronics* 22 (12): 2932–38. https://doi.org/10.1016/j.bios.2006.12.013.

Chen, Aicheng, and Sanghamitra Chatterjee. 2013. "Nanomaterials Based Electrochemical Sensors for Biomedical Applications." *Chemical Society Reviews* 42 (12): 5425. https://doi.org/10.1039/c3cs35518g.

Clark, Leland C., and Champ Lyons. 1962. "Electrode Systems for Continuous Monitoring in Cardiovascular Surgery." *Annals of the New York Academy of Sciences* 102 (1): 29–45. https://doi.org/10.1111/j.1749-6632.1962.tb13623.x.

Dai, Xuan, and Richard G. Compton. 2006. "Detection of As(Iii) via Oxidation to As(v) Using Platinum Nanoparticle Modified Glassy Carbon Electrodes: Arsenic Detection without Interference from Copper." *The Analyst* 131 (4): 516. https://doi.org/10.1039/b513686e.

Deng, Dehua, Yuanqiang Hao, Suling Yang, Qiuyang Han, Lin Liu, Yanrong Xiang, Fanbin Tu, and Ning Xia. 2019. "A Signal-on Electrochemical Biosensor for Evaluation of Caspase-3 Activity and Cell Apoptosis by the Generation of Molecular Electrocatalysts on Graphene Electrode Surface for Water Oxidation." *Sensors and Actuators B: Chemical* 286 (May): 415–20. https://doi.org/10.1016/j.snb.2019.01.137.

El-Gohary, Asmaa R.M., Ahmed Galal, and Nada F. Atta. 2024. "Novel Electrochemical Sensor for Simultaneous Determination of Amlodipine, Atorvastatin, and Acetylsalicylic Acid Based on Carbon Nanotubes-Ferrocene/Nickel Chromium Oxide Nanocomposite." *Microchemical Journal* 200 (May): 110221. https://doi.org/10.1016/j.microc.2024.110221.

El-Shibiny, A., S. El-Sahhar, and M. Adel. 2017. "Phage Applications for Improving Food Safety and Infection Control in Egypt." *Journal of Applied Microbiology* 123 (2): 556–67. https://doi.org/10.1111/jam.13500.

Fazio, Enza, Salvatore Spadaro, Carmelo Corsaro, Giulia Neri, Salvatore Gianluca Leonardi, Fortunato Neri, Nehru Lavanya, Chinnathambi Sekar, Nicola Donato, and Giovanni Neri. 2021. "Metal-Oxide Based Nanomaterials: Synthesis, Characterization and Their Applications in Electrical and Electrochemical Sensors." *Sensors* 21 (7): 2494. https://doi.org/10.3390/s21072494.

Flamm, H., J. Kieninger, A. Weltin, and G. A. Urban. 2015. "Superoxide Microsensor Integrated into a Sensing Cell Culture Flask Microsystem Using Direct Oxidation for Cell Culture Application." *Biosensors and Bioelectronics* 65 (March): 354–59. https://doi.org/10.1016/j.bios.2014.10.062.

Frączyk, Tomasz. 2021. "Phosphorylation Impacts Cu(II) Binding by ATCUN Motifs." *Inorganic Chemistry* 60 (12): 8447–50. https://doi.org/10.1021/acs.inorgchem.1c00939.

Fritea, Luminita, Florin Banica, Traian Costea, Liviu Moldovan, Luciana Dobjanschi, Mariana Muresan, and Simona Cavalu. 2021. "Metal Nanoparticles and Carbon-Based Nanomaterials for Improved Performances of Electrochemical (Bio)Sensors with Biomedical Applications." *Materials* 14 (21): 6319. https://doi.org/10.3390/ma14216319.

Gast, Manuela, Harald Sobek, and Boris Mizaikoff. 2019a. "Advances in Imprinting Strategies for Selective Virus Recognition a Review." *TrAC Trends in Analytical Chemistry* 114 (May): 218–32. https://doi.org/10.1016/j.trac.2019.03.010.

Gast, Manuela, Harald Sobek, and Boris Mizaikoff. 2019b. "Selective Virus Capture via Hexon Imprinting." *Materials Science and Engineering: C* 99 (June): 1099–1104. https://doi.org/10.1016/j.msec.2019.02.037.

Ghotekar, Suresh, C. R. Ravikumar, Ankush Chauhan, G. S. Hikku, Kun-Yi Andrew Lin, Abbas Rahdar, Louis Hitler, Majid S. Jabir, Abdolrazagh Marzban, and Rajeshwari Oza. 2024. "Eco-Friendly Fabrication of CdO Nanoparticles Using Polyalthia Longifolia Leaves Extract for Antibacterial and Electrochemical Sensing Studies." *Journal of Sol-Gel Science and Technology* 110 (1): 221–32. https://doi.org/10.1007/s10971-024-06352-6.

Gomes-Junior, Paulo Cardoso, Gustavo Patelli Longatto, Karen Kenlderi de Lima Augusto, Josias da Silveira Rocha, Evandro Piccin, and Orlando Fatibello-Filho. 2024. "Synthesis of Ultrasmall Cerium Oxide Nanoparticles in Deep Eutectic Solvent and Their Application in an Electrochemical Sensor to Detect Dopamine in Biological Fluid." *Microchimica Acta* 191 (7): 425. https://doi.org/10.1007/s00604-024-06480-4.

Graham, Simon P., Hazim F. El-Sharif, Sabha Hussain, Rieke Fruengel, Rebecca K. McLean, Philippa C. Hawes, Mark V. Sullivan, and Subrayal M. Reddy. 2019. "Evaluation of Molecularly Imprinted Polymers as Synthetic Virus Neutralizing Antibody Mimics." *Frontiers in Bioengineering and Biotechnology* 7 (May). https://doi.org/10.3389/fbioe.2019.00115.

Guntupalli, Rajesh, Iryna Sorokulova, Eric Olsen, Ludmila Globa, Oleg Pustovyy, and Vitaly Vodyanoy. 2013. "Biosensor for Detection of Antibiotic Resistant Staphylococcus Bacteria." *Journal of Visualized Experiments*, no. 75 (May). https://doi.org/10.3791/50474.

Haroon, Naila, and Keith J. Stine. 2023. "Electrochemical Detection of Hormones Using Nanostructured Electrodes." *Coatings* 13 (12): 2040. https://doi.org/10.3390/coatings13122040.

Huang, Lin, Jingjing Wan, Xiang Wei, Yu Liu, Jingyi Huang, Xuming Sun, Ru Zhang, et al. 2017. "Plasmonic Silver Nanoshells for Drug and Metabolite Detection." *Nature Communications* 8 (1): 220. https://doi.org/10.1038/s41467-017-00220-4.

Keum, Do Hee, Su-Kyoung Kim, Jahyun Koo, Geon-Hui Lee, Cheonhoo Jeon, Jee Won Mok, Beom Ho Mun, et al. 2020. "Wireless Smart Contact Lens for Diabetic Diagnosis and Therapy." *Science Advances* 6 (17). https://doi.org/10.1126/sciadv.aba3252.

Khalid, Shaimaa A., Rabeay Y. A. Hassan, Rasha Mohamed El Nashar, and Ibrahim M. El-Sherbiny. 2022. "Voltammetric Determination of Salmonella Typhimurium in Minced Beef Meat Using a Chip-Based Imprinted Sensor." *RSC Advances* 12 (6): 3445–53. https://doi.org/10.1039/D1RA08526C.

Kieninger, Jochen, Yaara Tamari, Barbara Enderle, Gerhard Jobst, Joe Sandvik, Erik Pettersen, and Gerald Urban. 2018. "Sensor Access to the Cellular Microenvironment Using the Sensing Cell Culture Flask." *Biosensors* 8 (2): 44. https://doi.org/10.3390/bios8020044.

Kimmel, Danielle W., Gabriel LeBlanc, Mika E. Meschievitz, and David E. Cliffel. 2012. "Electrochemical Sensors and Biosensors." *Analytical Chemistry* 84 (2): 685–707. https://doi.org/10.1021/ac202878q.

Kumar, Vineet, Praveen Guleria, and Surinder Kumar Mehta. 2017. "Nanosensors for Food Quality and Safety Assessment." *Environmental Chemistry Letters* 15 (2): 165–77. https://doi.org/10.1007/s10311-017-0616-4.

Kusumoputro, Sydney, Shannon Tseng, Jonathan Tse, Christian Au, Candice Lau, Xiang Wang, and Tian Xia. 2020. "Potential Nanoparticle Applications for Prevention, Diagnosis, and Treatment of COVID-19." *VIEW* 1 (4). https://doi.org/10.1002/VIW.20200105.

Leau, Sorina-Alexandra, Cecilia Lete, and Stelian Lupu. 2023. "Nanocomposite Materials Based on Metal Nanoparticles for the Electrochemical Sensing of Neurotransmitters." *Chemosensors* 11 (3): 179. https://doi.org/10.3390/chemosensors11030179.

Li, Xiang-Ling, Shu Shan, Meng Xiong, Xing-Hua Xia, Jing-juan Xu, and Hong-Yuan Chen. 2013. "On-Chip Selective Capture of Cancer Cells and Ultrasensitive Fluorescence Detection of Survivin MRNA in a Single Living Cell." *Lab on a Chip* 13 (19): 3868. https://doi.org/10.1039/c3lc50587a.

Liu, Xun, Lin Huang, and Kun Qian. 2021. "Nanomaterial-Based Electrochemical Sensors: Mechanism, Preparation, and Application in Biomedicine." *Advanced NanoBiomed Research* 1 (6). https://doi.org/10.1002/anbr.202000104.

Liu, Zhirong, Jinhui Nie, Bin Miao, Jiadong Li, Yuanbo Cui, Shu Wang, Xiaodi Zhang, et al. 2019. "Self-Powered Intracellular Drug Delivery by a Biomechanical Energy-Driven Triboelectric Nanogenerator." *Advanced Materials* 31 (12). https://doi.org/10.1002/adma.201807795.

Luo, Lianghui, Feng Zhang, Chunyan Chen, and Changqun Cai. 2020. "Molecular Imprinting Resonance Light Scattering Nanoprobes Based on PH-Responsive Metal-Organic Framework for Determination of Hepatitis A Virus." *Microchimica Acta* 187 (2): 140. https://doi.org/10.1007/s00604-020-4122-1.

Magar, Hend S., Ehab E. Abu-El Magd, Rabeay Y.A. Hassan, and Asmaa M. Fahim. 2022. "Rapid Impedimetric Detection of Cadmium Ions Using Nanocellulose/Ligand/Nanocomposite (CNT/Co3O4)." *Microchemical Journal* 182 (November): 107885. https://doi.org/10.1016/j.microc.2022.107885.

March, Gregory, Tuan Nguyen, and Benoit Piro. 2015. "Modified Electrodes Used for Electrochemical Detection of Metal Ions in Environmental Analysis." *Biosensors* 5 (2): 241–75. https://doi.org/10.3390/bios5020241.

Mousavi Nodoushan, Somayeh, Navid Nasirizadeh, Jafar Amani, Raheleh Halabian, and Abbas Ali Imani Fooladi. 2019. "An Electrochemical Aptasensor for Staphylococcal Enterotoxin B Detection Based on Reduced Graphene Oxide and Gold Nano-Urchins." *Biosensors and Bioelectronics* 127 (February): 221–28. https://doi.org/10.1016/j.bios.2018.12.021.

Oliveira, Micaela, Pedro Conceição, Krishna Kant, Alar Ainla, and Lorena Diéguez. 2021. "Electrochemical Sensing in 3D Cell Culture Models: New Tools for Developing Better Cancer Diagnostics and Treatments." *Cancers* 13 (6): 1381. https://doi.org/10.3390/cancers13061381.

Phuong, Thao Dao Vu, Nguyen Thi Kieu Trang, and Phuong Dinh Tam. 2024. "Electrochemical Behavior of Palladium Nanoparticles Deposition on Glassy Carbon Electrode from Reline and Ethaline and Its Application in DNA Sensors." *Journal of Solid State Electrochemistry* 28 (8): 2777–86. https://doi.org/10.1007/s10008-024-05844-3.

Piletska, Elena V., Eugeny Mirkes, Stanislav S. Piletsky, Hasan Abosoglu, Alfeshani Cassim, Edmund Chu, Simon Doughty, et al. 2020. "Combinatorial Screening of Polymer Nanoparticles for Their Ability to Recognize Epitopes of AAV-neutralizing Antibodies." *Journal of Molecular Recognition* 33 (4). https://doi.org/10.1002/jmr.2824.

Piñón-Balderrama, Claudia Ivone, Claudia Alejandra Hernández-Escobar, Simón Yobanni Reyes-López, Alain Salvador Conejo-Dávila, Anayansi Estrada-Monje, and Erasto Armando Zaragoza-Contreras. 2024. "Non-Enzymatic Electrochemical Sensing of Glucose with a Carbon Black/Polyaniline/Silver Nanoparticle Composite." *Chemosensors* 12 (2): 26. https://doi.org/10.3390/chemosensors12020026.

Pitiphattharabun, Siraprapa, Krittin Auewattanapun, Thura Lin Htet, Myo Myo Thu, Gasidit Panomsuwan, Ratchatee Techapiesancharoenkij, Jun Ohta, and Oratai Jongprateep. 2024. "Reduced Graphene Oxide/Zinc Oxide Composite as an Electrochemical Sensor for Acetylcholine Detection." *Scientific Reports* 14 (1): 14224. https://doi.org/10.1038/s41598-024-64238-7.

Rabbani, Gulam, Mohammad Ehtisham Khan, Anwar Ulla Khan, Syed Kashif Ali, Mazin A. Zamzami, Abrar Ahmad, Abdullateef H. Bashiri, and Waleed Zakri. 2024. "Label-Free and Ultrasensitive Electrochemical Transferrin Detection Biosensor Based on a Glassy Carbon Electrode and Gold Nanoparticles." *International Journal of Biological Macromolecules* 256 (January): 128312. https://doi.org/10.1016/j.ijbiomac.2023.128312.

Robinson, Jendai E., William R. Heineman, Laura B. Sagle, M. Meyyappan, and Jessica E. Koehne. 2016. "Carbon Nanofiber Electrode Array for the Detection of Lead." *Electrochemistry Communications* 73 (December): 89–93. https://doi.org/10.1016/j.elecom.2016.11.002.

Shamsipur, Mojtaba, Afshin Pashabadi, Fatemeh Molaabasi, and Saman Hosseinkhani. 2017. "Impedimetric Monitoring of Apoptosis Using Cytochrome-Aptamer Bioconjugated Silver Nanocluster." *Biosensors and Bioelectronics* 90 (April): 195–202. https://doi.org/10.1016/j.bios.2016.11.050.

Sharma, Arun, Vepa Kameswara Rao, Dev Vrat Kamboj, Ritu Gaur, Mahabul Shaik, and Anchal Roy Shrivastava. 2016. "Enzyme Free Detection of Staphylococcal Enterotoxin B (SEB) Using Ferrocene Carboxylic Acid Labeled Monoclonal Antibodies: An Electrochemical Approach." *New Journal of Chemistry* 40 (10): 8334–41. https://doi.org/10.1039/C5NJ03460D.

Sharma, Sarda, P. N. Sidhartha, and Karumbaiah N. Chappanda. 2024. "Black Titanium Oxide Nanotubes Deposited with Fe (III) Oxide Nanoparticles for Electrochemical Sensing of Dopamine." *Journal of Electronic Materials* 53 (2): 631–41. https://doi.org/10.1007/s11664-023-10819-7.

Su, Liang, Wenzhao Jia, Changjun Hou, and Yu Lei. 2011. "Microbial Biosensors: A Review." *Biosensors and Bioelectronics* 26 (5): 1788–99. https://doi.org/10.1016/j.bios.2010.09.005.

Tarlekar, Pravin, and Sanghamitra Chatterjee. 2024. "Dendritic Platinum Nanoparticles Decorated Electrochemical Sensor for Immensely Sensitive Determination of Antineoplastic Drug Methotrexate." *Diamond and Related Materials* 148 (October): 111417. https://doi.org/10.1016/j.diamond.2024.111417.

Tian, Yulan, Liping Du, Ping Zhu, Yating Chen, Wei Chen, Chunsheng Wu, and Ping Wang. 2021. "Recent Progress in Micro/Nano Biosensors for Shellfish Toxin Detection." *Biosensors and Bioelectronics* 176 (March): 112899. https://doi.org/10.1016/j.bios.2020.112899.

Tolba, Mona, Minhaz Uddin Ahmed, Chaker Tlili, Fritz Eichenseher, Martin J. Loessner, and Mohammed Zourob. 2012. "A Bacteriophage Endolysin-Based Electrochemical Impedance Biosensor for the Rapid Detection of Listeria Cells." *The Analyst* 137 (24): 5749. https://doi.org/10.1039/c2an35988j.

Wyantuti, Santhy, Safri Ishmayana, and Yeni Wahyuni Hartati. 2015. "Voltammetric Determination of Cr(VI) Using Gold Nanoparticles-Modified Glassy Carbon Electrode." *Procedia Chemistry* 16: 15–23. https://doi.org/10.1016/j.proche.2015.12.010.

Xu, Jianguo, Xinxin Wang, Chao Yan, and Wei Chen. 2019. "A Polyamidoamine Dendrimer-Based Electrochemical Immunosensor for Label-Free Determination of Epithelial Cell Adhesion Molecule- Expressing Cancer Cells." *Sensors* 19 (8): 1879. https://doi.org/10.3390/s19081879.

Xu, Meng, Ronghui Wang, and Yanbin Li. 2017. "Electrochemical Biosensors for Rapid Detection of Escherichia Coli O157:H7." *Talanta* 162 (January): 511–22. https://doi.org/10.1016/j.talanta.2016.10.050.

Yoon, Jinho, Joungpyo Lim, Minkyu Shin, Ji-Young Lee, and Jeong-Woo Choi. 2022. "Recent Progress in Nanomaterial-Based Bioelectronic Devices for Biocomputing System." *Biosensors and Bioelectronics* 212 (September): 114427. https://doi.org/10.1016/j.bios.2022.114427.

Yoshiga, Toyoshi, Vida P. Hernandez, Ann M. Fallon, and John H. Law. 1997. "Mosquito Transferrin, an Acute-Phase Protein That Is up-Regulated upon Infection." *Proceedings of the National Academy of Sciences* 94 (23): 12337–42. https://doi.org/10.1073/pnas.94.23.12337.

Yu, Ji, Anneng Yang, Naixiang Wang, Haifeng Ling, Jiajun Song, Xi Chen, Yadong Lian, Zhishan Zhang, Feng Yan, and Min Gu. 2021. "Highly Sensitive Detection of Caspase-3 Activity Based on Peptide-Modified Organic Electrochemical Transistor Biosensors." *Nanoscale* 13 (5): 2868–74. https://doi.org/10.1039/D0NR08453K.

Zhao, Zongmin, Anvay Ukidve, Jayoung Kim, and Samir Mitragotri. 2020. "Targeting Strategies for Tissue-Specific Drug Delivery." *Cell* 181 (1): 151–67. https://doi.org/10.1016/j.cell.2020.02.001.

Zhou, Shiwei, Yingying Wang, and Jun-Jie Zhu. 2016. "Simultaneous Detection of Tumor Cell Apoptosis Regulators Bcl-2 and Bax through a Dual-Signal-Marked Electrochemical Immunosensor." *ACS Applied Materials & Interfaces* 8 (12): 7674–82. https://doi.org/10.1021/acsami.6b01010.

Zhu, Chengzhou, Guohai Yang, He Li, Dan Du, and Yuehe Lin. 2015. "Electrochemical Sensors and Biosensors Based on Nanomaterials and Nanostructures." *Analytical Chemistry* 87 (1): 230–49. https://doi.org/10.1021/ac5039863.

CHAPTER 4

Fabrication of Electrochemical Sensors for Determination of Toxic Dyes

Vandna Sharma, Pankaj Kumar, and Vinayak Adimule

4.1 INTRODUCTION

The proliferation of synthetic dyes in modern industries has led to significant advancements in the quality and appeal of products ranging from textiles and cosmetics to pharmaceuticals and food items. However, the extensive use of these dyes, especially in food products, has raised considerable health and safety concerns. Synthetic dyes are often employed in food manufacturing to enhance the visual appeal, compensate for natural colour losses, and ensure product consistency. While some dyes are approved for use by regulatory agencies, the presence of non-permitted or excessive levels of dyes in food items poses serious risks to consumer health. Natural and synthetic food dyes in the form of colouring agents are generally used to enhance the plating and visual representation of the prepared food [1–3]. The sources of natural dyes are, for example, plants, animals, and minerals, and these are considered safe to be used in foods;

DOI: 10.1201/9781003512639-4

however, these dyes also lack the vitality and stability of their synthetic counterparts. In the case of synthetic dyes, such as tartrazine (Tz), sunset yellow (SY), Allura Red, etc., a varied range of bright colours can be obtained; also, these are more stable under various processing conditions [4, 5]. The synthetic dyes are more harmful because of their toxicity level and allergenic properties, which require rigorous observation and regulation. Therefore, the consumption of synthetic dyes in the form of food generates many health implications, such as hyperactivity in children, allergic reactions, and even carcinogenic effects, so it is now a subject of ongoing research and discussion [1, 6]. In most foods, azo dyes are used, which can be broken down by metabolism to form aromatic amines and are associated with mutagenic and carcinogenic activities [4, 5]. Consequently, there is high demand to overcome these problems by finding suitable methods to detect and quantify these dyes in food items to ensure consumer safety and compliance with food safety regulations. Several analytical methods, such as chromatography, spectrophotometry, and electrophoresis, have been introduced for the determination of synthetic dyes in food products. Chromatographic techniques, such as thin-layer chromatography (TLC), high-performance liquid chromatography (HPLC), and ionic chromatography, have been widely used for their high resolution and sensitivity. Spectrophotometric methods, often integrated with chemometric tools, provide robust quantitative analysis and are valuable for their simplicity and effectiveness. Electrophoretic techniques also offer precise separation and identification of dye compounds [1]. However, traditional methods required laborious sample preparation and the use of toxic substances, and quantification was possible only by using expensive instruments. Thus, attention has shifted towards developing methods that require simple procedures for sample preparation, the use of safe chemicals, and cost-effective usage of the instruments [7]. Among these advancements, electrochemical (EC) sensors have been developed for the detection of toxic dyes in food and beverages due to their high sensitivity, specificity, fast response times, low detection limits, cost-effectiveness, and the possibility of miniaturizing these EC sensors into portable devices for on-site analysis. In these sensors, the dyes are determined by converting the chemical information of the analyte (toxic dyes) into an electrical signal as well as from the presence of functional groups that can undergo redox reactions [8].

The fabrication of EC sensors includes the selection of a suitable material for the electrode to interact selectively with the specific dye molecule, the integration of advanced materials, such as nanomaterials and conducting polymers, into sensor design, and the optimization of performance parameters such as selectivity, sensitivity, and stability [9–11]. Improved detection of the amount of toxic dyes in food matrices can be obtained using these materials, as they provide large surface areas, better conductivity, and appropriate functionalization with specific elements.

This chapter aims to explore the recent advancements in fabrication techniques, selection of electrode materials, and surface modification approaches used to prepare EC sensors for the determination of toxic dyes. Further, we have explained the role of various key factors in the performance of EC sensors and highlighted their potential applications with future perspectives.

4.1.1 Synthetic Dyes and Their Physical/Chemical Properties

Synthetic dyes, originating from William Henry Perkin's accidental synthesis of mauve from coal tar in 1865, are organic compounds produced artificially using raw materials like coal tar and aniline dyes [12, 13]. Their vibrant colours stem from complex organic molecules containing chromophore groups (such as N=N, -NO_2, -NO, xanthine, and coumarin) and auxochromic groups (-OH, -NH_2, SO_3H), enhancing their solubility and affinity to substrates, rendering them electrochemically active [14]. Widely employed in the food and beverage industry for their stability to light, oxygen, and pH, as well as their uniform coloration and economic viability compared with natural dyes, synthetic dyes like erythrosine, Allura Red, Tz, SY, Brilliant Blue, and indigo carmine (IC) are commonly found in various food products from candy to baked goods [15]. These dyes are classified into indigoid, triphenylmethane, azo, and xanthene groups based on their chemical structures, further categorized into water-soluble and fat-soluble types, which dictate their applications across different food types [6, 13]. Their chemical and physical properties vary with different functional groups and influence their stability, reactivity, solubility, and volatility [16]. Azo dyes, characterized by their -N=N- bond, have gained attention due to their intense colour and ease of synthesis, while indigoid dyes exhibit complex molecular structures that provide chemical stability and resistance to biodegradation [17]. Triphenylmethane dyes, mainly soluble in water with sulfonic groups, and xanthene dyes, insoluble in acids but stable in alkaline media with strong fluorescence, show diverse properties within this category of synthetic colorants [18]. The extensive use of these dyes in beverages, confectionery, bakery products, dairy, seafood, meats, and pet food underscores their importance, necessitating advanced analytical methods such as EC techniques for accurate detection and quantification to confirm food safety monitoring and regulatory assessments [15].

Continuous monitoring of dye presence and concentration in various foods and beverages is essential, and several analytical methods are employed for this purpose. These include spectrophotometric [19], electrophoretic [20], and chromatographic techniques [21], with chromatography being the most commonly used. However, EC methods, particularly voltammetry, have gained prominence due to their numerous advantages, such as high sensitivity, selectivity, and reproducibility, and low investment and maintenance costs [22]. The following section will discuss the application of these EC methods, with a focus on the modification in their surface properties.

4.1.2 Basic Principles of EC Sensors

EC sensors for the detection of toxic dyes in food and beverages operate on the principle of EC reactions occurring at the interface between the electrode and the analyte. These sensors rely on the redox properties of dye molecules, where an applied potential causes the dye molecules to either lose or gain electrons, resulting in measurable electrical signals. Typically, the sensor setup includes a working electrode, a reference electrode, and a counter electrode. The transducer converts chemical energy into electrical energy, displaying the detected analyte's electrical signals for further analysis [23]. A pictorial illustration of the EC sensor is shown in Figure 4.1. The working electrode is made of materials like platinum, gold or carbon, fluorine doped tin oxide (FTO), indium doped tin oxide (ITO) and redox reaction of the dye occurs on this electrode. A fixed potential is maintained at the reference electrode, ensuring accurate measurement of the potential changes at the working electrode, while the counter electrode completes the electrical circuit. The silver–silver chloride (Ag/AgCl) electrode is commonly used as a reference electrode in electrochemical sensors. Other options include the saturated calomel electrode (SCE) and the reversible hydrogen electrode (RHE). For the auxiliary or counter electrode, materials such as platinum foil, platinum wire, platinum mesh, or graphite are typically employed [24, 25].

Further, a brief overview of electrode fabrication is shown in Figure 4.2. The detection properties of EC sensors primarily depend on the working electrode because it is the site where the redox reactions of the analyte

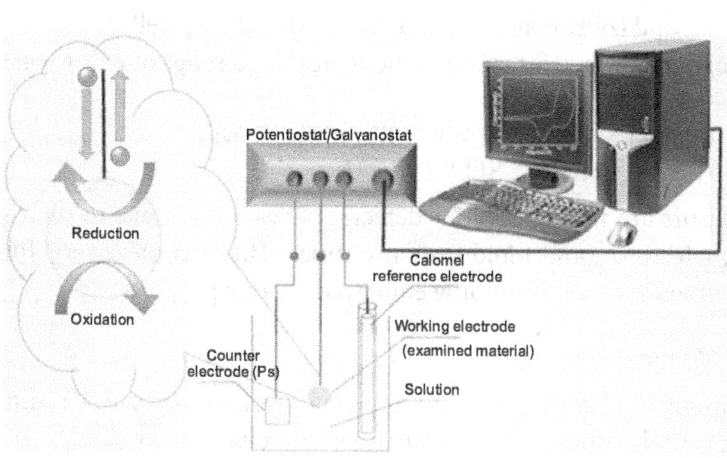

FIGURE 4.1 Pictorial illustration of the EC sensor. (Open Access Creative Common Attributions CC 4.0 from Bashir, S. et al., *Conducting Polymer Composites*, Central West Publishing, Australia, 2018.)

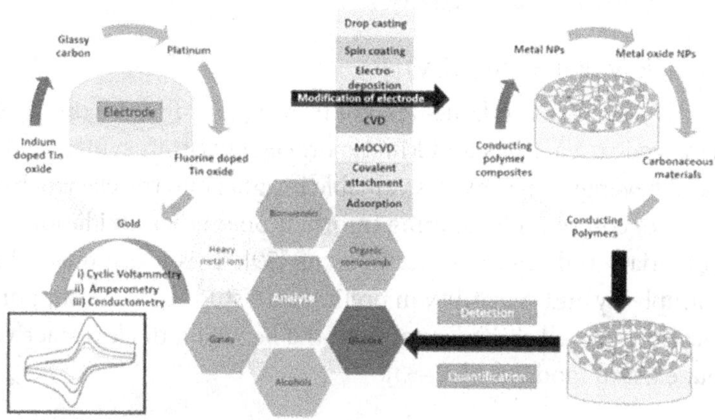

FIGURE 4.2 Brief overview of electrode fabrication. (Open Access Creative Common Attributions CC 4.0 from Bashir, S. et al., *Conducting Polymer Composites*, Central West Publishing, Australia, 2018.)

occur, directly influencing the sensitivity, selectivity, and overall performance of the sensor. Different electrode substrates such as glassy carbon, FTO, ITO, platinum, and gold are used in EC analysis due to their distinct EC properties, including conductivity, chemical stability, and surface characteristics [27, 28]. Glassy carbon is favoured for its wide potential window and low background currents, while FTO and ITO are transparent conductive oxides useful for optical–electrochemical applications.

Platinum and gold are noble metals known for their excellent conductivity and catalytic properties, making them ideal for a range of redox reactions.

4.2 DIFFERENT METHODS OF DETECTING DYES USING EC SENSORS

EC sensors are widely used for detecting dyes in food due to their sensitivity, selectivity, and rapid response times. This section covers different types of EC sensors commonly employed for this purpose.

4.2.1 Voltammetric Sensors

Voltammetric techniques, including cyclic voltammetry (CV), differential pulse voltammetry (DPV), square wave voltammetry (SWV), etc., are commonly employed for toxic dye determination. These sensors exploit the EC redox properties of the dye molecules, enabling their quantification based on the current response generated at different applied potentials. Various electrode materials and surface modifications can be utilized to enhance sensitivity and selectivity.

4.2.2 Cyclic Voltammetry (CV)

CV stands out as a fundamental EC technique utilized extensively for exploring redox reactions and kinetic processes within synthetic dyes in foods and beverages. It provides valuable insights into the electrochemical behaviour of compounds, offering information on their oxidation–reduction potentials and reaction mechanisms [29]. This technique is known for its simplicity and versatility in preliminary studies, and it supports the classification, as well as understanding the behavior, of electroactive species that exist in food dyes [30–33].

4.2.3 Square Wave Voltammetry (SWV) and Differential Pulse Voltammetry (DPV)

In SWV and DPV, potential pulses are applied to EC cells, and it is possible to analyse the results quantitatively due to their high sensitivity and ability to separate small amounts of synthetic dyes [34] in complex matrices like food and beverages. Therefore, this method provides robust systematic results, crucial for quality assurance and regulatory compliance [35].

4.2.4 Linear Sweep Voltammetry (LSV)

The rapid approach to EC analysis is linear sweep voltammetry (LSV); however, this is not an appropriate method to distinguish capacitive currents and trace analysis compared with the more sensitive methods like

DPV and SWV [36]. Despite these limitations, LSV is still a valuable method for preliminary testing and fast assessments of dye compositions in food items for testing [37].

4.2.5 Adsorptive Stripping Voltammetry (AdSV)

The adsorptive stripping voltammetry (AdSV) method is preferred to improve the sensitivity in trace analysis of synthetic dyes in food products. In this method, the electroactive species are deposited on the electrode surfaces of EC sensors, followed by an analysis of species during the redissolution phase, leveraging convection and diffusion processes to attain trace-level detection [38]. AdSV is specifically implemented in environments where conditions of high sensitivity and minimal interference are required to quantify the toxic dyes in food safety assessments [39, 40–42].

4.2.6 EC Biosensors

EC biosensors are fabricated to detect target dye analytes using biological elements, such as enzymes, antibodies, or nucleic acids, immobilized onto electrode surfaces. In the case of toxic dye detection, biosensors can be modified using specific enzymes or antibodies that are suitable to bind to the target dye molecules. The results are measured from the change in the EC signal obtained from the detected toxic dyes observed in complex sample matrices [26, 43, 44].

4.2.7 Impedimetric Sensors

Impedimetric sensors quantify the changes in the electrical impedance of the electrode–electrolyte interface upon the interaction with target dye molecules. These sensors utilize techniques such as EC impedance spectroscopy (EIS) to characterize the impedance response, which is influenced by factors such as dye concentration, charge transfer resistance, and solution conductivity. Impedimetric sensors offer label-free detection and can be highly sensitive to changes in dye concentration [45].

4.2.8 Conductometric Sensors

Conductometric sensors quantify the changes in the electrical conductivity of the solution resulting from the interaction between the target dye molecules and the electrode surface. These sensors typically utilize interdigitated electrode arrays or conductive polymers modified with specific recognition elements to enhance sensitivity and selectivity. Conductometric sensors provide fast response times and can detect toxic dyes in food products at low concentrations [46].

4.3 SURFACE MODIFICATION STRATEGIES

Surface modification approaches gained attention to enhance the sensitivity, selectivity, and stability of EC sensors fabricated for the determination of toxic dyes. One generally employed surface modification technique is the modification of electrode surfaces using the immobilization of molecular recognition elements, such as molecularly imprinted polymers (MIPs), aptamers, antibodies, or particular receptors to capture the toxic dye molecules. The synthetic polymer MIPs possess cavities with high affinity and selectivity towards target analyte molecules, representing the binding sites of natural receptors. Recent research work highlights the integration of MIPs with EC sensors for enhanced detection of azo dye with separation capability in complex food products [47, 48]. Further, EC sensors modified with carbonaceous materials, metal oxides, polymers, and nanoparticles (NPs) demonstrate increased sensitivity for azo dye detection [49]. MIP-based sensors utilize various sensing mechanisms, including EC, optical, and mass-sensitive approaches, providing efficient, sensitive, and low-cost detection methods [50]. These sensors overcome the limitations of traditional detection techniques, offering advantages such as rapidity, ease of use, and on-site analysis capabilities. Challenges remain in this approach, including electrode fouling and stability issues. Thus, ongoing research is focusing on exploring carbonaceous materials modified with metal oxides, polymers, and new nanomaterials–based EC sensors [47–49].

Various methods such as physical adsorption, self-assembled monolayers, and covalent coupling are implemented for the immobilization of antibodies or aptamers onto electrode surfaces, which is vital for the improvement of sensor performance [51, 52]. Aptamer-based biosensors use sandwich assays and conformational change–based approaches [52]. Recent developments include modifying a glassy carbon electrode (GCE) by adding calixarene and gold NPs, which revealed enhanced sensitivity and speedy detection of SY and Tz dyes in food products [53]. Various nanomaterials, such as metal NPs, carbon nanotubes, graphene oxide (GO), and quantum dots with unique characteristics including catalytic activity, kinetics of electron transfer, and higher surface area, are used for the fabrication of nanomaterial-based surface-modified electrodes. The functionalization of electrode surfaces with nanomaterials increases the adsorption of toxic dye molecules, which supports the electron transfer processes, resulting in improved signal amplification, biocompatibility, sensitivity, and resistance to fouling or interference from complex food products [54].

Thus, by the careful selection and optimization of surface modification strategies, researchers can modify the characteristics and performance of EC sensors for the sensitive and selective determination of toxic dyes and applicability in the monitoring and remediation of environmental samples.

4.4 DIFFERENT TYPES OF ELECTRODE-BASED EC SENSORS

EC sensors use several types of electrodes, which are applicable for specific applications and analytes. Procedures and steps performed in the development of different types of modified electrode–based EC sensors are shown in Figure 4.3. Carbon paste electrodes (CPEs) are versatile and cost-effective, commonly employed in the detection of heavy metals and organic pollutants. FTO electrodes and ITO electrodes are transparent and conductive, widely used in photoelectrochemical applications and

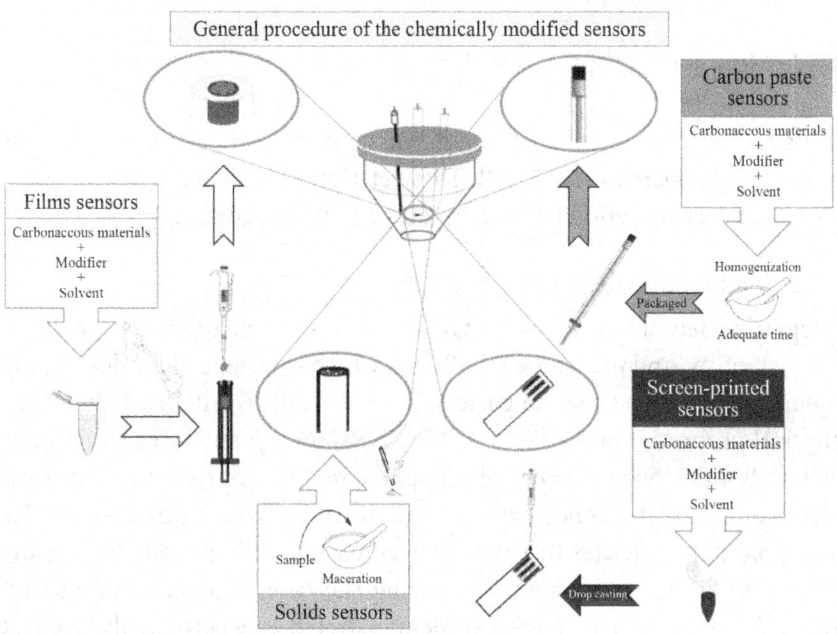

FIGURE 4.3 Steps involved and procedures followed in solid and modified EC sensors. (Adapted with kind permission from Springer Science+Business Media: *Food Analytical Methods*, Application of electrochemical sensors in the determination of synthetic dyes in foods or beverages and their toxicological effects on human health: A review, 15(9), 2022, 2394–2413, Christ-Ribeiro, A., Maciel, J.V., Bier, E.M., Pinto, J.S. and Dias, D.)

for sensors requiring optical transparency. GCEs are popular due to their broad potential range, chemical inertness, and ease of modification, making them ideal for detecting organic compounds and toxic dyes. Gold electrodes offer high conductivity and biocompatibility, and are often used in biosensors for detecting biomolecules such as glucose and DNA. Platinum electrodes are valued for their stability and catalytic properties, suitable for applications involving hydrogen and oxygen detection.

Screen-printed electrodes (SPEs) offer the advantage of mass production and disposability, used extensively in point-of-care testing and environmental monitoring. Modified electrodes, incorporating materials like NPs, graphene, and carbon nanotubes, significantly enhance sensitivity, selectivity, and stability for detecting a wide range of analytes in complex matrices. These various electrode materials and configurations enable electrochemical sensors to address diverse analytical challenges in fields such as environmental monitoring, food safety, clinical diagnostics, and industrial process control.

4.4.1 Modified Carbon Paste Electrode–Based Electrochemical Sensors

CPEs are prominent in electrochemical sensor applications due to their favourable properties [8, 55–59]. These attributes make CPEs highly suitable for detecting synthetic dyes in food and beverage samples.

4.4.2 Nanomaterials in Modified CPEs

Integrating nanomaterials with CPEs enhances their exchange rate (cations), stability, and mechanical stiffness, which is beneficial for developing chemically modified carbon paste sensors [55, 56]. The distinct chemical/physical properties of magnetic NPs (MNPs) incorporating carbonaceous materials have been successfully applied in the determination of dyes like Sudan II and SY. For example, modified nanostructured electrodes with gold nanoparticles (nAu) have been used to enhance the EC performance of CPEs, significantly improving the detection limits of SY and Tz [8]. The nanostructured sensor demonstrated an electrocatalytic effect for Sudan II with a detection limit of 0.0017 μM. The modification with nAu increased the effective electrode surface area, facilitating oxidation of SY and Tz and yielding detection limits of 3.0×10^{-8} and 2.0×10^{-9} mol l^{-1}, respectively. These values are comparable to those of other reported research works to detect SY using various methods such as electrochemical methods [60–65], including expanded graphite paste electrode [66], multi-walled carbon nanotubes (MWCNTs) and AuNPs–modified GCE

[67], poly(L-cysteine)-modified GCE [68], polyallylamine-modified electrode [69], B-doped diamond electrode [70], and GO and MWCNTs nanocomposite–modified electrode [71], as given in Table 4.1.

Furthermore, the determination of SY in food samples has been achieved using CPEs modified with bentonite clay particles and montmorillonite (MMT) nanoclay. The schematic of CPEs modified with nanoclay particles is shown in Figure 4.4. These modified electrodes exhibit improved resolution, sensitivity, and selectivity, with a detection limit of 2.6 nM compared with unmodified CPEs [55, 56]. Specifically, bentonite clay particle–modified CPE demonstrated a reversible, diffusion-controlled process with an equal number of protons and electrons, whereas MMT nanoclay–modified CPE showed a diffusion-controlled process with an unequal electron–proton transfer [55]. The electrochemical activity of SY was studied across various pH levels, with a pronounced peak observed at pH 6.0, as shown in Figure 4.4. The nanoclay-modified electrode exhibited remarkable electrocatalytic behaviour towards SY oxidation, demonstrating a strong linear response in the concentration range of 0.001 to 0.1 μM and achieving a low detection limit of 0.2 nM. These nanoclay-modified electrodes showcased excellent selectivity, stability, and sensitivity, making them highly effective for detecting SY at nanomolar levels in food and beverage samples.

4.4.3 Biocompatible Materials in CPEs

Another innovative approach involves using glucose combined with graphite powder to modify CPEs. The desirable properties of glucose, such as lower toxicity, chemical stability, biocompatibility, high mechanical strength, good adsorption in carbon powders, and high affinity towards organic molecules, make it an excellent modifier. This sensor was used to determine erythrosine in coated tablets, lake samples, and fruit juice by SWV, exhibiting sensitivity with a limit of detection (LOD) of 21.6 nM, accuracy (recovery from 82 to 98.5%), and a wide linear range (1.0×10^{-4} to 1.0×10^{-7} M) [59].

4.4.4 Redox Mediators in CPEs

The EC behaviour of indigo carmine has been studied using redox reagents to improve electron transfer capabilities. Redox reagents have shown good electrocatalytic activity and minimized surface fouling of the electrodes [57]. For instance, the use of 4-(4-nitrophenylazo) N-benzyl, N-ethylaniline as a mediator in carbon paste sensors improved the electrochemical behavior of IC, achieving a detection limit of 0.36 μmol l^{-1}

TABLE 4.1 Details of the Substrate, Modifier, Analyte, and Method Utilized for the Modification of Electrodes Used in Electrochemical Sensors and Comparison of Efficiencies

Substrate	Modifier	Analyte	Method	Detection limit ($\mu mol\ l^{-1}$)	Reference
GCE	Polyallylamine	Tz	SWV	1.8	Silva *et al.* (2007) [69]
GCE	MWCNT	SY	DPV	0.011	Song *et al.* (2010) [61]
Boron-Doped Diamond Electrode	–	SY	DPV	0.013	Medeiros *et al.* (2012) [70]
CPE	Gold nanoparticles	Tz	DPV	0.002	Ghoreishi *et al.* (2012) [8]
GCE	Graphene-wrapped phosphotungstic acid	SY	DPV	0.05	Gan *et al.* (2012) [62]
GPE	Expanded graphite paste electrode	SY	DPV	0.005	Zhang *et al.* (2013) [66]
GCE	Poly (L-cysteine)	SY	DPV	0.004	Zhang *et al.* (2013) [68]
	Multiwalled carbon nanotube/gold nanoparticles	SY	DPV	0.0403	Vikraman *et al.* (2014) [67]
GCE	Molecular imprinted ionic liquid polymer–ionic liquid functionalized graphene composite film (MIP-rGO-IL)	SY	DPV	0.004	Zhao *et al.* (2014) [63]
IL-EGPE	–	Ponceau 4R and Tz	SWSV	0.0014 and 0.003	Zhang *et al.* (2014) [75]
GCE	Hexadecyltrimethyl ammonium bromide functionalized graphene supported platinum nanoparticles composite (CTAB-Gr-Pt)	SY	DPV	0.0042	Yu *et al.* (2015) [64]
GCE	Gold nanoparticles and reduced graphene oxide (AuNP/RGO)	SY	DPV	0.002	Wang *et al.* (2015) [65]
GCE	IL-GO/MWCNT	Ponceau 4R and Tz	SWV	0.006 and 0.01	Wang *et al.* (2015) [93]
CPE	Al microfibre	Ponceau 4R and Tz	DPV	0.0008 and 0.002	Zhang *et al.* (2015) [77]

TABLE 4.1 Continued

Substrate	Modifier	Analyte	Method	Detection limit ($\mu mol\ l^{-1}$)	Reference
GCE	Porous carbon-2	Ponceau 4R and Tz	DPV	0.0035 and 0.0065	Cheng *et al.* (2015) [78]
	Graphene oxide and multiwalled carbon nanotubes (GO/MWCNTs) nanocomposite	SY	LSV	0.025	Qiu *et al.* (2016) [37]
CPE	Glucose combined with graphite powder	Erythrosine	SWV	0.0216	Nayak *et al.* (2016) [59]
CPE	4-(4-nitrophenylazo) N-benzyl, N-ethylaniline	Indigo carmine		0.36	Arvand *et al.* (2017) [57]
CPE	Montmorillonite (MMT) nanoclay	SY	SWV	0.0002	Shetti *et al.* (2018) [55]
CPE	ZnONPs	Sudan II	DPV	0.0017	Heydari *et al.* (2019) [58]
CPE	Bentonite clay particles	SY	DPV	0.0026	Moolya *et al.* (2019) [56]
CPE	Silica impregnated with cetylpyridinium chloride	Carmoisine	SWV	0.01	Chebotarev *et al.* (2020) [72]
GCE	TiO_2/electro-reduced graphene oxide (TiO_2/ErGO) nanocomposites	Ponceau 4R and Tz	AdSDPV	0.004 and 0.006	Qin *et al.* (2020) [74]
CPE	ZIF-8/g-C3N4/Co and 1-methyl-3-butylimidazolium bromide as an ionic liquid	D&C Red 33 and Patent Blue V	CV, DPV, and EIS	0.034	Maleh *et al.* (2022) [79]
GCE	Activated in phosphate buffersolution (PBS)	SY	LSV, DPV, SWV, CV	0.00167	Lu *et al.* (2022) [97]
CPE	Metallic glass modified (23Fe-21Cr-18Ni-20Ti-18Mn powders)	Methyl orange	CV	0.08	Rajendrachari *et al.* (2023) [80]
CPE	AuNPs	Congo Red dye	CV	0.07 μM in the 1–30 μM range and 0.7 μM in the 50–200 μM range	Ganash *et al.* (2023) [73]

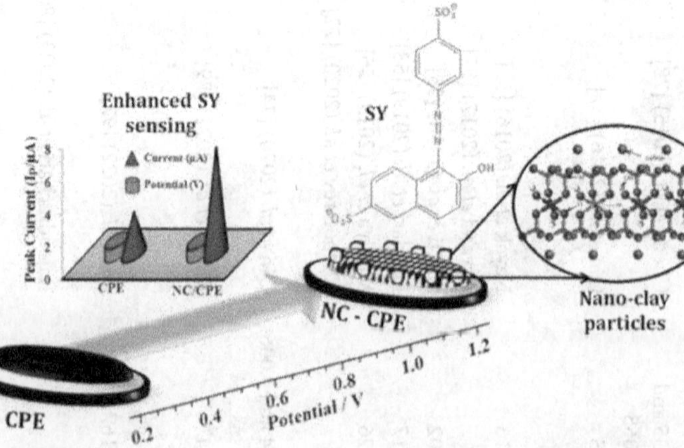

FIGURE 4.4 Schematic of nanoclay particles on carbon paste electrodes. Variation of peak current and potential for 0.01 mM SY in pH 6.0, phosphate buffer (I = 0.2 M); scan rate = 0.05 Vs^{-1}; accumulation time = 60 s. Chemical structure of Sunset yellow (SY). (Adapted from *Vacuum*, 155, Shetti, N.P., Nayak, D.S. and Malode, S.J., Electrochemical behavior of azo food dye at nanoclay modified carbon electrode-a nanomolar determination, 524–530, Copyright (2018),with permission from Elsevier.)

without pre-treatment of complex samples. This modified electrode was effectively applied for detecting IC in candy-coated chocolate, jelly powder, food dyes, and diazepam tablets.

4.4.5 CPEs with Silica Impregnated Sensors

CPEs modified with silica impregnated with cetylpyridinium chloride have been used to study the EC behaviour of Carmoisine. The sensor exhibited a single oxidation peak, indicating an irreversible process, with a wide linear range (0.08 to 1 µM), a low detection limit (0.01 µM), and high reliability, with a relative standard deviation (RSD) smaller than 5% [72]. Ganash *et al.* (2023) electrodeposited AuNPs on the CPE of an EC sensor and achieved high sensitivity and selectivity for the detection of Congo red (CR) dye. Further, there was significant reduction in the Rct value (896 Ω) of AuNPs/CPE compared with the bare CPE (1613 Ω), as observed from the Nyquist graphs shown in Figure 4.5 (A), with low detection limits of 0.07 µM in the 1–30 µM concentration range and 0.7 µM in the 50–200 µM range. CV measurements showed higher redox currents in the AuNP-modified CPE, exhibiting 0.017 and 0.018 mA cm^{-2} in the anodic and cathodic direction, respectively, compared with the bare CPE,

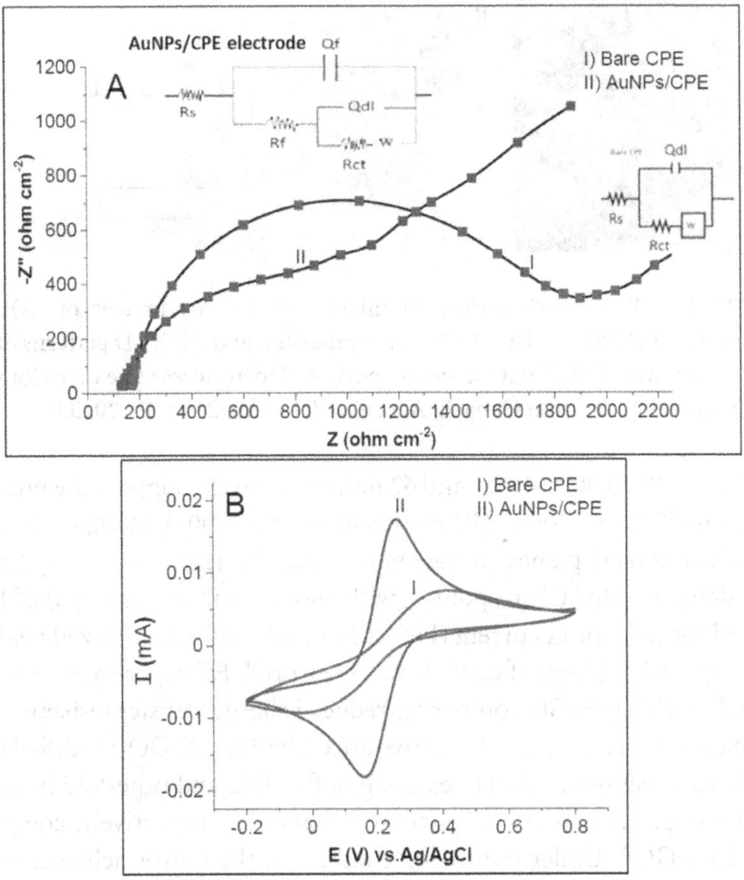

FIGURE 4.5 (A) Nyquist graphs of EIS measurements with the fitted equivalent circuits (insert images) and (B) CV graphs at 50 mV s^{-1} for (I) bare CPE and (II) AuNPs/CPE. (Open Access Creative Common Attributions CC 4.0 from Ganash, A. et al., *Molecules*, 28(1), 19, 2023.)

as shown in Figure 4.5 (B). The AuNPs/CPE sensor achieved an acceptable recovery range of 88–106% and was successfully applied to detect CR dye in different food samples [73].

Qin *et al.* (2020) [74] prepared a voltametric sensor using TiO_2/ErGO composites for the simultaneous detection of Ponceau 4R and Tz. TiO_2 NPs of cube-like structure with size ~30–35 nm were uniformly dispersed on GO nanoflakes, as shown in Figure 4.6 (A, B), creating an effective electrode interface for EC oxidation of the target compounds. Further, the purity of the TiO_2/ErGO composites was confirmed from the X-ray diffraction (XRD) data shown in Figure 4.6(C). Diffraction peaks at 25.26°,

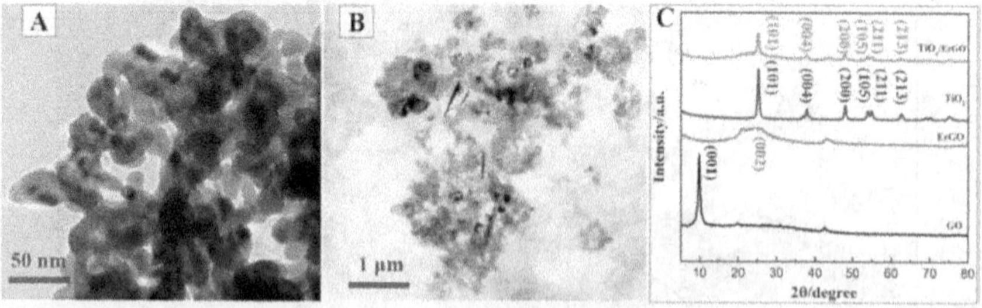

FIGURE 4.6 Transmission electron microscopy (TEM) images of (A) TiO_2 nanoparticles (NPs), (B) TiO_2/GO nanocomposites, and (C) XRD patterns of GO, ErGO, TiO_2, and TiO_2/ErGO nanocomposites. (Open Access Creative Common Attributions CC 4.0 from Ganash, A. et al., *Molecules*, 28(1), 19, 2023.)

36.92°, 47.94°, 53.98°, 54.90°, and 62.60°, respectively, support the presence of anatase TiO_2 NPs on ErGO substrates as (101), (004), (200), (105), (211), and (213) crystal planes (tetragonal, JCPDS 21-1272). The TiO_2/ErGO/GCE demonstrated CV responses, with anodic peak current (Ipa) of 148.2 μA and cathodic peak current (Ipc) of 144.1 μA, and a minimized peak-to-peak separation (ΔEp) of 0.097 V was measured. EC experiments demonstrated that TiO_2/ErGO composites reduced charge transfer resistance and significantly increased the EC active area. The TiO_2/ErGO-modified GCE exhibited synergistic effects, resulting in five-fold and nine-fold increases in anodic peak currents for ponceau 4R and Tz, respectively, compared with bare GCE. Under optimized conditions, the sensor achieved sensitive detection at nanomolar levels due to the straight line obtained in the Nyquist plot of TiO_2/ErGO/GCE, which reduced to the Rct value (44.9 Ω) and had low detection limits of 4.0 and 6.0 nM for Ponceau 4R and Tz, respectively, over a broad dynamic range (0.01–0.50 μM) [74].

In practical tests on orange juice samples, the sensor effectively detected both dyes at concentrations well below the maximum allowable limits, with satisfactory recovery rates and minimal interference from the juice matrix. The concentrations of Ponceau 4R and Tz in orange juice were 0.112 (67.7 μg l^{-1}) and 0.423 μM (226 μg l^{-1}), respectively, which are much lower than their maximum allowable dosages. The TiO_2/ErGO/GCE sensor is noted for its detection reliability, high sensitivity, selectivity, low cost, simple fabrication, and ease of operation, making it a promising tool for in-situ monitoring of azo colorants in food and beverage products, thus enhancing food safety and quality assurance processes [74–7576]. Compared with Al microfibre/CPE and porous C-2/GCE [77,

78], the fabrication of TiO_2/ErGO/GCE was simpler and more convenient. In short, the TiO_2/ErGO/GCE electrode possesses excellent sensing performance, low cost, and ease of fabrication for detection of toxic dyes. The sensing properties were comparable to or even exceeded the limits of previously reported electrodes (Table 4.1).

Maleh *et al.* (2022) [79] developed a highly sensitive and reproducible CPE modified with a ZIF-8/g-C_3N_4/Co nanocomposite (NC) and 1-methyl-3-butylimidazolium bromide as an ionic liquid (IL) for the first-time investigation of the EC behaviours of D&C Red 33 and Patent Blue V. The fabricated electrode demonstrated superior selectivity and sensitivity, achieving a wide linear concentration range of 0.08–10 μM (R^2 = 0.9906) and 10–900 μM (R^2 = 0.9932), with a low detection limit of 0.034 μM. The diffusion coefficient (D) and electron transfer coefficient (α) were calculated as 310×10^{-5} and 0.9, respectively. Using CV, DPV, and EIS, the sensor's performance was evaluated. Structural characterization of the NC was done using scanning electron microscopy (SEM), XRD, and Fourier transform infrared spectroscopy (FTIR). The sensor exhibited successful performance in determining D&C Red 33 and Patent Blue V in real samples, including mouthwashes and toothpaste, with recovery rates between 96% and 107%. This method paves the way for monitoring synthetic azo dyes in various products due to its high sensitivity and stability [79].

Rajendrachari *et al.* (2023) [80] synthesized nano-structured metallic glass powders with the composition 23Fe-21Cr-18Ni-20Ti-18Mn and modified a CPE (MG-MCPE) for detecting the carcinogenic dye Methyl Orange (MO) via CV. SEM images with energy-dispersive X-ray spectroscopy (EDS) analysis are shown in Figure 4.7. Panels (a–c) reveal that MG-MCPE had increased porosity, surface area, and roughness, resulting in a significantly higher current response (700 μA) compared with the bare CPE (BCPE, 85.12 μA). The optimal amount of the modifier was 6 mg, which showed strong electrocatalytic activity. The electrode surface area was determined using the Randles–Sevcik equation, yielding 0.4439 cm^2 for MG-MCPE versus 0.0546 cm^2 for BCPE. Scan rate studies indicated that MO oxidation involves both diffusion and adsorption-controlled processes. The pH studies showed MO's stability in a basic medium, with anodic peak currents decreasing as pH increased from 6 to 7.6. The limit of detection for MO using 6 mg MG-MCPE was 0.080 μM.

CPEs, especially when modified with nanomaterials, metal nanoparticles, redox mediators, and biocompatible materials, have proven to be

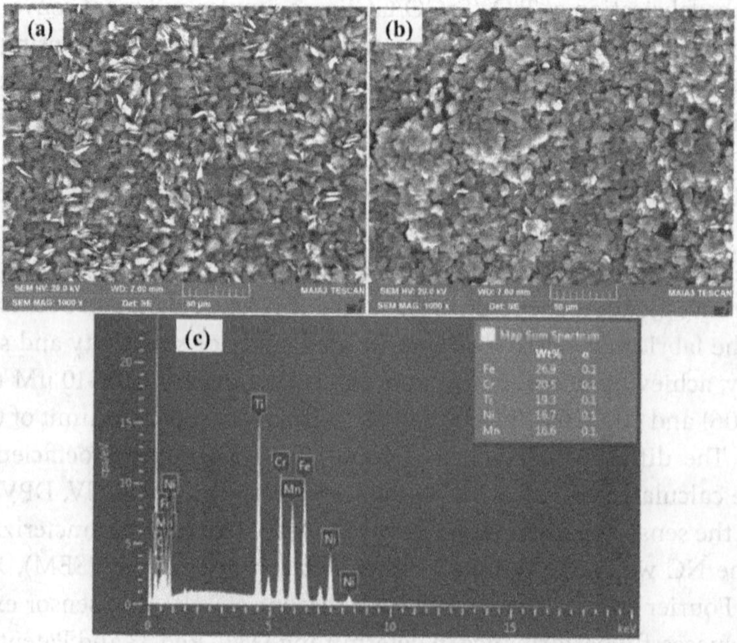

FIGURE 4.7 SEM images of the surface of (a) BCPE, (b) MG-MCPE (the presence of metallic glass powders on the surface of the carbon paste electrode is marked), and (c) EDS analysis of metallic glass powder. (Adapted from *Inorganic Chemistry Communications*, 155, Rajendrachari, S., Basavegowda, N., Vinaykumar, R., Narsimhachary, D., Somu, P. and Lee, M.J., Electrocatalytic determination of methyl orange dye using mechanically alloyed novel metallic glass modified carbon paste electrode by cyclic voltammetry, 111010, Copyright (2023), with permission from Elsevier.)

highly effective EC sensors for detecting toxic dyes in food and beverages. Their enhanced properties, including increased sensitivity, selectivity, and stability, and lower detection limits, make them invaluable tools for ensuring food safety and regulatory compliance.

4.4.6 Screen-Printed Electrode–Based EC Sensors

Screen printing is a widely used fabrication technique that involves depositing ink formulations containing electrode materials onto a substrate through a mesh screen. Screen printing has been found to be suitable for the mass production of EC sensors with flexibility in design of specific dimensions due to its unique characteristics, such as versatility, reproducibility, and scalability. Moreover, using screen printing, multiple electrode materials can be printed on a single substrate to prepare multi-sensor SPE arrays, which are suitable for simultaneous detection of different toxic dyes in food and beverages [81, 82].

4.4.7 Double-Stranded Copper (I) Helicate on SWCNT/SPCE and Rhenium NPs on Activated SPCE for the Detection of Amaranth, Tz, and SY

EC sensors were fabricated using a double-stranded copper (I) helicate (H) combined with a single-walled carbon nanotube–modified screen-printed carbon electrode (SWCNT/SPCE) by drop casting Cu (I) iodide and 1,3-bis(1H-benzotriazol-1-ylcarbonyl) benzene. Detection limits of 30.0 and 60.0 nM for amaranth and Tz dyes, respectively, were achieved using this EC sensor. These sensors were effectively used to examine dyes in synthetic orange juice and soft drinks [39]. In another technique, rhenium NPs were dispersed on activated carbon via drop casting onto the surface of SPCE. The performance of this sensor exhibited a detection limit of 16 nM for SY in multi-vitamin tablets with enhanced sensitivity and selectivity [42].

4.4.8 Boron-Doped Diamond Electrode for the Detection of Carmoisine E-122 Dye

The azo dye Carmoisine E-122 was detected in food and environmental samples using boron-doped diamond electrodes in the EC sensors. The anodic and cathodic processes in these electrodes showed linear analytical curves from 0.059 to 1.31 μmol l^{-1} and from 0.010 to 0.079 μmol l^{-1}, respectively. Additionally, the detection limits were 7.0 and 3.0 nmol l^{-1} for the anodic and cathodic processes, respectively, and the methods were observed to be reliable for analysing Carmoisine E-122 dye in the test samples [83].

4.4.9 Carbon Black–Polyethylene Composite Electrodes for the Detection of Tz and Brilliant Blue Dyes

A composite electrode composed of carbon black–polyethylene (CBPCE) with a renewable surface modified by carbon ink (CI) has been used for the voltammetric determination of Tz and Brilliant Blue (BB) in their mixtures. The method demonstrated a significant increase in the cathodic peak current for both dyes, allowing sensitive and selective determination by adjusting the pH of the medium. Detection limits were 0.019 μmol l^{-1} for Tz and 0.011 μmol l^{-1} for BB, with wide linear ranges of 0.037–1.38 μmol l^{-1} and 0.025–2.52 μmol l^{-1}, respectively [84].

4.4.10 Voltammetry of Immobilized Microparticles

An innovative screening method using voltammetry of immobilized microparticles on a simple graphite bar has been developed for identifying dyes such as SY, Tz, Brilliant Blue, indigotine, and erythrosine in culinary dye powders (yellow, blue, green, black, and red). This approach, which does not require prior sample preparation, elucidated the electrochemical behaviours of these compounds and offered a simpler, cost-effective alternative to traditional methods [85].

4.4.11 rGO-Methionine Modified SPCE for the Detection of Food Dyes

Food dyes are increasingly used in the food industry to enhance the visual appeal of products, necessitating stringent control over the levels of these dyes in food. To address this need, Akkapinyo *et al.* [86] constructed an rGO-methionine–modified SPCE for detecting the dyes amaranth, Tz, twilight yellow, and carminic acid. Figure 4.8 (a) shows a schematic illustration of the modification process of rGO-methionine/SPCE fabricated

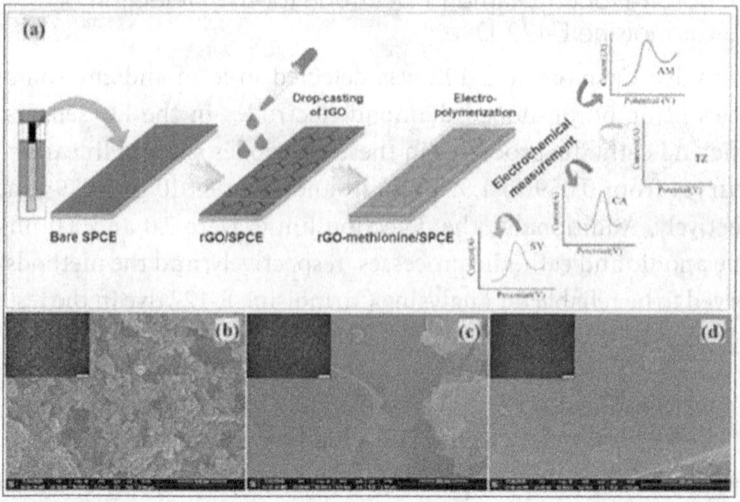

FIGURE 4.8 (a) Schematic representation of modification process of rGO-methionine/SPCE. SEM images of (b) bare SPCE, (c) rGO/SPCE, and (d) rGO-methionine/SPCE. The insets of (b–d) are the stereomicroscopic magnification at 10× of bare SPCE rGO/SPCE and rGO-methionine/SPCE, respectively. (Open Access Creative Common Attributions CC 4.0 from Akkapinyo, C. et al., *2019 12th biomedical engineering international conference (BMEiCON)* (pp. 1–4), IEEE, and Wu, J.H. and Lee, H.L., *Microchemical Journal*, 158, 105133, 2020.)

using two-steps drop-casting rGO and electropolymerizing a poly(L-methionine) film onto the SPCE. The poly(amino acid) component of the sensor enhanced its electrocatalytic activity and increased the number of active sites for dye interaction. rGO possesses a large surface area and has a 2D structure, serving as a robust conductive support for the dispersion of molecules. The performance of the rGO-methionine/SPCE was comparable to that of the classical UV-Vis analysis method. The sensor has similar recovery rates, indicating its viability and cost-effectiveness for the analysis of the dyes. Scanning electron microscopy showed the surface morphology of the modified electrode, as seen in Figure 4.8(b–d). The modified electrode exhibited significantly enhanced catalytic activity for the dyes, with optimal pH and accumulation times determined for each colorant. DPV revealed two linear detection ranges for each dye: 1 to 10 μM and 10 to 100 μM for amaranth, 1 to 10 μM and 10 to 85 μM for Tz, 1 to 10 μM and 10 to 50 μM for SY, and 1 to 20 μM and 20 to 60 μM for carminic acid. The LODs were calculated as 57 nM for amaranth, 41 nM for Tz, 48 nM for SY, and 36 nM for carminic acid. Additionally, the sensor showed high tolerance to interferants, repeatability, and enhanced performance in real sample analysis.

4.4.12 CuO/GO Nanocomposite–Modified SPCE for the Detection of Sudan I, Bisphenol A Dyes

Another study focused on the simultaneous detection of the food dye Sudan I and the additive Bisphenol A (found in food cans) using CuO/GO nanocomposites to modify an SPCE [88, 89]. This method demonstrated good sensitivity and excellent separation between the detection potentials of the analytes, indicating that the proposed sensor is simple and effective for analysing these substances.

Furthermore, solid electrodes have been applied for the determination and monitoring of dyes in food and beverages; however, their modifications tend to be more costly and labor-intensive compared with CPEs. Solid electrodes require thorough cleaning and polishing of their surfaces. Additionally, if film formation is involved, solvent evaporation is necessary. Depending on the immobilized material, sensor surface regeneration might be required between voltammetric measurements [32, 82]. Lipskikh *et al.* (2020) [83] created a carbon black–polyethylene composite electrode (CBPCE) featuring a renewable surface modified with carbon ink for the voltammetric analysis of Tz and Brilliant Blue FCF (BB) in mixtures. The EC characteristics of these dyes were examined using CV and linear sweep voltammetry (LSV) in a 0.1 mol l^{-1} Britton-Robinson (BR) buffer. Electrolytes with two different pH values, 2.0 for Tz and 10.0 for BB, were used to detect the two different dyes simultaneously. Linear concentration ranges of 0.037 to 1.38 μmol l^{-1} for Tz and 0.025 to 2.52 μmol l^{-1} for BB were achieved under optimal conditions of the experiment with LSV in the first-order derivative mode. The detection limits were found to be 0.019 and 0.011 μmol l^{-1} for Tz and BB, respectively. The novelty of this technique is the renewable surface of CBPCE and different pH levels of electrolytes for detecting the dyes simultaneously in candy and soft drink samples.

4.4.13 Film Sensors/GCE-based EC Sensors

In recent years, film-deposited EC sensors have been developed as advanced technology for the detection of toxic dyes in food and drinks. These sensors are typically fabricated by depositing thin films of active materials onto the electrodes' surface, enhancing their overall performance, including sensitivity, selectivity, reproducibility, robustness, etc. The following subsections include a systematic discussion on film-deposited EC sensors and their types, modifications, and applications.

4.4.13.1 Antimony Film Electrodes (SbFEs) for the Detection of Tz and Allura Red Dyes

Antimony film electrodes (SbFEs) can be used as alternative electrodes in the place of mercury electrodes due to their environment-friendly and low-toxicity nature. Serrano *et al.* (2016) [90] have observed the potential of SbFEs in "green analytical chemistry." Rodríguez *et al.* (2015) [91] observed that when SbFE was coupled with sequential injection, the analysis revealed the effective determination of dyes such as Tz and Allura Red in food testing samples, achieving a detection limit of 0.3 μmol l^{-1}. The advantage of SbFE electrodes over traditional electrodes is that they can be renewed easily online without disturbing analytical parameters, whereas traditional electrodes require extensive polishing.

4.4.13.2 Carbon Nanotube Films for the Detection of Tz, SY, and Carmoisine Dyes

Sierra-Rosales *et al.* (2017) [92] modified GCEs using MWCNTs combined with 1,3-dioxolane as a dispersing agent. The Tz, SY, and Carmoisine (CR) in commercial soft drinks were detected with the improved electrocatalytic activities, sensitivity, and stability of the EC sensors. The modified

MWCNT/GCE sensors showed higher current responses compared with GCEs without modification. The results were consistent with chromatographic methods but offered the advantage of shorter analysis times.

4.4.13.3 Metallic NP Films for the Detection of SY and Tz Dyes

Wang *et al.* (2015) [93] detected SY and Tz in drink samples simultaneously by modifying the GCEs using green synthesized ionic liquid 1-allyl-3-methylimidazolium chloride functionalized rGO supported by Au NPs (ILRGO-Au) composite films. Thirumalraj *et al.* (2018) [94] fabricated a Pd-Ru nanoparticle-incorporated carbon aerogel (Pd-Ru/CA) for the detection and catalytic degradation of SY. These techniques made these modified SPCEs valuable for their better recovery rates, high stability, accuracy, and robustness, making them suitable for practical applications in food quality assurance.

4.4.14 rGO and MNPs for the Detection of Orange II and Rhodamine B Dyes

Yun *et al.* (2015) [95] fabricated a sensor by modifying a GCE using rGO grafted with 5-amino-1,3,4-thiadiazole-2-thiol and Pt NPs (ERGO-ATDT-Pt) and obtained a sensor of good electrocatalytic activity with low detection limits for orange II dye. Shah *et al.* (2018) [96] also prepared a similar sensor using AgNPs functionalized with carbamazepine (Cbz-AgNPs) for the real-time determination of both orange II and rhodamine B dyes simultaneously. These sensors were recognized for their reproducibility, wide dynamic range, and high sensitivity, which make them highly favourable for detecting toxins in food items.

4.4.15 IL and Graphene Oxide (GO) Films for the Detection of Sudan Dyes (I–IV)

Chailapakul *et al.* (2008) [30] detected the Sudan dyes (I–IV) using a MWCNT ionic liquid gel–modified GCE (MWCNTs-IL-Gel/GCE) and attaching a high-performance liquid chromatograph (HPLC) with an EC detector. This method enhanced the EC performance of the electrode by influencing the properties of carbon nanotubes and ionic liquids. Stable amperometric measurements were obtained using MWCNTs-IL-Gel/GCE, and it was found to be very sensitive and effective to detect low levels of Sudan compounds. One important advantage of this technique is that measurements can be performed without blank interference, as the common problem observed in EC is that redox reactions occur due to non-target substances. The robustness of this method was confirmed by the coupling of flow-injection analysis (FIA) with amperometric detection. The sensitivity and selectivity of MWCNTs-IL-Gel/GCE were confirmed by examining the Sudan dyes in complex matrices, for example, soft drink samples. In many countries, Sudan dyes are banned because of their carcinogenic properties. Therefore, this technique is significant because of high-precision detection without significant interference from a food safety point of view. Moreover, based on this study, this technique could be applicable for monitoring and estimating Sudan dyes in different food products. Also, it will improve the overall safety and quality declaration procedures in the food industry. Qiu *et al.* (2016) [37] also used graphene oxide (GO) with MWCNTs for the fabrication of EC sensors and detected SY and Tz dyes in orange juice. The dispersant GO behaves as a signal amplifier in the EC sensor and shows high sensitivity, a wide linear range, and a low detection limit.

Lu *et al.* (2022) [97] detected SY by SWV using the fabricated EC-activated GCE (AGCE) as shown in Figure 4.9 (a). The enhancement of oxidation peak current for SY was obtained due to the good conductivity of AGCE compared with non-activated GCE. Further, it was reported that when the SY concentration (0.005 to 1.0 μM) increased, a linear increase in current was obtained, and the detection limit was 0.00167 μM (S/N = 3). The method was also implemented to detect SY in beverage samples and attained recovery rates between 96.19 and 103.47%, which is appropriate for real-world applications, confirming food safety and quality due to better catalytic activity, selectivity, and repeatability for SY detection using an AGCE sensor.

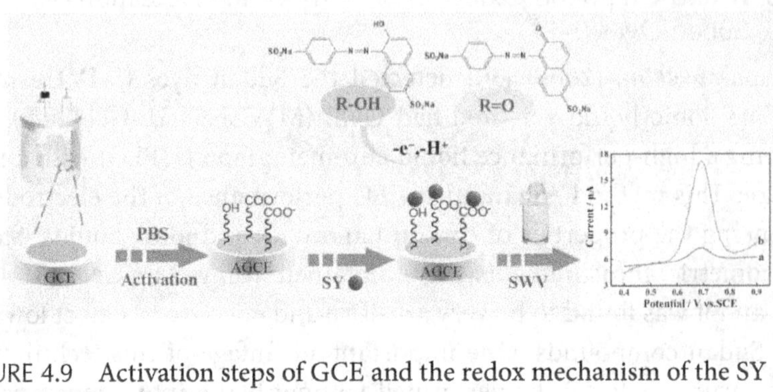

FIGURE 4.9 Activation steps of GCE and the redox mechanism of the SY at the AGCE. (Open Access Creative Common Attributions CC 4.0 from Lu, Y. et al., *Molecules*, 27(16), 5221, 2022.)

4.5 RECENT ADVANCEMENTS AND FUTURE PERSPECTIVES

Recent developments in the fabrication of EC sensors for the determination of toxic dyes have been explored. These enhanced the performance of EC sensors by integrating nanomaterials such as carbon-based materials, MNPs, and NC into electrode surfaces. The enhancement in the detection of sensitivity and selectivity of toxic dyes using these sensors was obtained due to the high surface area, catalytic activity, and biocompatibility of the nanomaterials used. Furthermore, for the detection of toxic dyes in complex food and beverage samples, other recognized elements were also introduced, such as MIPs, aptamers, and biomimetic receptors to immobilize onto the electrodes for surface modification, which provided specific binding sites for target analyte molecules. Another advancement is the miniaturization of EC sensors for making these sensors portable to perform on-site analysis of toxic dyes in food products and environmental monitoring applications in resource-limited conditions with rapid response, low detection limits, and ease of operation.

Looking ahead, the future perspectives in the field of detecting toxic dyes in food products using EC sensors include the fabrication of multifunctional EC sensors proficient in detecting multiple analytes simultaneously; the investigation of advanced methods for data analysis, such as artificial intelligence and machine learning, for enhanced quantification and identification of toxic dyes in food and beverages; and the coupling of renewable energy sources, such as solar cells or microbial fuel cells, into EC sensors to make the sensors self-powered for their applicability in remote or off-grid areas. Continuous research efforts and interdisciplinary collaborations are crucial for developing these advancements into impactful technologies that will contribute to environmental sustainability and human health.

REFERENCES

1. Kucharska, M. and Grabka, J., 2010. A review of chromatographic methods for determination of synthetic food dyes. *Talanta*, *80*(3), pp.1045–1051.
2. Wu, D. and Sun, D.W., 2013. Advanced applications of hyperspectral imaging technology for food quality and safety analysis and assessment: A review—Part I: Fundamentals. *Innovative Food Science & Emerging Technologies*, *19*, pp.1–14.
3. Christ-Ribeiro, A., Maciel, J.V., Bier, E.M., Pinto, J.S. and Dias, D., 2022. Application of electrochemical sensors in the determination of synthetic dyes in foods or beverages and their toxicological effects on human health: A review. *Food Analytical Methods*, *15*(9), pp.2394–2413.

4. Chung, K.T. and Cerniglia, C.E., 1992. Mutagenicity of azo dyes: Structure-activity relationships. *Mutation Research/Reviews in Genetic Toxicology, 277*(3), pp.201–220.
5. Calam, T.T., 2024. A new, fast and selective electrochemical sensor based on Bismuthiol-I for the voltammetric detection of tartrazine, a synthetic food dye, in soft drinks. *Microchemical Journal, 197*, p.109907.
6. König, J., 2015. Food colour additives of synthetic origin. In *Colour additives for foods and beverages* (pp. 35–60). Woodhead Publishing.
7. Komissarchik, S. and Nyanikova, G., 2014. Test systems and a method for express detection of synthetic food dyes in drinks. *LWT-Food Science and Technology, 58*(2), pp.315–320.
8. Ghoreishi, S.M., Behpour, M. and Golestaneh, M., 2012. Simultaneous determination of sunset yellow and tartrazine in soft drinks using gold nanoparticles carbon paste electrode. *Food Chemistry, 132*(1), pp.637–641.
9. Şişman, E. and Ağın, F., 2024. Voltammetric methods for determination of Allura Red AC in foods and beverages. *Turkish Journal of Analytical Chemistry, 6*(1), pp.40–49.
10. Chaudhari, S.S., Patil, P.O., Bari, S.B. and Khan, Z.G., 2024. A comprehensive exploration of tartrazine detection in food products: Leveraging fluorescence nanomaterials and electrochemical sensors: Recent progress and future trends. *Food Chemistry, 433*, p.137425.
11. Santra, S., Datta, D., Biswas, S. and Das, B., 2024. Role of Silica, Carbon, and ZnO Nanomaterials in the fabrication of electrochemical sensors for the detection of water contaminants and food dye. *Particle & Particle Systems Characterization, 41*, p.2400020.
12. Oliveira, A.R.M.D. and Szczerbowski, D., 2009. Quinina: 470 anos de história, controvérsias e desenvolvimento. *Química Nova, 32*, pp.1971–1974.
13. Rovina, K., Prabakaran, P.P., Siddiquee, S. and Shaarani, S.M., 2016. Methods for the analysis of sunset yellow FCF (E110) in food and beverage products-a review. *TrAC Trends in Analytical Chemistry, 85*, pp.47–56.
14. Bessegato, G.G., Brugnera, M.F. and Zanoni, M.V.B., 2019. Electroanalytical sensing of dyes and colorants. *Current Opinion in Electrochemistry, 16*, pp.134–142.
15. Dey, S. and Nagababu, B.H., 2022. Applications of food color and bio-preservatives in the food and its effect on the human health. *Food Chemistry Advances, 1*, p.100019.
16. Zanoni, M.V.B. and Yamanaka, H.I.D.E.K.O., 2016. *Caracterização química, toxicológica, métodos de detecção e tratamento*. Cultura Acadêmica.
17. Chaves, K.O., Monteiro, C.R.L., Muniz, C.R., Gomes, R.B. and Buarque, H.L.D.B., 2008. Adsorção de índigo carmim em biomassas mortas de aspergillus niger. *Engenharia sanitária e ambiental, 13*, pp.351–355.
18. Prado, M.A. and Godoy, H.T., 2003. Corantes artificiais em alimentos. *Alimentos e Nutrição, 14*(2), pp.237–250.
19. Di Anibal, C.V., Ruisánchez, I., Fernández, M., Forteza, R., Cerdà, V. and Callao, M.P., 2012. Standardization of UV-visible data in a food adulteration classification problem. *Food Chemistry, 134*(4), pp.2326–2331.

20. da Silva Neto, G.F., de Andrade Rodrigues, M.L. and Fonseca, A., 2021. A new quantitative gel electrophoresis method with image-based detection for the determination of food dyes and metallic ions. *Talanta, 221*, p.121602.
21. Vlase, L., Muntean, D., Cobzac, S.C. and Filip, L., 2014. Development and validation of an HPLC-UV method for determination of synthetic food colorants. *Revue Roumaine de Chimie, 59*(9), pp.719–725.
22. Durigon, A.M.M., da Silveira, G.D., Sokal, F.R., Pires, R.A.C.V. and Dias, D., 2020. Food dyes screening using electrochemistry approach in solid state: The case of sunset yellow dye electrochemical behavior. *Journal of Solid State Electrochemistry, 24*, pp.2907–2921.
23. Stetter, J.R., Penrose, W.R. and Yao, S., 2003. Sensors, chemical sensors, electrochemical sensors, and ECS. *Journal of The Electrochemical Society, 150*(2), p.S11.
24. Stradiotto, N.R., Yamanaka, H. and Zanoni, M.V.B., 2003. *Electrochemical sensors: A powerful tool in analytical chemistry.* Sociedade Brasileira de Química.
25. Hitchman, M.L. and Berlouis, L.E.A., 1995. Electrochemical methods. In *Process analytical chemistry* (pp. 217–258). Springer Netherlands.
26. Bashir, S., Ramesh, S., Ramesh, K., Numan, A. and Iqbalc, J., 2018. Conducting polymer composites in electrochemical sensors. *Conducting Polymer Composites Central West Publishing, Australia*, pp.41–68.
27. Yang, C., Denno, M.E., Pyakurel, P. and Venton, B.J., 2015. Recent trends in carbon nanomaterial-based electrochemical sensors for biomolecules: A review. *Analytica Chimica Acta, 887*, pp.17–37.
28. Omar, F.S., Numan, A., Duraisamy, N., Bashir, S., Ramesh, K. and Ramesh, S., 2017. A promising binary nanocomposite of zinc cobaltite intercalated with polyaniline for supercapacitor and hydrazine sensor. *Journal of Alloys and Compounds, 716*, pp.96–105.
29. Brett, C.M. and Brett, O., 1993. Principles, methods, and applications. *Electrochemistry, 67*(2), p.444.
30. Chailapakul, O., Wonsawat, W., Siangproh, W., Grudpan, K., Zhao, Y. and Zhu, Z., 2008. Analysis of sudan I, sudan II, sudan III, and sudan IV in food by HPLC with electrochemical detection: Comparison of glassy carbon electrode with carbon nanotube-ionic liquid gel modified electrode. *Food Chemistry, 109*(4), pp.876–882.
31. Yun, M., Choe, J.E., You, J.M., Ahmed, M.S., Lee, K., Üstündağ, Z. and Jeon, S., 2015. High catalytic activity of electrochemically reduced graphene composite toward electrochemical sensing of Orange II. *Food Chemistry, 169*, pp.114–119.
32. Dossi, N., Piccin, E., Bontempelli, G., Carrilho, E. and Wang, J., 2007. Rapid analysis of azo-dyes in food by microchip electrophoresis with electrochemical detection. *Electrophoresis, 28*(22), pp.4240–4246.
33. Yilmaz, U.T., Ergun, F. and Yilmaz, H., 2014. Determination of the food dye carmine in milk and candy products by differential pulse polarography. *Journal of Food and Drug Analysis, 22*(3), pp.329–335.

34. Mirceski, V., Gulaboski, R., Lovric, M., Bogeski, I., Kappl, R. and Hoth, M., 2013. Square-wave voltammetry: A review on the recent progress. *Electroanalysis, 25*(11), pp.2411–2422.
35. Wang, M. and Zhao, J., 2015. Facile synthesis of Au supported on ionic liquid functionalized reduced graphene oxide for simultaneous determination of Sunset yellow and Tartrazine in drinks. *Sensors and Actuators B: Chemical, 216*, pp.578–585.
36. Bard, A.J. and Faulkner, L.R., 1983. Electrochemical methods: Fundamentals and applications. *Surf Technologies, 20*(1), pp.91–92.
37. Qiu, X., Lu, L., Leng, J., Yu, Y., Wang, W., Jiang, M. and Bai, L., 2016. An enhanced electrochemical platform based on graphene oxide and multi-walled carbon nanotubes nanocomposite for sensitive determination of Sunset Yellow and Tartrazine. *Food Chemistry, 190*, pp.889–895.
38. Kalvoda, R. and Kopanica, M., 1989. Adsorptive stripping voltammetry in trace analysis. *Pure and Applied Chemistry, 61*(1), pp.97–112.
39. Nuñez-Dallos, N., Macías, M.A., García-Beltrán, O., Calderón, J.A., Nagles, E. and Hurtado, J., 2018. Voltammetric determination of amaranth and tartrazine with a new double-stranded copper (I) helicate-single-walled carbon nanotube modified screen printed electrode. *Journal of Electroanalytical Chemistry, 822*, pp.95–104.
40. Gómez, M., Arancibia, V., Aliaga, M., Núñez, C. and Rojas-Romo, C., 2016. Determination of Sudan I in drinks containing Sunset yellow by adsorptive stripping voltammetry. *Food Chemistry, 212*, pp.807–813.
41. Prabakaran, E. and Pandian, K., 2015. Amperometric detection of Sudan I in red chili powder samples using Ag nanoparticles decorated graphene oxide modified glassy carbon electrode. *Food Chemistry, 166*, pp.198–205.
42. Veerakumar, P., Rajkumar, C., Chen, S.M., Thirumalraj, B. and Lin, K.C., 2018. Activated porous carbon supported rhenium composites as electrode materials for electrocatalytic and supercapacitor applications. *Electrochimica Acta, 271*, pp.433–447.
43. Zare, E.N., Motahari, A. and Sillanpää, M., 2018. Nanoadsorbents based on conducting polymer nanocomposites with main focus on polyaniline and its derivatives for removal of heavy metal ions/dyes: A review. *Environmental Research, 162*, pp.173–195.
44. Cardenas-Riojas, A.A., Calderon-Zavaleta, S.L., Quiroz-Aguinaga, U., Muedas-Taipe, G., Carhuayal-Alvarez, S.M., Ascencio-Flores, Y.F., Ponce-Vargas, M. and Baena-Moncada, A.M., 2024. Modified electrochemical sensors for the detection of selected food azo dyes: A review. *ChemElectroChem, 11*(4), p.e202300490.
45. Stanković, D.M., Ognjanović, M., Espinosa, A., del Puerto Morales, M., Bessais, L., Zehani, K., Antić, B. and Dojcinović, B., 2019. Iron oxide nanoflower–based screen print electrode for enhancement removal of organic dye using electrochemical approach. *Electrocatalysis, 10*, pp.663–671.
46. Majeed, S., Naqvi, S.T.R., ul Haq, M.N. and Ashiq, M.N., 2022. Electroanalytical techniques in biosciences: Conductometry, coulometry, voltammetry, and electrochemical sensors. In *Analytical Techniques in Biosciences* (pp. 157–178). Academic Press.

47. Zhou, S., Liu, C., Lin, J., Zhu, Z., Hu, B. and Wu, L., 2022. Towards development of molecularly imprinted electrochemical sensors for food and drug safety: Progress and trends. *Biosensors, 12*(6), p.369.
48. Elfadil, D., Lamaoui, A., Della Pelle, F., Amine, A. and Compagnone, D., 2021. Molecularly imprinted polymers combined with electrochemical sensors for food contaminants analysis. *Molecules, 26*(15), p.4607.
49. Cardenas-Riojas, A.A., Calderon-Zavaleta, S.L., Quiroz-Aguinaga, U., López, E.O., Ponce-Vargas, M. and Baena-Moncada, A.M., 2023. Evaluation of an electrochemical sensor based on gold nanoparticles supported on carbon nanofibers for detection of tartrazine dye. *Journal of Solid State Electrochemistry, 27*(8), pp.1969–1982.
50. Cao, Y., Feng, T., Xu, J. and Xue, C., 2019. Recent advances of molecularly imprinted polymer-based sensors in the detection of food safety hazard factors. *Biosensors and Bioelectronics, 141*, p.111447.
51. Qiao, L., Jia, H., Sun, X. and Wang, X., 2014. Recent advance of electrochemical immunosensor for pesticide residues detection. *Sensors & Transducers, 164*(2), p.6.
52. Radi, A.E., 2011. Electrochemical aptamer-based biosensors: Recent advances and perspectives. *International Journal of Electrochemistry, 2011*(1), p.863196.
53. Jawahar, C.P. and Jana, A.K., 2023. Sensing of harmful dyes in food particles using nanomaterial-based electrochemical analysis. *Migration Letters, 20*(S13), pp.188–195.
54. Tarannum, N., Gautam, A., Chauhan, T. and Kumar, D., 2024. Nanomaterial based sensors for detection of food contaminants: A prospect. *Sensing Technology, 2*(1), p.2373196.
55. Shetti, N.P., Nayak, D.S. and Malode, S.J., 2018. Electrochemical behavior of azo food dye at nanoclay modified carbon electrode-a nanomolar determination. *Vacuum, 155*, pp.524–530.
56. Moolya, C.V., Shetti, N.P. and Nayak, D.S., 2019. Clay coated carbon electrode sensor for a food dye sunset yellow. *Materials Today: Proceedings, 18*, pp.1116–1123.
57. Arvand, M., Saberi, M., Ardaki, M.S. and Mohammadi, A., 2017. Mediated electrochemical method for the determination of indigo carmine levels in food products. *Talanta, 173*, pp.60–68.
58. Heydari, M., Ghoreishi, S.M. and Khoobi, A., 2019. Novel electrochemical procedure for sensitive determination of Sudan II based on nanostructured modified electrode and multivariate optimization. *Measurement, 142*, pp.105–112.
59. Nayak, D.S. and Shetti, N.P., 2016. A novel sensor for a food dye erythrosine at glucose modified electrode. *Sensors and Actuators B: Chemical, 230*, pp.140–148.
60. Gómez, M., Arancibia, V., Rojas, C. and Nagles, E., 2012. Adsorptive stripping voltammetric determination of tartrazine and sunset yellow in gelatins and soft drink powder in the presence of cetylpyridinium bromide. *International Journal of Electrochemical Science, 7*(8), pp.7493–7502.

61. Song, Y.Z., 2010. Electrochemical reduction of sunset yellow at a multi-walled carbon nanotube (MWCNT)-modified glassy carbon electrode and its analytical application. *Canadian Journal of Chemistry, 88*(7), pp.676–681.
62. Gan, T., Sun, J., Cao, S., Gao, F., Zhang, Y. and Yang, Y., 2012. One-step electrochemical approach for the preparation of graphene wrapped-phosphotungstic acid hybrid and its application for simultaneous determination of sunset yellow and tartrazine. *Electrochimica Acta, 74*, pp.151–157.
63. Zhao, L., Zhao, F. and Zeng, B., 2014. Preparation and application of sunset yellow imprinted ionic liquid polymer– ionic liquid functionalized graphene composite film coated glassy carbon electrodes. *Electrochimica Acta, 115*, pp.247–254.
64. Yu, L., Shi, M., Yue, X. and Qu, L., 2015. A novel and sensitive hexadecyltrimethyl ammonium bromide functionalized graphene supported platinum nanoparticles composite modified glassy carbon electrode for determination of sunset yellow in soft drinks. *Sensors and Actuators B: Chemical, 209*, pp.1–8.
65. Wang, J., Yang, B., Wang, H., Yang, P. and Du, Y., 2015. Highly sensitive electrochemical determination of Sunset Yellow based on gold nanoparticles/graphene electrode. *Analytica Chimica Acta, 893*, pp.41–48.
66. Zhang, J., Zhu, H., Wang, M., Wang, W. and Chen, Z., 2013. Electrochemical determination of sunset yellow based on an expanded graphite paste electrode. *Journal of The Electrochemical Society, 160*(8), p.H459.
67. Vikraman, A.E., Thomas, D., Cyriac, S.T. and Kumar, K.G., 2014. Kinetic and thermodynamic approach in the development of a voltammetric sensor for sunset yellow. *Journal of the Electrochemical Society, 161*(14), p.B305.
68. Zhang, K., Luo, P., Wu, J., Wang, W. and Ye, B., 2013. Highly sensitive determination of Sunset Yellow in drink using a poly (l-cysteine) modified glassy carbon electrode. *Analytical Methods, 5*(19), pp.5044–5050.
69. Silva, M.L.S., Garcia, M.B.Q., Lima, J.L. and Barrado, E., 2007. Voltammetric determination of food colorants using a polyallylamine modified tubular electrode in a multicommutated flow system. *Talanta, 72*(1), pp.282–288.
70. Medeiros, R.A., Lourencao, B.C., Rocha-Filho, R.C. and Fatibello-Filho, O., 2012. Simultaneous voltammetric determination of synthetic colorants in food using a cathodically pretreated boron-doped diamond electrode. *Talanta, 97*, pp.291–297.
71. Qiu, X., Lu, L., Leng, J., Yu, Y., Wang, W., Jiang, M. and Bai, L., 2016. An enhanced electrochemical platform based on graphene oxide and multi-walled carbon nanotubes nanocomposite for sensitive determination of Sunset Yellow and Tartrazine. *Food Chemistry, 190*, pp.889–895.
72. Chebotarev, A.N., Pliuta, K.V. and Snigur, D.V., 2020. Determination of carmoisine onto carbon-paste electrode modified by silica impregnated with cetylpyridinium chloride. *ChemistrySelect, 5*(12), pp.3688–3693.
73. Ganash, A., Alshammari, S. and Ganash, E., 2023. Development of a novel electrochemical sensor based on gold nanoparticle-modified carbon-paste electrode for the detection of congo red dye. *Molecules, 28*(1), p.19.

74. Qin, Z., Zhang, J., Liu, Y., Wu, J., Li, G., Liu, J. and He, Q., 2020. A simple but efficient voltammetric sensor for simultaneous detection of tartrazine and ponceau 4R based on TiO_2/electro-reduced graphene oxide nanocomposite. *Chemosensors*, *8*(3), p.70.
75. Zhang, J., Wang, X., Zhang, S., Wang, W., Hojo, M. and Chen, Z., 2014. An electrochemical sensor for simultaneous determination of ponceau 4R and tartrazine based on an ionic liquid modified expanded graphite paste electrode. *Journal of the Electrochemical Society*, *161*(9), p.H453.
76. Wang, M. and Zhao, J., 2015. A facile method used for simultaneous determination of ponceau 4R, allura red and tartrazine in alcoholic beverages. *Journal of the Electrochemical Society*, *162*(6), p.H321.
77. Zhang, Y., Hu, L., Liu, X., Liu, B. and Wu, K., 2015. Highly-sensitive and rapid detection of ponceau 4R and tartrazine in drinks using alumina microfibers-based electrochemical sensor. *Food Chemistry*, *166*, pp.352–357.
78. Cheng, Q., Xia, S., Tong, J. and Wu, K., 2015. Highly-sensitive electrochemical sensing platforms for food colourants based on the property-tuning of porous carbon. *Analytica Chimica Acta*, *887*, pp.75–81.
79. Karimi-Maleh, H., Darabi, R., Shabani-Nooshabadi, M., Baghayeri, M., Karimi, F., Rouhi, J., Alizadeh, M., Karaman, O., Vasseghian, Y. and Karaman, C., 2022. Determination of D&C Red 33 and patent blue V Azo dyes using an impressive electrochemical sensor based on carbon paste electrode modified with ZIF–8/g-C3N4/Co and ionic liquid in mouthwash and toothpaste as real samples. *Food and Chemical Toxicology*, *162*, p.112907.
80. Rajendrachari, S., Basavegowda, N., Vinaykumar, R., Narsimhachary, D., Somu, P. and Lee, M.J., 2023. Electrocatalytic determination of methyl orange dye using mechanically alloyed novel metallic glass modified carbon paste electrode by cyclic voltammetry. *Inorganic Chemistry Communications*, *155*, p.111010.
81. Negahdary, M., Ameku, W.A., Santos, B.G., dos Santos Lima, I., de Oliveira, T.G., França, M.C. and Angnes, L., 2023. Recent electrochemical sensors and biosensors for toxic agents based on screen-printed electrodes equipped with nanomaterials. *Microchemical Journal*, *185*, p.108281.
82. Silva, R.M., da Silva, A.D., Camargo, J.R., de Castro, B.S., Meireles, L.M., Silva, P.S., Janegitz, B.C. and Silva, T.A., 2023. Carbon nanomaterials-based screen-printed electrodes for sensing applications. *Biosensors*, *13*(4), p.453.
83. Micheletti, L., Coldibeli, B., Salamanca-Neto, C.A.R., Almeida, L.C. and Sartori, E.R., 2020. Assessment of the use of boron-doped diamond electrode for highly sensitive voltammetric determination of the azo-dye carmoisine E– 122 in food and environmental matrices. *Talanta*, *220*, p.121417.
84. Lipskikh, O.I., Korotkova, E.I., Barek, J., Vyskocil, V., Saqib, M. and Khristunova, E.P., 2020. Simultaneous voltammetric determination of brilliant blue FCF and Tartrazine for food quality control. *Talanta*, *218*, p.121136.
85. Durigon, A.M.M., da Silveira, G.D., Sokal, F.R., Pires, R.A.C.V. and Dias, D., 2020. Food dyes screening using electrochemistry approach in solid state: The case of sunset yellow dye electrochemical behavior. *Journal of Solid State Electrochemistry*, *24*, pp.2907–2921.

86. Akkapinyo, C., Ngamkham, K., Plongthongkum, N. and Poo-arporn, R.P., 2019, November. Reduced graphene oxide and poly (L-methionine) modified screen-printed carbon electrode for determination of yellow azo dye. In *2019 12th biomedical engineering international conference (BMEiCON)* (pp. 1–4). IEEE.
87. Wu, J.H. and Lee, H.L., 2020. Determination of sunset yellow and tartrazine in drinks using screen-printed carbon electrodes modified with reduced graphene oxide and NiBTC frameworks. *Microchemical Journal*, *158*, p.105133.
88. Ebrahimi-Tazangi, F., Beitollahi, H., Hekmatara, H. and Seyed-Yazdi, J., 2021. Design of a new electrochemical sensor based on the CuO/GO nanocomposites: Simultaneous determination of Sudan I and bisphenol A. *Journal of the Iranian Chemical Society*, *18*(1), pp.191–199.
89. Yahya, R., Shah, A., Kokab, T., Ullah, N., Hakeem, M.K., Hayat, M., Haleem, A. and Shah, I., 2022. Electrochemical Sensor for detection and degradation studies of ethyl violet dye. *ACS omega*, *7*(38), pp.34154–34165.
90. Serrano i Plana, N., Díaz Cruz, J.M., Ariño Blasco, C. and Esteban i Cortada, M., 2016. Antimony-based electrodes for analytical determinations. *Trac-Trends in Analytical Chemistry*, *77*, pp. 203–213.
91. Rodríguez, J.A., Juárez, M.G., Galán-Vidal, C.A., Miranda, J.M. and Barrado, E., 2015. Determination of allura red and tartrazine in food samples by sequential injection analysis combined with voltammetric detection at antimony film electrode. *Electroanalysis*, *27*(10), pp.2329–2334.
92. Sierra-Rosales, P., Toledo-Neira, C. and Squella, J.A., 2017. Electrochemical determination of food colorants in soft drinks using MWCNT-modified GCEs. *Sensors and Actuators B: Chemical*, *240*, pp.1257–1264.
93. Wang, M. and Zhao, J., 2015. Facile synthesis of Au supported on ionic liquid functionalized reduced graphene oxide for simultaneous determination of Sunset yellow and Tartrazine in drinks. *Sensors and Actuators B: Chemical*, *216*, pp.578–585.
94. Thirumalraj, B., Rajkumar, C., Chen, S.M., Veerakumar, P., Perumal, P. and Liu, S.B., 2018. Carbon aerogel supported palladium-ruthenium nanoparticles for electrochemical sensing and catalytic reduction of food dye. *Sensors and Actuators B: Chemical*, *257*, pp.48–59.
95. Yun, M., Choe, J.E., You, J.M., Ahmed, M.S., Lee, K., Üstündağ, Z. and Jeon, S., 2015. High catalytic activity of electrochemically reduced graphene composite toward electrochemical sensing of Orange II. *Food Chemistry*, *169*, pp.114–119.
96. Shah, A., Malik, M.S., Zahid, A., Iftikhar, F.J., Anwar, A., Akhter, M.S., Shah, M.R., Zia, M.A., Ashiq, M.N. and Shah, A.H., 2018. Carbamazepine coated silver nanoparticles for the simultaneous electrochemical sensing of specific food toxins. *Electrochimica Acta*, *274*, pp.131–142.
97. Lu, Y., Bao, C., Zou, J., Xiao, J., Zhong, W. and Gao, Y., 2022. Highly sensitive electrochemical sensor for sunset yellow based on electrochemically activated glassy carbon electrode. *Molecules*, *27*(16), p.5221.

CHAPTER 5

Advanced Electrochemical Sensing Devices For Analysis of Pesticides, Insecticides, and Other Toxins in Food Products

Nikhila N Pai, Shashanka Rajendrachari, and Akshatha R Salian

5.1 INTRODUCTION

Ensuring the quality and safety of food products has consistently been essential for human health and well-being. However, in recent decades, the rapid growth of global population and the accompanying demand for increased food production have led to widespread use of chemical agents such as pesticides, insecticides, and other agrochemicals in agriculture. In addition, various other food contaminants, like veterinary drugs, unauthorized food additives, pathogens, mycotoxins, industrial chemicals, and heavy metals significantly impact the quality of food products.[1–4] These

DOI: 10.1201/9781003512639-5

contaminants can give rise to significant public health concerns, as their residues can persist in food products, posing potential risks such as endocrine disruption, neurological disorders, and cancer to consumers, if they are not adequately controlled and monitored.[5, 6]

The need for continuous, rapid, and efficient methods for detection and quantification of these harmful substances has thus become a critical issue, driving the development of advanced analytical techniques. Among these, electrochemical sensing devices have emerged as a powerful tool in the analysis of contaminants in food products.[7]

Traditional methods for detecting toxins in food, such as chromatography and spectrometry, though accurate and reliable, are often expensive, require specialized equipment, complex sample preparation, and lengthy analysis time, and demand professional manpower, making them inappropriate for regular sensing. In contrast, the electrochemical sensing devices are generally highly versatile, cost-effective, and rapid, and can be miniaturized for real-time automatic on-site analysis.[1, 7–10].

At the outset, this chapter outlines various food contaminants, including pesticides, veterinary drugs, mycotoxins, food additives, heavy metals, and pathogens. It then analyses electrochemical methods such as conductometry, potentiometry, amperometry, voltammetry, and electrochemical impedance spectroscopy (EIS), focusing on their key parameters for pollutant detection. The chapter also provides a detailed examination of how nanomaterials (NMs) and their composites have been applied in sensing food contaminants in recent years and concludes by discussing the current challenges and future directions in the field (Figure 5.1).

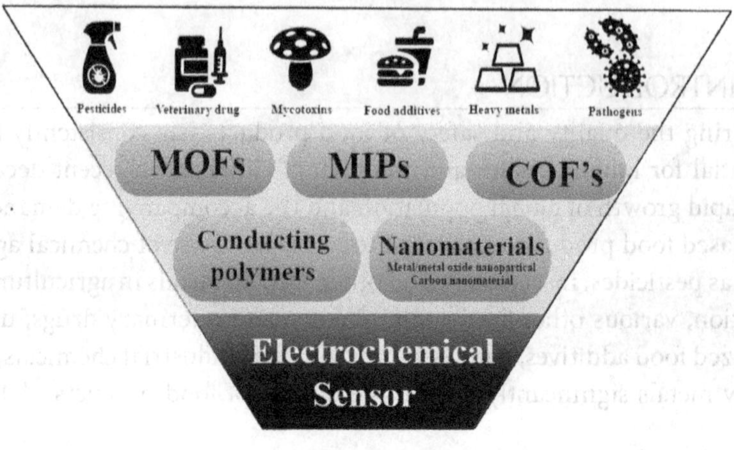

FIGURE 5.1 Electrode modifiers and their role in sensing food contaminants.

5.2 FOOD CONTAMINANTS

5.2.1 Pesticides

Pesticides are utilized to maintain crop health and yields by preventing diseases and infestations.[11, 12] These products encompass insecticides, herbicides, fungicides, plant growth regulators, and repellents. Pesticides are chemically categorized into organophosphates, carbamates, organochlorines, pyrethrin, pyrethroids, etc. Different classes of pesticides may have various impacts on their intended targets, including potential effects on humans.[13–15]

Organophosphorus pesticides are widely utilized across the globe to enhance agricultural yields because of their potent effectiveness against insect pests and parasitic weeds.[15–18] Carbamates and organophosphates are recognized for their toxic properties, primarily due to their capacity to block the enzyme acetylcholinesterase within the nervous system of insects, mammals, and humans. This leads to the build-up of acetylcholine, potentially causing severe health issues such as respiratory and heart malfunctions.[19–21] Some of the most commonly used organophosphates include malathion, parathion, methyl parathion, fenthion, diazinon, and chlorpyrifos. Carbaryl, carbosulfan, and fenobucarb are the carbamate compounds widely used in agriculture due to their cost-effectiveness and strong efficacy against a variety of insects.[15, 22, 23]

Organochlorine pesticides have been observed to stimulate the nervous system and act as hormone disruptors, leading to both immediate and long-term harmful effects. The commonly used organochlorine pesticides include aldrin, dieldrin, endrin, chloridazon, chlordane, toxaphene, heptachlor, chlordecone, dichloro-diphenyl-trichloroethane, mirex, lindane, hexachlorobenzene, dicofol, pentachlorophenol, pentachlorobenzene, and hexachlorocyclohexanes.[24, 25] Pyrethroid pesticides lead to neuronal hyperexcitation, which results in continuous synaptic firing and sustained depolarization.[17, 26–29]

High food production often requires pesticide use, leading governments to regulate it and set maximum residue levels for safety. Despite these measures, misuse of pesticides persists, especially in developing countries. Reliable tools are needed for rapid pesticide detection to ensure proper regulation and monitoring.[30]

5.2.2 Veterinary Drugs

Anabolic drugs are used in veterinary care for treatment and prevention of diseases and to increase breeding efficiency but are often illegally used to promote growth and improve feed efficiency in animals.[31] Pharmaceuticals used in veterinary medicine could be antibiotics, hormones, anti-inflammatory agents, analgesics, antiepileptics, antidepressants, etc. A significant quantity of these substances used in animals leads to the presence of their residues in food products like meat, poultry, eggs, honey, milk, and fish.[32–40]

The excessive use of antibiotics promotes their accumulation in animals and food, which can result in allergies, antibiotic resistance, and superinfections in humans. Various antibiotics, such as penicillin, tetracyclines, fluoroquinolones, β-lactams, macrolides, amphenicols, aminoglycosides, glycopeptides, lincosamides, and sulfonamides, have been recognized as potential environmental pollutants. These antibiotics have been targeted for detection in different matrices, including food, environmental, and milk samples.[41–46]

Strictly controlling veterinary drug residues and improving detection methods, particularly those that are fast, sensitive, accurate, and easy to use, are crucial for protecting ecological ecosystems and safeguarding human health.[47, 48]

5.2.3 Mycotoxins

Mycotoxins, produced by filamentous fungi, are widely studied biological toxins that contaminate food even at low concentrations, posing significant risks to human health and food safety. They can contaminate various agricultural products during preparation, packaging, and distribution. These toxins are difficult to eliminate during food processing due to their resistance to heat, chemical, and physical treatments. Consumption of mycotoxin-contaminated food can lead to acute or chronic ~~storage~~poisoning with serious health consequences.[49–51]

They have the potential to contaminate various foods, such as dairy products, groundnuts, soybeans, maize, sorghum, cereals, nuts, spices, and other food commodities and feed crops, at any stage of processing of harvest.[52] The most common mycotoxins are patulin, aflatoxins, fumonisins, trichothecenes, zearalenones, and ochratoxins, which are generated by specific plant-pathogenic species of *Penicillium*, *Aspergillus*, and *Fusarium*.[53, 54]

The modification of mycotoxins into different structural forms can make conventional detection methods inadequate for identifying these various mycotoxin types. Therefore, there is a need for detection techniques that are not only sensitive, efficient, and precise but also user-friendly.[55]

5.2.4 Food Additives

In the current world, food additives play an essential role in the composition of food products.[56] Food additives are substances incorporated into food during processing, packing, and transporting processes to prolong its shelf life and enhance its physical, chemical, sensory, and microbiological characteristics.[57] They aim to enhance product quality, stability, and durability as well as to alter odour, colour, appearance of various food products, and flavour characteristics. Common food additives include colorants, sweeteners, flavouring agents, preservatives, and antioxidants. However, some additives, like Sudan dyes, dimethyl yellow, and ethoxyquin, are banned or restricted for use in food for human consumption but may still be present in contaminated products.[58, 59]

Certain additives, such as acidity regulators, colorants, antioxidants, and preservatives, can increase toxicity for consumers and negatively affect the flavour of food. Repeated consumption of food additives has been linked to an increased risk of allergies, diabetes, obesity, and metabolic issues.[60] The shift from natural to synthetic food additives has raised concerns due to overuse, misuse, and potential toxicity. Although additives improve sensory appeal and offer commercial advantages, they also pose health risks, making quantitative assessments essential.[61]

5.2.5 Heavy Metal Ions

Heavy metal contamination in fruits and vegetables is a critical concern for food quality. Elements like Pb, As, Cr, Cu, Hg, Co, Ni, Zn, and Cd in our diet can be toxic even at low concentrations. They can disrupt various biochemical processes in the body, potentially leading to a higher risk of chronic diseases.[62–65] The primary source of toxic metal contamination in agricultural lands is irrigation with sewage and industrial wastewater.[66, 67]

Heavy metal compounds are absorbed by plants along with essential nutrients and are subsequently distributed to various tissues within the plant, as well as throughout the food chain.[68] Eating fruits and vegetables that contain heavy metals is a primary way humans are exposed to these toxins through the food chain. Since the conventional heavy metal detection methods are unsuitable for rapid on-site use, there is a need to improve current technologies for more cost-effective solutions.[69–75]

5.2.6 Pathogens

Pathogens like bacteria, fungi, viruses, yeast, and parasites often cause foodborne illnesses. The pathogens or their toxins can contaminate various food products (fresh, raw, or processed) throughout their lifecycle, from growth to consumption. Once ingested, these pathogens can cause a range of foodborne diseases.[76, 77] Foodborne illnesses pose serious health risks. Major pathogens include *Escherichia coli*, *Bacillus cereus*, *Clostridium botulinum*, *Campylobacter* spp., *Clostridium perfringens*, *Salmonella* spp., *Staphylococcus aureus*, *Vibrio* spp., and *Shigella* spp. Foodborne viruses of concern include Rotavirus, Enterovirus, Norovirus, Hepatitis A and E, Adenovirus, Achivirus, Sapovirus, and Astrovirus.[78, 79]

The rising popularity of street foods and the growing demand for minimally processed, ready-to-eat products have raised concerns among public health agencies regarding food safety. Foodborne pathogens can be found in various items, including fruits, vegetables, and ready-to-eat products, which are consumed without additional processing.[77]

There is no current method that provides selective and sensitive binding to specific microorganisms while being fast, cost-effective, and adaptable. Nevertheless, electrochemical sensors have gained attention because of their label-free and highly sensitive detection ability to identify pathogens in water and food.[80]

5.3 ELECTROCHEMICAL DEVICES

As a fundamental element in electrochemical detection, the properties of the working electrode are vital to the detection capability of the sensor. Electrochemical sensing involves quantifying electrical properties like charge, current, or potential, which arise from the interaction between the analyte of interest and the sensing electrode, along with a reference electrode. Depending on their operating principles, electrochemical devices can be classified into categories such as conductometry, potentiometry, amperometry, voltammetry, and electrochemical impedance spectroscopy (EIS).[6, 81, 82]

5.3.1 Conductometric Devices

Conductometric sensors measure changes in the electrical conductivity of a sample solution between two electrodes, where the conductivity of the analyte varies as the electrochemical reaction occurs. The conductivity value represents the total concentration of all ions present in the analyte solution being tested. These sensors are commonly employed to observe metabolic processes in living biological systems.[81–83]

Conductometry can study reactions that alter the concentration of charged species in a solution. However, the ionic species generated by the electrode reaction must significantly affect the total ionic strength for accurate results; otherwise, fluctuations in ionic strength may cause interference. Additionally, conductometry has lower sensitivity compared with other electrochemical techniques.[84, 85]

5.3.2 Potentiometric Devices

Potentiometric devices measure the build-up of electrical potential at the working electrode in relation to the reference electrode within an electrochemical cell when no substantial current passes between them.[86] This depends on the potential shift caused by a change in the ratio between the oxidized and reduced forms of an indicator redox species. Potential limitations in ion-selective potentiometry may include deviations from the Nernst equation, which can result from fluctuations in ion activity or temperature.[87] The development of potentiometric and conductometric sensors is very limited in the field of sensors.[88]

5.3.3 Amperometric Devices

Amperometric devices use a constant voltage applied between a reference electrode and a working electrode to trigger an electrochemical reaction. Adjusting this voltage to a specific level facilitates the desired redox reaction of the analyte. The current generated from this process is then measured to determine the concentration of the analyte.[89–91] Amperometric measurements are often employed as a highly accurate and sensitive analytical technique.[92]

5.3.4 Voltammetric Devices

A voltammetric measurement involves scanning the potential difference across an electrochemical cell from one predetermined value to another while recording the cell current as a function of the applied potential.[93] The peak current value observed over a linear potential range is directly proportional to the concentration of the analyte, or the electroactive species. Although this sensing method can be indirect, these devices are reported to have better sensitivity compared with potentiometric devices.[94–98] Voltammetry includes various electrochemical techniques such as cyclic voltammetry (CV), square wave voltammetry (SWV), linear sweep voltammetry (LSV), and differential pulse voltammetry (DPV).[99]

5.3.5 Electrochemical Impedance Spectroscopy

EIS assesses intrinsic material properties affecting conductivity, resistivity, or capacitance in electrochemical systems. In electrochemical sensing, impedance methods track changes in electrical properties due to biorecognition events on modified electrode surfaces.[100] Impedimetric sensors operate on the principle that changes in conductance and capacitance on the working electrode's surface are affected by receptor activity due to the presence of the target analyte.[101]

EIS is highly adept at examining the interface between the electrode and electrolyte, offering crucial insights into the characteristics and activities at this junction. This technique enables the determination of analytes by measuring changes in the charge transfer resistance of the electrode surface.[100, 102]

For an electrochemical sensor, sample analysis involves recognizing the analyte with electrode modification materials and converting its concentration into an electrical signal. Bare electrodes have limited performance; hence, modification materials are crucial for enhancing the sensitivity and selectivity.[103]

5.4 ELECTRODE MATERIALS

At present, enhancing the performance of electrochemical devices for contaminant detection is primarily achieved through the modification of electrodes.[103] The materials employed for modifications consist of non-metal materials, metal materials, and conducting polymers (CPs). Among these, non-metal and metal NMs are the most commonly used in research studies.[104] The incorporation of NMs into electrochemical sensing has significantly advanced the progress of electrochemical sensors and biosensors.[105, 106] Products derived from nanotechnology have provided a diverse array of materials that enhance the stability, sensitivity, and selectivity of electrochemical sensors.[107]

5.4.1 Nanomaterials

NMs enhance electrochemical sensor performance due to their distinct physical, chemical, and electrocatalytic characteristics. These include a large surface-to-volume ratio, excellent electrical conductivity, good biocompatibility, catalytic activity, and the ability to be easily tailored with functional groups.[108–111] Non-enzymatic electrochemical sensors that use NMs offer benefits such as compact size, ease of use, affordability, rapid performance, and superior sensitivity.[112] NMs are typically modified or combined with various organic and inorganic functional materials to create nanocomposites for practical use.[105] Metal, metal oxide, and carbon-based NMs are commonly employed for modifying sensing electrodes in the electrochemical detection of food contaminants.[113]

5.4.1.1 Metal/Metal Oxide Nanoparticles

Metal nanoparticles possess distinctive physical and chemical characteristics, like extensive surface areas, consistent particle sizes, and excellent electrical, optical, and catalytic properties.[112]. Recent research shows that transition metals and their oxide nanoparticles, including single-metal and multi-metal composites, are commonly used for electrode modification. Most studies focus on Cu, Ag, Au, and Pt, with Ni and Pd also being significant. Transition metal carbides and nitrides (MXene) are emerging as promising two-dimensional (2D) NMs.[104, 114]

A Co–Ag bimetallic nanoparticle-modified electrode stabilized in poly(vinylpyrrolidone) was developed by Gerent *et al.* to monitor a herbicide, butralin, in honey and apple jam samples. A linear calibration curve was obtained using SWV within the concentration range of 0.1–1.0 μmol L^{-1}, detection range and quantification limits of 32 and 106 nmol L^{-1}, respectively, due to the synergy of the electrocatalytic activity of Co–Ag bimetallic nanoparticles.[115] In another study, a highly efficient electrochemical sensor using Mn/Zn/V nanocomposite was developed for detecting vanillin, a common flavouring agent in food, ice creams, beverages, candies, etc. The sensor's performance was evaluated using various voltammetry techniques (Figure 5.2).[116]

Compared with traditional costly metal catalysts, metal oxides are gaining attention because of their electrocatalytic activity, adsorption properties, and electron transport characteristics. They are also easy to control in terms of morphology and size. Metal oxide-based electrochemical sensors are advantageous due to their simple construction, low cost, and ease of operation.[117–120] A polarizable electrode with Cu/Fe/NiO nanocomposite was created by Kumari *et al.* to evaluate the electrocatalytic behaviour of a synthetic colorant, tartrazine dye, in soft drinks using CV, LSV, and DPV techniques. This sensor proved to be cost-effective and reliable for detecting food preservatives and colouring agents in food.[121] A sensitive analytical method using Fe_3O_4@SiO_2-based extraction combined with direct electrochemical detection was developed by Zhang *et al.* to analyse heavy metal ions like Cu^{2+}, Pb^{2+}, Cd^{2+}, and Hg^{2+} simultaneously in milk with detection limits in the nanomolar range.[122] Anh *et al.* investigated how crystallinity, phase ratio, and heterojunction formation influence the performance of ZnO/$ZnFe_2O_4$ nanocomposite electrochemical sensors in

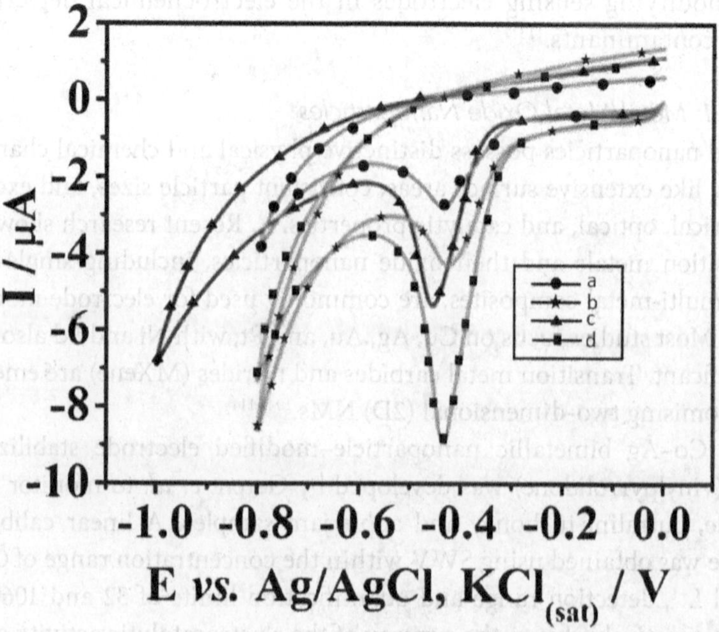

FIGURE 5.2 Cyclic voltammograms for 50.0 μmol L^{-1} butralin at (a) GCE, (b) Co NPs-PVP/GCE, (c) Ag NPs-PVP/GCE, and (d) Co-Ag BMNPs-PVP/GCE. Parameters: $E_{acc} = -0.15$ V, $t_{acc} = 120$ s, $\nu = 100$ mV s^{-1}. (Reprinted from *Food Chemistry*, 343, Gerent GG, Santana ER, Martins EC, and Spinelli A, A non-mercury electrode for the voltammetric determination of butralin in foods, 128419, Copyright (2021), with permission from Elsevier.)

detecting an antibiotic drug, furazolidone, in shrimp and pork samples (Figure 5.3).[123]

Metal oxides are often paired with conductive materials like three-dimensional (3D) graphene (Gr), multi-walled carbon nanotubes (MWCNTs), and organometallic frameworks (MOFs) to enhance their conductivity, forming layered nanostructures with larger surface areas and more active sites.[112] Shukla *et al.* developed a reduced graphene oxide/tin oxide (rGO/SnO_2) composite for electrochemical detection of a fungal contaminant, patulin, in contaminated apple juice with a DPV response, removing the need for biological or chemical receptors.[124]

5.4.1.2 Non-Metallic Nanomaterials

Non-metallic NMs mainly include carbon-based and silicon-based types. Carbon-based materials are the most common, while silicon-based materials are usually combined with other substances.[104] Carbon-based NMs

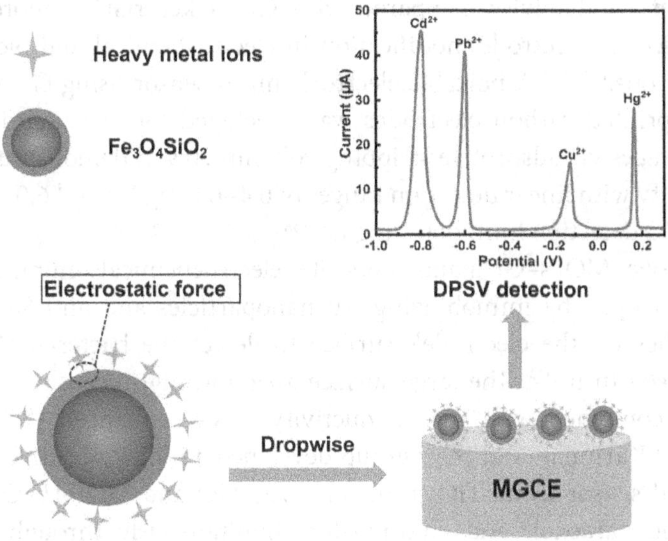

FIGURE 5.3 The working principle of Fe_3O_4@SiO_2 sensors for the direct electrochemical detection of heavy metal ions. (Reprinted from *Food Chemistry*, 406, Zhang M and Guo W, Simultaneous electrochemical detection of multiple heavy metal ions in milk based on silica-modified magnetic nanoparticles, 135034, Copyright (2023), with permission from Elsevier.)

like single-walled carbon nanotubes (SWCNTs), MWCNTs, carbon nanohorns (CNHs), buckypaper, Gr and its derivatives, fullerenes, and graphitic carbon nitride (gC_3N_4) exhibit high surface-to-volume ratios, excellent electrical conductivity, chemical stability, and high mechanical strength. These properties make them highly valuable in sensing technology.[125–133]

Tan and colleagues developed a fluorinated Gr and Au nanocage nanocomposite (FGr/Au) electrode that could simultaneously detect Cd^{2+}, Pb^{2+}, Hg^{2+}, Cu^{2+}, and Zn^{2+} in peanut and tea samples, with a linear detection range of 4–7000 μg L^{-1} and limit of detection below the recommended threshold.[134] Veenuttranon *et al.* developed a novel electrochemical sensor by combining nickel oxide (NiO) with carboxylated carbon nanotubes (MWCNT(–COOH)) that without using biomolecules, can detect the mycotoxins zearalenone and ochratoxin A simultaneously in corn flour and wheat flour samples. The chemical functionalization of MWCNTs facilitated their interaction with metal oxide nanoparticles, thereby showing excellent synergy in magnifying the sensor performance.[135]

Graphene oxide (GO) disperses well in aqueous solutions due to its hydrophilic functional groups (OH, COOH, and epoxides). However, its lower conductivity compared with Gr makes rGO a more popular choice for electrode modification in electrochemical and biosensing applications.[136, 137] A portable electrochemical sensor using Gr-modified screen-printed carbon electrodes was developed for xylazine detection in beverages via adsorptive stripping voltammetry. It demonstrated high sensitivity with linear detection ranges of 0.4–6.0 mg L^{-1} and 6.0–80.0 mg L^{-1}, and a detection limit of 0.1 mg L^{-1}.[138]

A CoFe–MOFs–Gr nanocomposite electrochemical immunosensor was developed by immobilizing Au nanoparticles and anti-*Salmonella* antibodies on the electrode's surface to detect the bacterial pathogen *Salmonella* in milk. The large surface area and high porosity of CoFe-MOFs, combined with the conductivity of Gr, enhanced the current signal.[139] Karimian and their group developed a highly sensitive electrochemical sensor using TiO_2-functionalized GO and MOF (UiO-66) for detecting paraoxon and chlorpyrifos simultaneously through impedance and CV, with optimized conditions leading to successful application in monitoring pesticide residues in vegetables and water.[140] Malode *et al.* established nanostructured WO_3/rGO composite–modified sensors for the detection of triclosan, a bactericide, in fruits, vegetables, soil, and water with a lower detection limit.[141]

Carbon nanotubes (CNTs) can boost electrochemical analysis by facilitating the immobilization of additional chemical species, such as organic molecules and metallic nanoparticles. Their chemical stability, high conductivity, and large surface area drive their increasing use in habitat rehabilitation.[142, 143] CNTs come in two types: SWCNT, with a diameter of about 1 nm, and MWCNT, with diameters over 100 nm. They have high porosity and a large surface area, and are easily modified. MWCNTs have metal-like conductivity, while SWCNTs' conductivity varies based on their chiral vector, ranging from metallic to semiconducting or insulating.[125, 144]

Appaturi and colleagues designed a biosensor utilizing a nanocomposite made of reduced graphene oxide and carbon nanotubes (rGO/CNT) for precise and swift label-free electrochemical detection of pathogenic bacteria, specifically *S. typhimurium*, in real food samples. The nanocomposite was drop-cast onto a glassy carbon electrode (GCE) and further enhanced with an amino-modified DNA aptamer. The resulting ssDNA/rGO/CNT aptasensor was then employed to detect the bacteria using the DPV technique.[145] Chen *et al.* developed an Au/MWCNT-modified GCE for detecting dichlorvos, a pesticide, in agricultural food. Using CV and

DPV, they found that the sensor had a detection limit of 5 nM and a linear range of 1–120 μM.[146]

MXenes are a new class of 2D transition metal–carbon/nitride materials similar to Gr but with a richer composition and varied bonds, offering greater customization for specific uses. Combining MXenes with other NMs enhances their properties, boosting the selectivity and sensitivity of MXene-based electrochemical sensors.[147–150] Tu *et al.* designed a nanoarchitectured composite comprising MXene, CNHs, and β-cyclodextrin MOFs (β-CD-MOFs) for the detection of carbendazim pesticide in tomato samples. The β-CD--MOFs offered remarkable adsorption capacity for carbendazim, attributed to their host–guest recognition ability and high porosity. Meanwhile, the MXene/CNHs component improved mass transport and catalytic efficiency due to its extensive surface area and superior conductivity.[151] Further, a highly sensitive electrochemical enzyme biosensor was developed using Ti_3C_2 MXene/MoS_2/Au nanoparticles/acetylcholine esterase for detecting an organothiophosphate insecticide, phoxim, in fruits. The two-dimensional Ti_3C_2 MXene structure offers supports for MoS_2, increasing surface area and preventing re-stacking. The synergistic effect of Au nanoparticles with MoS_2 enhanced the conductivity.[152]

5.4.2 Metal–Organic Framework

MOFs represent an emerging class of NMs, characterized by crystalline porous polymers formed through the coordination of organic ligands with metal particles.[153, 154] The wide variety of organic ligands and metal ions enables the creation of MOFs with customizable porosity, exceptionally high surface areas, versatile and attractive morphologies, and adaptable surface properties, making them highly valuable in sensor design.[155–160]

The porous nature of MOFs allows guest molecules to diffuse into their pores, and certain specialized MOF materials can be tailored to match the morphology and size of the target substance, enhancing detector concentration, reducing interference from impurities, and improving the selectivity, sensitivity, and accuracy of detection methods. Commonly used metal ions in the fabrication of chemically stable MOFs include Al, Fe, and Zr.[161, 162] The binding strength between analytes and MOFs can be enhanced through the diversity of pore sizes, the hydrophilicity of the pores, the presence of specific functional groups, or functionalization with biomolecules that increase affinity between analytes and functional groups or open metal sites.[163–165]

Electrochemical sensors incorporating electrodes modified with MOF composites are capable of effectively detecting drug molecules, organic compounds, metallic ions, and biological molecules, thereby enhancing the detection sensitivity for specific targets.[166–170] Nevertheless, the direct electrochemical application of single MOFs is limited due to drawbacks such as low conductivity, poor water stability, and limited mechanical strength. To overcome these limitations, composites with additional functionalities have been developed by combining MOFs with carbon materials, metallic nanoparticles, metal oxides, and polymers.[154, 171] Compared with single MOFs and traditional porous materials, modified MOF composites feature unique framework structures and diverse functional groups, along with distinct electrical, magnetic, catalytic, and optical properties.[172]

Methyl parathion, a common organophosphorus pesticide, was detected using a Zr-based MOF combined with rGO. The Zr-MOF/rGO nanocomposite, with its Zr-OH groups, selectively interacts with methyl parathion's phosphate groups, enhancing adsorption and lowering detection limits due to rGO's large surface area and excellent electron transport properties.[173] Zhan and co-workers developed a new electrochemical immunosensor for rapid and precise detection of *Salmonella typhimurium* in milk due to the combined effects of its various components, utilizing Pt nanoparticles, a Co/Zn-MOF, and carboxylic MWCNT.[174]

A novel electrochemical sensor using a samarium metal–organic framework (Sm-MOF) and gC_3N_4 composite was prepared by Srinivasan *et al.* for selective detection of propyl gallate, an antioxidant used to preserve food. The Sm-MOF and gC_3N_4 combination enhances current response and lowers oxidation potential, improving the sensing of propyl gallate (an antioxidant).[175] Kaur and team introduced a novel electrochemical biosensor created from a multifunctional composite by combining molybdenum-based 2D MOF with Ti_3C_2 MXene on a screen-printed electrode. Anti-Aflatoxin B1 antibodies (Ab) are covalently bonded to the composite, and the resulting Ab-MOF/MXene sensor is evaluated for its effectiveness in detecting Aflatoxin B1 using EIS.[176]

5.4.3 Conducting Polymers

Conducting polymers (CPs) have emerged as significant materials in the field of electrochemical sensors due to their favourable characteristics, such as high sensitivity and good redox properties. They can be used as sensitive compounds that can change their conductivity or their electrochemical behaviour in the presence of the analyte or substrate to immobilize specific molecules that can recognize and detect the analyte.[177–179]

The key and unique characteristic of a conducting polymer is the presence of alternating single and conjugated double bonds along its main chain. These π-bonds are highly delocalized, polarized, and rich in electrons, which give the polymer its electrical and optical properties.[180] CPs such as polypyrroles, polyanilines (PANI), and polythiophenes have attracted considerable attention due to their good film-forming property, electrical semiconductivity, high transparency in the visible region, and excellent thermal and environmental stability.[181, 182]

CPs are highly valued and widely employed materials for sensor modification due to their distinctive physical and chemical characteristics. These include customizable architecture, flexibility, adaptability, ambient stability, and heightened sensitivity to changes in their electrochemical behaviour with minor surface alterations. They also offer semiconductivity, structural versatility, various synthetic approaches, and stability under environmental conditions. Sensitivity enhancement in electrochemical sensors has been recently reported by using conjugated microporous polymers synthesized from multifunctional monomers rather than from bifunctional precursor monomers.[177, 183–187] Furthermore, the electrical conductivity of CPs can be controlled by doping processes (p-doping or n-doping). The characteristic electrical and optical properties of CPs can be fine-tuned through controlling the doping states of CPs.[186]

Wang *et al.* explored an electrochemically assisted approach to detect Pb^{2+} ions from beverage samples using a PANI nanofiber array on GO (PANI/GO). The integration of GO's flexibility with the short ionic-diffusion path provided by the vertical PANI nanofibres improved the sensor's efficiency, achieving a detection limit of 0.04 μg L^{-1}.[188] Ranjith Kumar *et al.* modified a GCE using a poly(4-aminodiphenylamine)-4-aminothiophenol, Au nanocomposite (p-ADPA-4-ATP/Au) to detect the food preservative sodium sulfite (SO_3^{2-}) in beverages. The composite allowed detection across a broad linear range from 5 to 160 μM.[189]

A modified copper electrode coated with a PANI/GO conductive nanocomposite was designed by Sereshti *et al.* to detect the antibiotics oxytetracycline, tetracycline, and doxycycline in milk samples. This method was used to analyse analytes in milk and water samples with varying fat levels.[190] In the research work carried out by Pangajam and their group, a specialized electrochemical biosensor featuring a nanoassembly of carbon dots, ZnO nanorods, and PANI was created to detect Cd^{2+}. It demonstrated

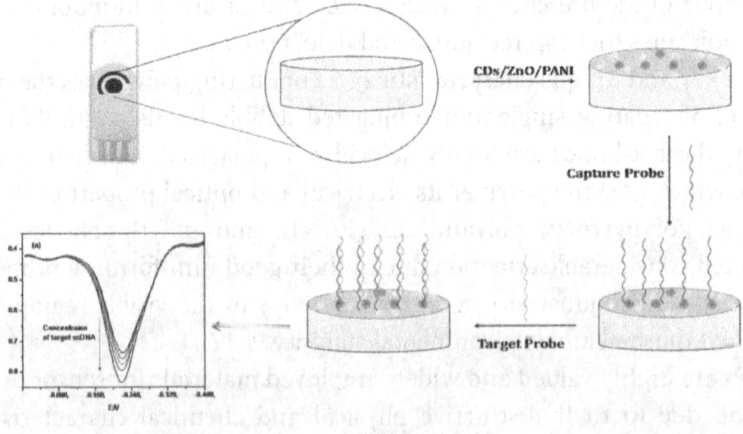

FIGURE 5.4 The working principle of the label-free electrochemical sensor, utilizing carbon dots/ZnO/PANI, designed for the detection of *E. coli*. (Reprinted from *Sensing and Bio-Sensing Research*, 29, Pangajam A, Theyagarajan K, and Dinakaran K, Highly sensitive electrochemical detection of *E. coli* O157:H7 using conductive carbon dot/ZnO nanorod/PANI composite electrode, 100317, Copyright (2020), with permission from Elsevier.)

high selectivity and could identify *E. coli* O157 in water with a detection limit of 1.3×10^{-18} M (Figure 5.4).[191]

5.4.4 Molecularly Imprinted Polymers

Electrochemical sensors utilizing molecularly imprinted polymers (MIPs) are an emerging category of synthetic biomimetic molecular receptors, known for their strong affinity towards specific target molecules and high chemical stability. MIPs are particularly appealing in electrochemistry because they enable selective recognition of target molecules over closely related compounds based on their shape, size, and functional groups. In the molecular imprinting technique, the target or analyte molecule serves as a template, and crosslinked monomers are organized around it. These monomers then polymerize to form a template-like layer. Once polymerization is complete, the target or analyte molecule is removed, leaving binding sites that are complementary in size, shape, and functional group placement to the target molecule.[192–195]

As an innovative material and biomimetic molecular recognition element, MIPs offer unique benefits over traditional enzymes, antibodies, and receptors, such as superior stability, high selectivity, ease of preparation, stability in organic solvents, and broad compatibility across various scientific fields.[196, 197] Along with their outstanding identification features,

MIPs offer the benefits of low cost, resistance to heat and pressure, storage stability, and suitability for use in harsh chemical environments.[198]

However, the electrically insulative nature of MIPs can hinder electrochemical detection. To overcome this limitation, MIPs constructed from conducting polymers are preferred due to their controllable electrochemical properties. Among conductive polymers, PANI stands out as a promising choice due to its ease of preparation, good conductivity, environmental stability, and customizable electrical properties. When combined with carbon materials, PANI-based composites significantly enhance the electrochemical performance of electrodes compared with carbon materials alone.[199–201] These binding sites endow MIPs with affinity and selectivity similar to those of natural receptors like antibodies or enzymes. MIPs boast several advantages, including thermal and chemical stability, nearly unlimited shelf-life, and an ethical, cost-effective, and straightforward synthesis process.[202, 203]

These devices are characterized by excellent chemical and physical stability, low production costs, good selectivity, and rapid response times. Research indicates that incorporating nanoparticles into these sensors enhances their analytical properties, particularly sensitivity.[197] The integration of MIPs with electrochemical sensors paves the way for creating low-cost, user-friendly devices with high sensitivity and selectivity, rapid response, excellent chemical and mechanical stability, miniaturization, automation, reusability, and in-situ detection of target analytes.[204]

Ma and colleagues developed an electrochemical sensor for detecting endosulfan in food samples, incorporating a MIP and g-C_3N_4 nanostructure on a GCE. Electrochemical measurements conducted using CV and DPV demonstrated the sensor's high sensitivity and selectivity for endosulfan, with a linear detection range of 0–96 μM and a detection limit of 20 nM.[205] A reliable electrochemical MIP sensor was created utilizing MIP-coated Au nanoparticles combined with black phosphorus nanocomposites for the selective detection of pefloxacin (a broad-spectrum antibacterial drug) in milk and orange juice. This sensor demonstrated high sensitivity to the drug, with MIP/BP/Au offering a wide linear detection range (0.005–10 μM) and a low detection limit of 0.80 nM.[206]

Yang *et al.* developed a novel portable electrochemical sensor for the quick detection of the antibiotic tetracycline in food samples utilizing Fe_3O_4/COF/MIP. The sensor exhibited strong magnetic properties, high adsorption capacity, and enhanced detection efficiency, with a linear detection range from 1×10^{-10} to 1×10^{-4} g mL^{-1}.[207] A highly sensitive electrochemical sensor utilizing nano-MIPs was developed for the detection of Fumonisin B_1, a carcinogenic mycotoxin that contaminates food and animal feed. This sensor employed EIS and DPV for signal analysis, enabling the detection of the contaminant in maize samples.[208]

Işık and team fabricated molecularly imprinted kanamycin (KAN) electrodes by electrochemically polymerizing pyrrole (Py) onto a GCE coated with GO. The kanamycin template was then removed using HCl, resulting in GO-pPy-KAN, which was used to detect kanamycin (an aminoglycoside antibiotic commonly used to treat infections in farm animals) in milk samples.[209] Mugo *et al.* developed a portable sensor using MIP technology for electrochemical detection of imidacloprid. The sensor was created via a layer-by-layer coating process using a composite of carbon nanotubes and cellulose nanocrystals (CNC). It featured an imidacloprid-imprinted PANI layer co-polymerized with imidazole-functionalized CNCs. The resulting MIP/CNT/CNC sensor showed excellent electrochemical performance in detecting imidacloprid in honey samples.[210] Further, Zeb and colleagues created a highly sensitive and selective biomimetic sensor utilizing magnetic nanoparticles combined with an MIP-based electrochemical sensor to measure tetracycline levels in both commercial and natural cow milk samples.[211]

5.4.5 Covalent Organic Frameworks

Covalent organic frameworks (COFs) represent a new class of porous organic polymers composed of light elements like carbon, hydrogen, nitrogen, and boron connected via covalent bonds. These materials have garnered increasing attention due to their low density, chemical stability, low toxicity, durability, thermal stability, structural diversity, and a range of tunable porous structures, all of which make them excellent candidates for constructing sensing interfaces.[212–217]

The ordered network configuration and numerous active sites within COFs provide a large adhesion surface, surpassing that of 2D Gr nanosheets. Unlike MOFs, COFs exhibit superior thermal and chemical stability due to their covalent bond structure.[218, 219] Recently, COFs have gained traction in electrochemical sensing because of their ability to enhance the selectivity, sensitivity, and speed of electrochemical sensors. However, the inherent poor conductivity and weak electrocatalytic properties of most COFs limit their application.[216, 220, 221] To address this, combining COFs with materials that possess good conductivity, such as carbon materials, metal nanoparticles, metallic oxides, and conducting polymers, is a promising approach to improving the electrochemical performance of COFs.[212] Additionally, modifying COFs with different functional groups or metal ions can lead to the development of highly specific and targeted sensors.[222]

COFs also exhibit unmatched biocompatibility and chemical stability, along with high surface areas, high porosities, and ease of functionalization, similarly to MOFs and zeolitic imidazolate frameworks. Their advantages lie in providing numerous binding sites and π-π stacking interactions, which facilitate charge transfer and enhance the electrochemical performance of sensors.[223] A ternary composite of CoCu nanoparticles integrated into a COF/SWCNT matrix (CoCu/COF/SWCNT) has been employed as an electrochemical sensor for detecting carbendazim, a commonly used fungicide in crop protection. The composite's superior performance is attributed to its enhanced charge transfer, larger active surface area, and high electrocatalytic activity resulting from the synergy of its components. Under optimal conditions, the proposed sensor demonstrated a detection range of 0.001 to 10 μM and a detection limit of 0.65 nM for carbendazim in water and pear samples.[224]

Yang *et al.* developed a sensor utilizing gold nanoparticles embedded in COFs (Au nanoparticles/COFs) for the detection of ractopamine, a feed additive. This sensor demonstrated excellent sensitivity and reliability, with a detection range spanning 1.2 to 1600 μmol L^{-1} and a low detection limit in pork and chicken samples.[225] In the research study carried out by Luo *et al.*, a novel electrochemical sensor was developed using an ultrafine $ZnO/ZnNi_2O_4$/porous carbon/COF composite for the detection of paracetamol (PA) and tert-butylhydroquinone (TBHQ) in food samples. This sensor demonstrated broad linear detection ranges, spanning from 48.5 nM to 130 μM for PA and from 47.85 nM to 130 μM for TBHQ.[226]

Further, a highly conductive composite, composed of COFs and Au nanoparticles (3,5-tris(4-aminophenyl)benzene-p-phthalaldehyde-COFs/Au nanoparticles), was developed by Lu *et al.* for the detection of the antibiotic enrofloxacin using square wave stripping voltammetry. The sensor demonstrated two linear detection ranges: 0.05–10 μmol L^{-1} and 10–120 μmol L^{-1}, with a detection limit of 0.041 μmol L^{-1},[227] and a COF/Au nanoparticles/Gr composite was employed by Yu *et al.* for the simultaneous detection of heavy metal ions in Baijiu. This electrode modifier's high porosity and numerous adsorption sites, combined with the exceptional electrical conductivity of Au nanoparticles and Gr, along with the large specific surface area and synergy between COF, Au nanoparticles, and Gr, enabled the proposed sensor to deliver outstanding performance in detecting Cd^{2+}, Pb^{2+}, and Cu^{2+} ions (Table 5.1).[228]

TABLE 5.1 Voltammetric Determination of Food Contaminants Using Various Nanomaterials/Nanocomposites as Electrode Modifiers

Electrode modifier	Analyte	Food sample	Linear range	Detection limit	Ref.
Co-Ag/PVP	Butralin (herbicide)	Honey and apple jam	0.1–1.0 μmol L^{-1}	32 nmol L^{-1}	115
Cu/Fe/NiO	Tartrazine dye (synthetic colorant)	Soft drink	0–500 μM	0.010 μM	121
$ZnO/ZnFe_2O_4$	Furazolidone (antibiotic drug)	Pork and shrimp	1–100 μM	0.65 μM	123
rGO/SnO_2	Patulin (fungal contaminant)	Apple juice	50–600 nM	0.6635 nM	124
FGr/Au	Zn^{2+}	Peanut and tea	6–7000 μg L^{-1}	0.08 μg L^{-1}	134
	Cd^{2+}		4–6000 μg L^{-1}	0.09 μg L^{-1}	
	Pb^{2+}		6–5000 μg L^{-1}	0.05 μg L^{-1}	
	Cu^{2+}		4–4000 μg L^{-1}	0.19 μg L^{-1}	
	Hg^{2+}		6–5000 μg L^{-1}	0.01 μg L^{-1}	
	(Heavy metal ions)				
NiO/MWCNT(–COOH)	Zearalenone and Ochratoxin A (mycotoxins)	Corn flour and wheat flour	0.01–10.24 μg mL^{-} 0.04–10.24 μg mL^{-1}	6 ng mL^{-1} 15 ng mL^{-1}	135
Gr	Xylazine (analgesic)	Beverages	0.4–80.0 mg L^{-1}	0.1 mg L^{-1}	138
CoFe/MOFs/Gr	*Salmonella* (pathogen)	Milk	2.4×10^2 –2.4×10^8 CFU mL^{-1}	1.2×10^2 CFU mL^{-1}	139
TiO_2/GO/MOF(UiO-66)	Paraoxon and Chlorpyrifos (pesticides)	Vegetables and water	1.0–100.0 nM 5.0–300.0 nM	0.2 nM 1.0 nM	140
Au/MWCNT	Dichlorvos (pesticide)	Agricultural food	1–120 μM	5 nM	146
Mxene/CNH/β-CD/MOF	Carbendazim (pesticide)	Tomato	3.0 nM–10.0 μM	1.0 nM	151
Ti_3C_2 MXene/MoS_2/Au	Phoxim (Insecticide)	Fruits	1×10^{-1}–1×10^{-7} M	5.29×10^{-15} M	152
PtNP/Co/Zn-MOF/C-MWCNT	*Salmonella typhimurium* (pathogen)	Milk	1.3×10^2–1.3×10^8 CFU mL^{-1}	9.4×10^1 CFU mL^{-1}	174
Sm-MOF/gC_3N_4	Propyl gallate (antioxidant)	Food	0.02–60 μM	2.05 nM	175

(*Continued*)

TABLE 5.1 Continued

Electrode modifier	Analyte	Food sample	Linear range	Detection limit	Ref.
Ab-MOF/MXene	Aflatoxin B_1 ~~(antibiotic)~~ (mycotoxin)	Spiked samples of pistachio	0.06–50 ng mL^{-1}	8 pg mL^{-1}	176
PANI/GO	Pb^{2+} (heavy metal ion)	Beverage	–	0.04 μg L^{-1}	188
p-ADPA-4-ATP/Au	(SO_3^{2-}) (preservative)	Beverages	5–160 μM	1.5 μM	189
PANI/GO	Tetracycline (antibiotic)	Milk	8.0–750 μg L^{-1}	2.42–7.59 μg L^{-1}	190
Carbon dot/ZnO/PANI	*E. coli* (pathogen)	Water	–	1.3×10^{-18} M	191
MIP/g-C_3N_4	Endosulfan (insecticide and acaricide)	Celery samples	0–96 μM	20 nM	205
MIP/BP/Au	Pefloxacin (antibacterial drug)	Milk and orange juice	0.005–10 μM	0.80 nM	206
Fe_3O_4/COF/MIP	Tetracycline (antibiotic)	Food samples	1×10^{-10}–1×10^{-4} g mL^{-1}	–	207
NanoMIP	Fumonisin B_1 (mycotoxin)	Food and animal feed	1 fM–10 pM	0.7 fM	208
GO/pPy-KAN	Kanamycin (antibiotic)	Milk	5 nM–1 μM	5 nM	209
MIP/CNT/CNC	Imidacloprid (Insecticide)	Honey	CV: 2.0–99 μM DPV: 0.20–92 μM	CV: 0.35 μM DPV: 0.06 μM	210
CoCu/COF/SWCNT	Carbendazim (fungicide)	Water and pear	0.001–10 μM	0.65 nM	224
Au/COF	Ractopamine (feed additive)	Pork and chicken	1.2–1600 μmol L^{-1}	–	225
COF/Au/Gr	Cd^{2+} Pb^{2+} Cu^{2+} (Heavy metal ions)	Baijiu	0.5–18 μM 0.5–6.0 μM 0.5–6.0 μM	3.025 nM 5.849 nM 9.993 nM	228

PVP: poly(vinylpyrrolidone), FGr: fluorinated graphene, MWCNT–COOH: carboxylated carbon nanotubes, β-CD: β-cyclodextrin, C-MWCNT: carboxylic MWCNT, Ab: antibodies, p-ADPA-4-ATP: poly(4-aminodiphenylamine)-4-aminothiophenol, BP: black phosphorus, KAN: Kanamycin, pPy: poly pyrrole, CNC: cellulose nanocrystals.

5.5 CONCLUSION

This chapter has explored the critical developments in sensor technology, from the fundamental principles of electrochemical detection to the integration of novel materials that enhance sensor performance. The advancements in electrochemical sensing devices have significantly enhanced the ability to detect pesticides, insecticides, and other toxins in food products with high sensitivity and specificity, addressing the growing global concern over food safety. The integration of electrochemical sensors with digital platforms has further revolutionized the field, enabling real-time monitoring and data analysis. However, despite the progress made, challenges such as multi-analyte detection, detection of ultra-low levels of contaminants, reducing interference from complex food matrices, and ensuring the long-term stability and reproducibility of sensor performance persist, and these shortcomings will pave the path for continued research and innovation in this field.

The future of electrochemical sensing devices for detecting chemical toxicants in food products holds significant promise, driven by ongoing advancements in materials science. The development of highly selective, sensitive, miniaturized, portable, and user-friendly sensors that can operate in complex food matrices with minimal sample preparation will be a key focus. Moreover, these sensors are expected to evolve into smart devices, integrated with digital platforms for seamless monitoring and data management, playing a crucial role in enhancing food safety on a global scale.

Abbreviations

NMs	Nanomaterials
EIS	Electrochemical impedance spectroscopy
CV	Cyclic voltammetry
SWV	Square wave voltammetry
DPV	Differential pulse voltammetry
Gr	Graphene
GO	Graphene oxide
rGO	Reduced graphene oxide
CNTs	Carbon nanotubes
SWCNTs	Single-walled carbon nanotubes
MWCNTs	Multi-walled carbon nanotubes
CNHs	Carbon nanohorns
gC_3N_4	Graphitic carbon nitride

MXene	Transition metal carbides and nitrides
MOFs	Organometallic frameworks
MIPs	Molecularly imprinted polymers
CPs	Conducting polymers
PANI	Polyanilines
COFs	Covalent organic frameworks
CFU	Colony-forming unit
GCE	Glassy carbon electrode

REFERENCES

1. Umapathi R, Ghoreishian SM. Review-emerging trends in the development of electrochemical devices for the on-site detection of food contaminants. *ECS Sensors Plus*. 2022;1:044601.
2. Carvalho FP. Pesticides, environment, and food safety. *Food Energy Secur.* 2017;6(2):48–60.
3. Fyles H, Madramootoo C. Key drivers of food insecurity. In: *Emerging technologies for promoting food security*, ed. Madramootoo C. 2016. p. 1–19. Woodhead Publishing.
4. Rosenfeld PE, Feng LGH. 11 - Pesticides. In: *Risks of hazardous wastes*. 2011. p. 127–54. William Andrew Applied Science Publishers.
5. Mugo SM, Lu W, Robertson SV. Molecularly imprinted polymer-modified microneedle sensor for the detection of imidacloprid pesticides in food samples. *Sensors (Basel)*. 2022;22(21):8492.
6. Noori JS, Mortensen J, Geto A. Recent development on the electrochemical detection of selected pesticides: A focused review. *Sensors (Basel)*. 2020;20:2221.
7. Wang J. Real-time electrochemical monitoring: Toward green analytical chemistry. *Acc Chem Res*. 2002;35(9):811–6.
8. Seymour I, Narayan T, Creedon N, Kennedy K, Murphy A, et al. Advanced solid state nano-electrochemical sensors and system for Agri 4.0 applications. *Sensors (Basel)*. 2021;21:3149.
9. Sumathi C, Venkateswara Raju C, Muthukumaran P, Wilson J, Ravi G. Au–Pd bimetallic nanoparticles anchored on α-Fe2O3 nonenzymatic hybrid nanoelectrocatalyst for simultaneous electrochemical detection of dopamine and uric acid in the presence of ascorbic acid. *J Mater Chem B*. 2016;4(15):2561–9.
10. Ghosh S, AlKafaas SS, Bornman C, Apollon W, Hussien AM, et al. The application of rapid test paper technology for pesticide detection in horticulture crops: A comprehensive review. *Beni-Suef Univ J Basic Appl Sci*. 2022;11:73.
11. Liu X, Zhang H, Ma Y, Wu X, Meng L, et al. Graphene-coated silica as a highly efficient sorbent for residual organophosphorus pesticides in water. *J Mater Chem A*. 2013;1(5):1875–84.

12. Yang Q, Wang J, Zhang W, Liu F, Yue X, et al. Interface engineering of metal organic framework on graphene oxide with enhanced adsorption capacity for organophosphorus pesticide. *Chem Eng J.* 2017;313:19–26.
13. Yadav IC, Devi NL. Pesticides classification and its impact on environment. *Environ Sci Eng.* 2017;6:140–58.
14. Hara TO, Singh B. Electrochemical biosensors for detection of pesticides and heavy metal toxicants in water: Recent trends and progress. *ACS ES&T Water.* 2021;1(6):1462–78.
15. Hossain MI, Hasnat MA. Recent advancements in non-enzymatic electrochemical sensor development for the detection of organophosphorus pesticides in food and environment. *Heliyon.* 2023;9(9):e19299.
16. Kaur N, Prabhakar N. Current scenario in organophosphates detection using electrochemical biosensors. *TrAC Trends Anal Chem.* 2017;92:62–85.
17. Bolat G, Abaci S. Non-enzymatic electrochemical sensing of malathion pesticide in tomato and apple samples based on gold nanoparticles-chitosan-ionic liquid hybrid nanocomposite. *Sensors (Basel).* 2018;18:773.
18. Tang W, Yang J, Wang F, Wang J, Li Z. Thiocholine-triggered reaction in personal glucose meters for portable quantitative detection of organophosphorus pesticide. *Anal Chim Acta.* 2019;1069:10–17.
19. Lin JN, Lin CL, Lin MC, Lai CH, Lin HH, et al. Increased risk of dementia in patients with acute organophosphate and carbamate poisoning: A nationwide population-based cohort study. *Medicine (Baltimore).* 2015;94:e1187.
20. Elfadil D, Lamaoui A, Della Pelle F, Amine A, Compagnone D. Molecularly imprinted polymers combined with electrochemical sensors for food contaminants analysis. *Molecules.* 2021;26:4607.
21. Pundir CS, Chauhan N. Acetylcholinesterase inhibition-based biosensors for pesticide determination: A review. *Anal Biochem.* 2012;429:19–31.
22. Thanos CAA, Tomuka D, Mallo NTS. Livor mortis on organophosphate insecticide poisoning in rabbits. *E-CliniC.* 2016;4:1.
23. Kunpatee K, Kaewdorn K, Duangtong J, Chaiyo S, Chailapakul O, et al. A new disposable electrochemical sensor for the individual and simultaneous determination of carbamate pesticides using a nanocomposite modified screen-printed electrode. *Microchem J.* 2022;177:107318.
24. Alnahdi HS, Mousa RMA, El-Said WA. Development of organochlorine pesticide electrochemical sensor based on Fe3O4 nanoparticles@indium tin oxide electrode. *Electroanalysis.* 2023;35:e202100659.
25. Moldovan R, Iacob BC, Farcău C, Bodoki E, Oprean R. Strategies for SERS detection of organochlorine pesticides. *Nanomaterials (Basel).* 2021;11(2):304.
26. Tsagkaris AS, Pulkrabova J, Hajslova J. Optical screening methods for pesticide residue detection in food matrices. *Adv Emerg Anal Trends.* 2021;10:88.
27. Santoni G, De Sousa J, De La Mora E, Dias J, Jean L, et al. Structure-based optimization of nonquaternary reactivators of acetylcholinesterase inhibited by organophosphorus nerve agents. *J Med Chem.* 2018;61(17):7630–9.
28. Gazzi EN, Sorodoc V, Jaba IM, Lionte C, Bologa C, et al. Profile of adult acute cholinesterase inhibitors substances poisoning: A 30 years analysis. *Open Med (Wars).* 2015;10(1):278–284.

29. Hassani S, Momtaz S, Vakhshiteh F, Maghsoudi AS, Ganjali MR, et al. Biosensors and their applications in detection of organophosphorus pesticides in the environment. *Arch Toxicol.* 2017;91:109–30.
30. Khairy M, Ayoub HA, Banks CE. Non-enzymatic electrochemical platform for parathion pesticide sensing based on nanometer-sized nickel oxide modified screen-printed electrodes. *Food Chem.* 2018;255:104–11.
31. Toldrá F, Reig M. Methods for rapid detection of chemical and veterinary drug residues in animal foods. *Trends Food Sci Technol.* 2006;17(9):482–9.
32. Marx ÍMG. Emerging trends of electrochemical sensors in food analysis. *Electrochem.* 2023;4:42–6.
33. López R, Khan S, Wong A, Sotomayor M del PT, Picasso G. Development of a new electrochemical sensor based on Mag-MIP selective toward amoxicillin in different samples. *Front Chem.* 2021;9:1-9.
34. Wang Q, Xue Q, Chen T, Li J, Liu Y, Shan X, et al. Recent advances in electrochemical sensors for antibiotics and their applications. *Chin Chem Lett.* 2021;32(2):609–19.
35. Arsène MMJ, Davares AKL, Viktorovna PI, Andreevna SL, Sarra S, Khelifi I, Sergueïevna DM. The public health issue of antibiotic residues in food and feed: Causes, consequences, and potential solutions. *Vet World.* 2022;15(3):662–71.
36. Majdinasab M, Mishra RK, Tang X, Marty JL. Detection of antibiotics in food: New achievements in the development of biosensors. *TrAC Trends Anal Chem.* 2020;127:115883.
37. Liang JF, Peng C, Li P, Ye QX, Wang Y, Yi YT, et al. A review of detection of antibiotic residues in food by surface-enhanced Raman spectroscopy. *Bioinorg Chem Appl.* 2021:8180154.
38. Khan MZH. Recent biosensors for detection of antibiotics in animal derived food. *Crit Rev Anal Chem.* 2020;52(4):780–90.
39. Stolker AAM, Brinkman UAT. Analytical strategies for residue analysis of veterinary drugs and growth-promoting agents in food-producing animals: A review. *J Chromatogr A.* 2005;1067:15–53.
40. Mikkelsen KH, Allin KH, Knop FK. Effect of antibiotics on gut microbiota, glucose metabolism and body weight regulation: A review of the literature. *Diabetes Obes Metab.* 2016;18:444–53.
41. Hutchings M, Truman A, Wilkinson B. Antibiotics: Past, present and future. *Curr Opin Microbiol.* 2019;51:72–80.
42. Mboya EA, Sanga LA, Ngocho JS. Irrational use of antibiotics in the Moshi Municipality Northern Tanzania: A cross-sectional study. *Pan Afr Med J.* 2018;31:165.
43. Sachi S, Ferdous J, Sikder MH, Hussani SMKA. Antibiotic residues in milk: Past, present, and future. *J Adv Vet Anim Res.* 2019;6:315–32.
44. Singh B, Bhat A, Dutta L, Pati KR, Korpan Y, Dahiya I. Electrochemical biosensors for the detection of antibiotics in milk: Recent trends and future perspectives. *Biosensors.* 2023;13:867.
45. Adane WD, Chandravanshi BS, Tessema M. A novel electrochemical sensor for the detection of metronidazole residues in food samples. *Chemosphere.* 2024;359:142279.

46. Aihaiti A, Li Z, Qin Y, Meng F, Li X, Huangfu Z, et al. Construction of electrochemical sensors for antibiotic detection based on carbon nanocomposites. *Nanomaterials (Basel)*. 2022;12:2789.
47. Kim C, Ryu HD, Chung EG, Kim Y, Lee KJ. A review of analytical procedures for the simultaneous determination of medically important veterinary antibiotics in environmental water: Sample preparation, liquid chromatography, and mass spectrometry. *J Environ Manag*. 2018;217:629–45.
48. Yang YF, Zhang H, Zhou GY, Zhang SW, Chen JH, et al. Risk assessment of veterinary drug residues in pork on the market in the People's Republic of China. *J Food Prot*. 2022;85:815–27.
49. Boshra MH, El-Housseiny GS, Farag MMS, Aboshanab KM. Innovative approaches for mycotoxin detection in various food categories. *AMB Express*. 2024;14(1):7.
50. Nawaz M, Rauf S, Hayat A, Catanante G, Raza R, Marty J. Determination of mycotoxins in food. In: *Analysis of food toxins and toxicants*, ed. Wong Y-c, Lewis RJ. 2017. p. 137–68. Wiley-Blackwell.
51. Agriopoulou S, Stamatelopoulou E, Varzakas T. Advances in analysis and detection of major mycotoxins in foods. *Foods*. 2020;9(4):518.
52. Kharayat BS, Singh Y. Chapter 13 - Mycotoxins in foods: Mycotoxicoses, detection, and management. In: *Handbook of food bioengineering*, ed. Holban AM, and Grumezescu AM. 2018. pp. 395–421. Academic Press.
53. Inglis A, Parnell AC, Subramani N, Doohan FM. Machine learning applied to the detection of mycotoxin in food: A systematic review. *Toxins (Basel)*. 2024;16(6):268.
54. Martinez L, He L. Detection of mycotoxins in food using surface-enhanced Raman spectroscopy: A review. *ACS Appl Bio Mater*. 2021;4(1):295–310.
55. Freire L, Sant'ana AS. Modified mycotoxins: An updated review on their formation, detection, occurrence, and toxic effects. *Food Chem Toxicol*. 2018;111:189–205.
56. PS A, Thadathil DA, George L, Varghese A. Food additives and evolved methods of detection: A review. *Crit Rev Anal Chem*. 2024:1–20.
57. Onaolapo AY, Onaolapo OJ. Food additives, food and the concept of 'food addiction': Is stimulation of the brain reward circuit by food sufficient to trigger addiction? *Pathophysiology*. 2018;25:263–76.
58. Hsu CL, Fang M, Tseng HC, Lin YT, Tseng SH, Wang DY, et al. Comprehensive detection of 120 additives in food using nontargeted MS data acquisition. *J Food Drug Anal*. 2021;29(3):419–32.
59. Mahale RS, Shashanka R, Vasanth S, Vinaykumar R. Voltammetric determination of various food azo dyes using different modified carbon paste electrodes. *Biointerface Res Appl Chem*. 2022;12(4):4557–66.
60. Gao Y. Research of food additives and their detection technology. *BIO Web Conf*. 2024;111.
61. Gosadi II, Alamri AK, Saleh RA, Almutairi BA, Batais MA, Alamri NK. Preference of sweeteners among Saudi diabetes patients from a tertiary health care centre in Riyadh, Saudi Arabia. *Saudi J Biol Sci*. 2020;27(3):947–52.

62. Nisa K, Samiullah S, Khan N, Kakar A. Detection of heavy metals in fruits and vegetables available in the market of Quetta City. *Al-Nahrain J Sci.* 2020;23:47–56.
63. Liang G, Gong W, Li B, Zuo J, Pan L, Liu X. Analysis of heavy metals in foodstuffs and an assessment of the health risks to the general public via consumption in Beijing, China. *Int J Environ Res Public Health.* 2019;16(6):909.
64. Koch W, Czop M, Iłowiecka K, Nawrocka A, Wiącek D. Dietary intake of toxic heavy metals with major groups of food products: Results of analytical determinations. *Nutrients.* 2022;14(8):1626.
65. Sulthana SF, Iqbal UM, Suseela SB, Anbazhagan R, Chinthaginjala R, Chitathuru D, et al. Electrochemical sensors for heavy metal ion detection in aqueous medium: A systematic review. *ACS Omega.* 2024;9(24):25493–512.
66. Liang G, Gong W, Li B, Zuo J, Pan L, Liu X. Analysis of heavy metals in foodstuffs and an assessment of the health risks to the general public via consumption in Beijing, China. *Int J Environ Res Public Health.* 2019;16(6):909.
67. Nisa K, Samiullah S, Khan N, Kakar A. Detection of heavy metals in fruits and vegetables available in the market of Quetta City. *Al-Nahrain J Sci.* 2020;23:47–56.
68. Nisa K, Samiullah S, Khan N, Kakar A. Detection of heavy metals in fruits and vegetables available in the market of Quetta City. *Al-Nahrain J Sci.* 2020;23:47–56.
69. Scutarașu EC, Trincă LC. Heavy metals in foods and beverages: Global situation, health risks and reduction methods. *Foods.* 2023;12(18):3340.
70. Qin G, Niu Z, Yu J, Li Z, Ma J, Xiang P. Soil heavy metal pollution and food safety in China: Effects, sources and removing technology. *Chemosphere.* 2021;267:128909.
71. Rai PK, Lee SS, Zhang M, Tsang YF, Kim KH. Heavy metals in food crops: Health risks, fate, mechanisms, and management. *Environ Int.* 2019;125:365–85.
72. Verger PJ, Boobis AR. Reevaluate pesticides for food security and safety. *Science.* 2013;341:717–8.
73. Mattison RL, Bowyer AA, New EJ. Small molecule optical sensors for nickel: The quest for a universal nickel receptor. *Coord Chem Rev.* 2020;425:213522.
74. Verma N, Singh MA. *Bacillus sphaericus*-based biosensor for monitoring nickel ions in industrial effluents and foods. *J Autom Methods Manag Chem.* 2006:1–4.
75. Rajendrachari S, Kenchappa Somashekharappa K, Shrinivas Mahale R, Vasanth S, Peramenahalli Chikkegouda S. A review on cyclic voltammetric investigation of toxic heavy metals. *Front Voltammetry.* 2023:1–19.
76. Paparella A, Maggio F. Detection and control of foodborne pathogens. *Foods.* 2023;12(19):3521.
77. Law JW, Ab Mutalib NS, Chan KG, Lee LH. Rapid methods for the detection of foodborne bacterial pathogens: Principles, applications, advantages and limitations. *Front Microbiol.* 2015;5:770.
78. Paparella A, Maggio F. Detection and control of foodborne pathogens. *Foods.* 2023;12:3521.

79. Priyanka B, Patil RK, Dwarakanath S. A review on detection methods used for foodborne pathogens. *Indian J Med Res.* 2016;144(3):327–38.
80. Banerjee P, Bhunia AK. Cell-based biosensor for rapid screening of pathogens and toxins. *Biosens Bioelectron.* 2010;26(1):99–106.
81. Curulli A. Electrochemical biosensors in food safety: Challenges and perspectives. *Molecules.* 2021;26(10):2940.
82. Naresh V, Lee N. Review on biosensors and recent development of nanostructured materials-enabled biosensors. *Sensors.* 2021;21:1109.
83. Chen ZG. Conductometric immunosensors for the detection of staphylococcal enterotoxin B based bio-electrocalytic reaction on micro-comb electrodes. *Bioprocess Biosyst Eng.* 2008;31:345–50.
84. Shanbhag MM, Manasa G, Mascarenhas RJ, Mondal K, Shetti NP. Fundamentals of bio-electrochemical sensing. *Chem Eng J Adv.* 2023;16:100516.
85. Jaffrezic-Renault N, Dzyadevych SV. Conductometric microbiosensors for environmental monitoring. *Sensors.* 2008;8(4):2569–88.
86. Velusamy V, Arshak K, Korostynska O, Oliwa K, Adley C. An overview of foodborne pathogen detection: In the perspective of biosensors. *Biotechnol Adv.* 2010;28:232–54.
87. Demir E, Mercan Silah H, Aydogdu Ozdogan N. Electrochemical analysis of pesticide in food samples. In: *Medicine & health.* 2021:213–34.
88. Rostiyanti H, Mubarok AZ. *IOP Conf Ser Earth Environ Sci.* 2021;924:012005.
89. Grieshaber D, MacKenzie R, Vörös J, Reimhult E. Electrochemical biosensors: Sensor principles and architectures. *Sensors (Basel).* 2008;8(3):1400–58.
90. Kumala AS, Sulmartiwi L, Lutfiyah L. A review development of electrochemical sensors for detection of organophosphate pesticides in fruits and vegetables. *IOP Conf Ser Earth Environ Sci.* 2021;924(1):012005.
91. Maheshwaran S, Chen WH, Lin SL, Ghorbani M, Hoang AT. Metal oxide-based electrochemical sensors for pesticide detection in water and food samples: A review. *Environ Sci Adv.* 2024;3(2):154–76.
92. Hu J, Sang G, Zeng N, Lv C, Xu C. Amperometric sensor for the detection of hydrogen stable isotopes based on Pt nanoparticles confined within single-walled carbon nanotubes (SWNTs). *Sens Actuators B Chem.* 2022;356:131344.
93. Stradiotto NR, Yamanaka H, Zanoni MVB. Electrochemical sensors: A powerful tool in analytical chemistry. *J Braz Chem Soc.* 2003;14(2):159–73.
94. Mishra RK, Hubble LJ, Martín A, Kumar R, Barfidokht A, Kim J, et al. Wearable flexible and stretchable glove biosensor for on-site detection of organophosphorus chemical threats. *ACS Sensors.* 2017;2:553–61.
95. Thanos CAA, Tomuka D, Mallo NTS. Livor mortis on organophosphate insecticide poisoning in rabbits. *E-CliniC.* 2016;4:1.
96. Saiya A. Analysis of clorpirifos residual in cabbage with HPLC method in some traditional markets in North Sulawesi. *EKSAKTA Berk Ilm Bid MIPA.* 2017;18:77–85.
97. Harsojo H, Chairul SM. Pathogenic microbes, organophosphate insecticide residues and heavy metals content in vegetables. *J Ecolab.* 2011;5:89–95.

98. Carvalho FP. Pesticides, environment, and food safety. *Food Energy Secur.* 2017;6:48–60.
99. Jebril S, Fredj Z, Saeed AA, Gonçalves AM, Kaur M, Kumar A, et al. Nanomaterial-based electrochemical chemo(bio)sensors for the detection of nanoplastic residues: Trends and future prospects. *RSC Sustain.* 2024;2(4):832–51.
100. Ribeiro BV, Ferreira LF, Franco DL. Advances in biosensor development for the determination of antibiotics in cow's milk: A review. *Talanta Open.* 2022;6:100145.
101. Zamfir LG, Puiu M, Bala C. Advances in electrochemical impedance spectroscopy detection of endocrine disruptors. *Sensors.* 2020;20(22):6443.
102. Barreiros dos Santos M, Agusil JP, Prieto-Simón B, Sporer C, Teixeira V, et al. Highly sensitive detection of pathogen Escherichia coli O157:H7 by electrochemical impedance spectroscopy. *Biosens Bioelectron.* 2013;45:174–80.
103. Shi Z, Xia L, Li G. Recent progress of electrochemical sensors in food analysis. *Chemosensors.* 2023;11:478.
104. Wang W, Wang X, Cheng N, Luo Y, Lin Y, Xu W, et al. Recent advances in nanomaterials-based electrochemical (bio)sensors for pesticides detection. *TrAC Trends Anal Chem.* 2020;132:116041.
105. Lu L, Hu X, Zhu Z, Li D, Tian S, Chen Z. Review: Electrochemical sensors and biosensors modified with binary nanocomposite for food safety. *J Electrochem Soc.* 2020;167:037512.
106. Mahale RS, Vasanth S, Krishna H, Shashanka R, Sharath PC, Sreekanth NV. Electrochemical sensor applications of nanoparticle modified carbon paste electrodes to detect various neurotransmitters: A review. *Appl Mech Mater.* 2022;908:69–88.
107. Raja IS, Vedhanayagam M, Preeth DR, Kim C, Lee JH, Han DW. Development of two-dimensional nanomaterials based electrochemical biosensors on enhancing the analysis of food toxicants. *Int J Mol Sci.* 2021;22:3277.
108. Zhao F, Wu J, Ying Y, She Y, Wang J, Ping J. Carbon nanomaterial-enabled pesticide biosensors: Design strategy, biosensing mechanism, and practical application. *TrAC Trends Anal Chem.* 2018;106:62–83.
109. Adeniji TM, Stine KJ. Nanostructure modified electrodes for electrochemical detection of contaminants of emerging concern. *Coatings.* 2023;13:381.
110. Mustafa F, Andreescu S. Sensing and packaging applications. *RSC Adv.* 2020;10:19309–36.
111. Shashanka R, Kumara Swamy BE. Simultaneous electro-generation and electro-deposition of copper oxide nanoparticles on glassy carbon electrode and its sensor application. *SN Appl Sci.* 2020;2(5):1–10.
112. Zhou C, Feng J, Tian Y, Wu Y, He Q, et al. Non-enzymatic electrochemical sensors based on nanomaterials for detection of organophosphorus pesticide residues. *Environ Sci Adv.* 2023;2(7):933–56.
113. Meng R, Zhu Q, Long T, He X, Luo Z, Gu R. The innovative and accurate detection of heavy metals in foods: A critical review on electrochemical sensors. *Food Control.* 2023;150:109743.

114. Adeniji TM, Stine KJ. Nanostructure modified electrodes for electrochemical detection of contaminants of emerging concern. *Coatings*. 2023;13(2):381.
115. Gerent GG, Santana ER, Martins EC, Spinelli A. A non-mercury electrode for the voltammetric determination of butralin in foods. *Food Chem*. 2021;343:128419.
116. Kumari R, Kumar H, Yadav A, Sharma R, Kumari A, Kumar A. Metal nanocomposites-based electrochemical sensor for the detection of vanillin (food additives): Experimental and theoretical approach. *Food Biosci*. 2023;52:102464.
117. Chakraborty P, Dhar S, Deka N, Debnath K, Mondal SP. Non-enzymatic salivary glucose detection using porous CuO nanostructures. *Sens Actuators B Chem*. 2020;302:127134.
118. Akanji SP, Ama OM, Ray SS, Osifo PO. Metal oxide nanomaterials for electrochemical detection of heavy metals in water. In: *Nanostructured metal-oxide electrode materials for water purification: Fabrication, electrochemistry and applications*, ed. Ama OM, Ray SS. 2020. pp.113–26. Springer.
119. Ensafi AA, Sohrabi M, Jafari-Asl M, Rezaei B. Selective and sensitive furazolidone biosensor based on DNA-modified TiO2-reduced graphene oxide. *Appl Surf Sci*. 2015;356:301–7.
120. Wang M, Sun Y, Yang X, Zhao J. Sensitive determination of amaranth in drinks by highly dispersed CNT in graphene oxide "water" with the aid of small amounts of ionic liquid. *Food Chem*. 2015;179:318–24.
121. Kumari R, Kumar H, Sharma R, Kumar G, Tundwal A, Dhayal A, et al. Highly efficient and reliable voltammetry food sensor for tartrazine dye using a nanocomposite reformed electrode. *Microchem J*. 2024;196:109583.
122. Zhang M, Guo W. Simultaneous electrochemical detection of multiple heavy metal ions in milk based on silica-modified magnetic nanoparticles. *Food Chem*. 2023;406:135034.
123. Anh NT, Huyen NN, Dinh NX, Vinh LK, Tung LM, Vinh NT, et al. ZnO/ZnFe2O4 nanocomposite-based electrochemical nanosensors for the detection of furazolidone in pork and shrimp samples: Exploring the role of crystallinity, phase ratio, and heterojunction formation. *New J Chem*. 2022;46(15):7090–102.
124. Shukla S, Haldorai Y, Khan I, Kang SM, Kwak CH, Gandhi S, et al. Bioreceptor-free, sensitive and rapid electrochemical detection of patulin fungal toxin, using a reduced graphene oxide@SnO2 nanocomposite. *Mater Sci Eng C*. 2020;113:110916.
125. Rajendrachari S, Basavegowda N, Adimule VM, Avar B, Somu P, et al. Assessing the food quality using carbon nanomaterial based electrodes by voltammetric techniques. *Biosensors*. 2022;12:1173.
126. Yu D, Nagelli E, Du F, Dai LM. Metal-free carbon nanomaterials become more active than metal catalysts and last longer. *J Phys Chem Lett*. 2010;1:2165–73.
127. He Q, Liu J, Feng J, Wu Y, Tian Y, et al. Sensitive voltammetric sensor for tryptophan detection by using polyvinylpyrrolidone functionalized graphene/GCE. *Nanomaterials*. 2020;10:125.

128. Karimi-Maleh H, Arotiba OA. Simultaneous determination of cholesterol, ascorbic acid and uric acid as three essential biological compounds at a carbon paste electrode modified with copper oxide decorated reduced graphene oxide nanocomposite and ionic liquid. *J Colloid Interface Sci.* 2020;560:208–12.
129. Mishra K, Kumar Thakur V, Singh Siwal S. Graphitic carbon nitride based palladium nanoparticles: A homemade anode electrode catalyst for efficient direct methanol fuel cells application. *Mater Today Proc.* 2022;56:107–11.
130. Porto LS, Silva N, de Oliveira AEF, Pereira AC, Borges KB. Carbon nanomaterials: Synthesis and applications to development of electrochemical sensors in determination of drugs and compounds of clinical interest. *Rev Anal Chem.* 2019;38:20190017.
131. Bounegru AV, Apetrei C. Carbonaceous nanomaterials employed in the development of electrochemical sensors based on screen-printing technique—A review. *Catalysts.* 2020;10:680.
132. Kour R, Arya S, Young SJ, Gupta V, Bandhoria P, et al. Review—Recent advances in carbon nanomaterials as electrochemical biosensors. *J Electrochem Soc.* 2020;16:037555.
133. Bobrinetskiy II, Knezevic NZ. Graphene-based biosensors for on-site detection of contaminants in food. *Anal Methods.* 2018;10:5061–71.
134. Tan Z, Wu W, Feng C, Wu H, Zhang Z. Simultaneous determination of heavy metals by an electrochemical method based on a nanocomposite consisting of fluorinated graphene and gold nanocage. *Microchim Acta.* 2020;187:1–9.
135. Veenuttranon K, Lu X, Chen J. Ultrasensitive electrochemical sensing for simultaneous rapid detection of zearalenone and ochratoxin A in feedstuffs and foodstuffs. *Chem Eng J.* 2024;497:154807.
136. Curulli A. Nanomaterials in electrochemical sensing area: Applications and challenges in food analysis. *Molecules.* 2020;25:5759.
137. Kaur H, Singh S, Chauhan G, Kumar A, Kumari A, Kumar V. Recent advances in electrochemical-based sensors amplified with carbon-based nanomaterials (CNMs) for sensing pharmaceutical and food pollutants. *Chemosphere.* 2022;304:135182.
138. Saisahas K, Soleh A, Promsuwan K, Phonchai A, Mohamed Sadiq NS, Teoh WK, et al. A portable electrochemical sensor for detection of the veterinary drug xylazine in beverage samples. *J Pharm Biomed Anal.* 2021;198:113958.
139. Feng K, Li T, Ye C, Gao X, Yang T, et al. A label-free electrochemical immunosensor for rapid detection of salmonella in milk by using CoFe-MOFs-graphene modified electrode. *Food Control.* 2021;130:108357.
140. Karimian N, Fakhri H, Amidi S, Hajian A, Arduini F, et al. A novel sensing layer based on metal–organic framework UiO-66 modified with TiO2–graphene oxide: Application to rapid, sensitive and simultaneous determination of paraoxon and chlorpyrifos. *New J Chem.* 2019;43(6):2600–9.
141. Malode SJ, Prabhu K, Pollet BG, Kalanur SS, Shetti NP. Preparation and performance of WO3/rGO modified carbon sensor for enhanced electrochemical detection of triclosan. *Electrochim Acta.* 2022;429:141010.

142. Dhull V. Fabrication of AChE/SnO2-cMWCNTs/Cu nanocomposite-based sensor electrode for detection of methyl parathion in water. *Int J Anal Chem.* 2018;2874059.
143. Sireesha M, Jagadeesh Babu V, Kranthi Kiran AS, Ramakrishna S. A review on carbon nanotubes in biosensor devices and their applications in medicine. *Nanocomposites.* 2018;4:36–57.
144. Rafiquea I, Anwara AKZ, Muhammad B. Exploration of epoxy resins, hardening systems, and epoxy/carbon nanotube composite designed for high performance materials: A review. *Polym Plast Technol Eng.* 2016;55:312–33.
145. Appaturi JN, Pulingam T, Thong KL, Muniandy S, Ahmad N, Leo BF. Rapid and sensitive detection of Salmonella with reduced graphene oxide-carbon nanotube based electrochemical aptasensor. *Anal Biochem.* 2020;589:113489.
146. Chen Y, He K, Sun F, Wei D, Li H. Efficient electrochemical sensor based on multi-walled carbon nanotubes/gold nanocomposite for detection of dichlorvos pesticide in agricultural food. *Int J Electrochem Sci.* 2021;16(3):210321.
147. Lukatskaya MR, Mashtalir O, Ren CE, Dall'Agnese Y, Rozier P, et al. Cation intercalation and high volumetric capacitance of two-dimensional titanium carbide. *Science.* 2013;341:1502–5.
148. Khazaei M, Arai M, Sasaki T, Chung CY, Venkataramanan NS, et al. Novel electronic and magnetic properties of two-dimensional transition metal carbides and nitrides. *Adv Funct Mater.* 2013;23:2185–92.
149. Zhang H. Ultrathin two-dimensional nanomaterials. *ACS Nano.* 2015;9:9451–69.
150. He LD, Wu J, Zhu YZ, Wang YM, Mei Y. Covalent immobilization of black phosphorus quantum dots on MXene for enhanced electrocatalytic nitrogen reduction. *Ind Eng Chem Res.* 2021;60:5443–50.
151. Tu X, Gao F, Ma X, Zou J, Yu Y, et al. MXene/carbon nanohorn/β-cyclodextrin-metal-organic frameworks as high-performance electrochemical sensing platform for sensitive detection of carbendazim pesticide. *J Hazard Mater.* 2020;396:122776.
152. Zhang J, Guo X, Zhang J, Guo X, Xu Y, et al. Ti3C2 MXene/MoS2@AuNPs ternary nanocomposite for highly sensitive electrochemical detection of phoxim residues in fruits. *Food Chem.* 2025;462:140939.
153. Sohrabi H, Salahshour Sani P, Zolfaghari R, Majidi MR, Yoon Y, Khataee A. MOF-based mycotoxin nanosensors for food quality and safety assessment through electrochemical and optical methods. *Molecules.* 2022;27(21):7511.
154. Yang Y. Research progress on electrochemical sensors based on metal-organic framework materials for the detection of mycotoxins in food. *Acad J Sci Technol.* 2024;11(2):261–81.
155. Kitagawa S. Metal–organic frameworks (MOFs). *Chem Soc Rev.* 2014;43:5415–8.
156. Zhang W, Li X, Ding X, Hua K, Sun A, et al. Progress and opportunities for metal-organic framework composites in electrochemical sensors. *RSC Adv.* 2023;13(16):10800–17.

157. Reinsch H, Van der Veen MA, Gil B, Marszalek B, Verbiest T, et al. Structures, sorption characteristics, and nonlinear optical properties of a new series of highly stable aluminium MOFs. *Chem Mater.* 2013;25(1):17–26.
158. Liu L, Zhou Y, Liu S, Xu M. The applications of metal-organic frameworks in electrochemical sensors. *Chem Electro Chem.* 2018;5(1):6–19.
159. Lan BL, Shao B, Yang FJ, Pang W, Guo Z, et al. Metal-organic frameworks-derived nickel–iron oxyhydroxide with highly active edge sites for electrochemical oxygen evolution. *Small Struct.* 2022;3:2200085.
160. Li MY, Wang F, Zhang J. Synthesis, structure, and proton conductivity of a metal–organic framework with rich hydrogen-bonds between the layers. *Inorg Chem Commun.* 2017;79:37–40.
161. Jiao L, RuSeow JY, ScottSkinner W, Wang ZU, Jiang HL. Metal-organic frameworks: Structures and functional applications. *Mater Today.* 2019;27:43–68.
162. Yuan S, Feng L, Wang K, Pang J, Bosch M, et al. Stable metal-organic frameworks: design, synthesis, and applications. *Adv Mater.* 2018;30(37):1704303.
163. Hosseinzadeh B, Rodriguez-Mendez ML. Electrochemical sensor for food monitoring using metal-organic framework materials. *Chemosensors.* 2023;11:357.
164. Nong W, Liu X, Wang Q, Wu J, Guan Y. Metal-organic framework-based materials: synthesis, stability, and applications in food safety and preservation. *ES Food Agrofor.* 2020;1:11–40.
165. Stavila V, Talin AA, Allendorf MD. MOF-based electronic and opto-electronic devices. *Chem Soc Rev.* 2014;43:5994–6010.
166. Du T, Wang J, Zhang L, Wang S, Yang C, et al. Missing-linker engineering of Eu(III)-doped UiO-MOF for enhanced detection of heavy metal ions. *Chem Eng J.* 2022;431:134050.
167. Zhang W, Li X, Ding X, Hua K, Sun A, et al. Progress and opportunities for metal-organic framework composites in electrochemical sensors. *RSC Adv.* 2023;13(16):10800–17.
168. Kajal N, Singh V, Gupta R, Gautam S. Metal-organic frameworks for electrochemical sensor applications: a review. *Environ Res.* 2022;204:112320.
169. Song J, Huang M, Lin X, Li S, Jiang N, et al. Novel Fe-based metal-organic framework (MOF) modified carbon nanofiber as a highly selective and sensitive electrochemical sensor for tetracycline detection. *Chem Eng J.* 2021;427:130913.
170. Umesh NM, Antolin Jesila J, Wang SF, Vishnu N, Yang YJ. Novel voltammetric detection of norfloxacin in urine and blood serum using a flexible Ni foam based Ni-Co-MOF ultrathin nanosheets derived from Ni-Co-LDH. *Microchem J.* 2021;160:105747.
171. Zhao J, Kan Y, Chen Z, Li H, Zhang W. MOFs-modified electrochemical sensors and the application in the detection of opioids. *Biosensors.* 2023;13:284.
172. Jahangiri-Dehaghani F, Zare HR, Shekari Z, Benvidi A. Development of an electrochemical aptasensor based on Au nanoparticles decorated on metal-organic framework nanosheets and p-biphenol electroactive label for the measurement of aflatoxin B1 in a rice flour sample. *Anal Bioanal Chem.* 2022;414(5):1973–85.

173. Gao N, Runan T, Zhiwei C, Zhao H, Chang G, He Y. A novel electrochemical sensor via Zr-based metal-organic framework–graphene for pesticide detection. *J Mater Sci.* 2021;56:19060–19074.
174. Zhan K, Chen L, Li S, Yu Q, Zhao Z, et al. A novel metal-organic framework based electrochemical immunosensor for the rapid detection of *Salmonella typhimurium* in milk. *Food Chem.* 2024;444:138672.
175. Srinivasan P, Gopalakrishnan Sethuraman M, Chinnapayan S, Govindasamy M, Albaqami MD, et al. Electrocatalytic sensing of propyl gallate in real-time food samples using a synergistic Sm-MOF@GCN composite modified glassy carbon electrode. *Microchem J.* 2024;201:110647.
176. Kaur G, Sharma S, Bhardwaj N, Nayak MK, Deep A. Highly sensitive and selective electrochemical detection of aflatoxin B1 in water and pistachio samples with MOF/MXene composite based sensor. *Food Control.* 2024;165:110694.
177. Simões FR, Henrique L, Mattoso C, Manoel C, Vaz P. Conducting polymers as sensor materials for the electrochemical detection of pesticides. *Sensor Lett.* 2006;4(3):319–24.
178. Kros A, Sommerdijk NAJM, Nolte RJM. Poly(pyrrole) versus poly(3,4-ethylenedioxythiophene): implications for biosensor applications. *Sensors Actuators B Chem.* 2005;106(1):289–95.
179. Gerard M, Chaubey A, Malhotra BD. Application of conducting polymers to biosensors. *Biosens Bioelectron.* 2002;17(5):345–59.
180. Thanh HL, Yukyung K, Hyeonseok Y. Electrical and electrochemical properties of conducting polymers. *Polymers.* 2017;9:150.
181. Rattan S, Singhal P, Verma A. Synthesis of PEDOT: PSS (poly(3,4-ethylenedioxythiophene))/poly(4-styrene sulfonate)/ngps (nanographitic platelets) nanocomposites as chemiresistive sensors for detection of nitroaromatics. *Polym Eng Sci.* 2013;53:2045–52.
182. Nur S, Mohd A, Adriana A, Adnan C, Isa I, et al. Conducting polymer functionalized graphene-based electrochemical sensors for sensing pollutants in water. *J. Electrochem. Sci. Eng.* 2023;13(2):251–74.
183. Terán-Alcocer Á, Bravo-Plascencia F, Cevallos-Morillo C, Palma-Cando A. Electrochemical sensors based on conducting polymers for the aqueous detection of biologically relevant molecules. *Nanomaterials.* 2021;11(1):252.
184. Terán-Alcocer Á, Bravo-Plascencia F, Cevallos-Morillo C, Palma-Cando A. Electrochemical sensors based on conducting polymers for the aqueous detection of biologically relevant molecules. *Nanomaterials.* 2021;1:252.
185. Samanta S, Roy P, Kar P. Sensing of ethanol and other alcohol contaminated ethanol by conducting functional poly(ophenylenediamine). *Mater Sci Eng B.* 2020;256:114541.
186. Lin CH, Lin JH, Chen CF, Ito Y, Luo SC. Conducting polymer-based sensors for food and drug analysis. *J Food Drug Anal.* 2021;29(4):544–58.
187. Lakard B. Electrochemical biosensors based on conducting polymers: A review. *Appl Sci.* 2020;10:6614.
188. Wang J, Zhu W, Zhang T, Zhang L, Du T, et al. Conductive polyaniline-graphene oxide sorbent for electrochemically assisted solid-phase extraction of lead ions in aqueous food samples. *Anal Chim Acta.* 2020;1100:57–65.

189. Ranjith Kumar D, Dhakal G, Muhammed Shafi P, Saad Sayed M, Lee J, et al. Sulfite food additive electrochemical determination by nucleophilic addition on poly(4-aminodiphenylamine)-4-aminothiophenol-Au composite electrode. *Microchem J.* 2022;181:107635.
190. Sereshti H, Karami F, Nouri N, Farahani A. Electrochemically controlled solid phase microextraction based on a conductive polyaniline-graphene oxide nanocomposite for extraction of tetracyclines in milk and water. *J Sci Food Agric.* 2021;101(6):2304–11.
191. Pangajam A, Theyagarajan K, Dinakaran K. Highly sensitive electrochemical detection of *E. coli* O157:H7 using conductive carbon dot/ZnO nanorod/PANI composite electrode. *Sens Bio-Sensing Res.* 2020;29:100317.
192. Do MH, Florea A, Farre C, Bonhomme A, Bessueille F, et al. Molecularly imprinted polymer-based electrochemical sensor for the sensitive detection of glyphosate herbicide. *Int J Environ Anal Chem.* 2015;95:1489–1501.
193. Anirudhan TS, Alexander S. Design and fabrication of molecularly imprinted polymer-based potentiometric sensor from the surface modified multiwalled carbon nanotube for the determination of lindane (-hexachlorocyclohexane), an organochlorine pesticide. *Biosens Bioelectron.* 2015;64:586–93.
194. Kanoun O, Lazarević-Pašti T, Pašti I, Nasraoui S, Talbi M, et al. A review of nanocomposite-modified electrochemical sensors for water quality monitoring. *Sensors.* 2021;21(12):1–50.
195. Elfadil D, Lamaoui A, Pelle F Della, Amine A, Compagnone D. Molecularly imprinted polymers combined with electrochemical sensors for food contaminants analysis. *Molecules.* 2021;26(15):1–41.
196. Beyazit S, Tse Sum Bui JB, Haupt K, Gonzato C. Molecularly imprinted polymer nanomaterials and nanocomposites by controlled/living radical polymerization. *Prog Polym Sci.* 2016;62:1–21.
197. Kamel AH, Abd-Rabboh HSM, Hefnawy A. Molecularly imprinted polymer-based electrochemical sensors for monitoring the persistent organic pollutants chlorophenols. *RSC Adv.* 2024;14(28):20163–81.
198. Peng S, Wang A, Lian Y, Jia J, Ji X, et al. Technology for rapid detection of cyromazine residues in fruits and vegetables: molecularly imprinted electrochemical sensors. *Biosensors.* 2022;12(6):414.
199. Yuan Z, Yang Y, Yang Y, Wang G, Huo Y. Preparation and electrochemical properties of carbon/PANI composite mesh electrode materials. *J Electron Mater.* 2022;51:2289–97.
200. Mugo SM, Alberkant J. Flexible molecularly imprinted electrochemical sensor for cortisol monitoring in sweat. *Anal Bioanal Chem.* 2020;412:1825–33.
201. Wood M, Mugo SM. A MIP-enabled stainless-steel hypodermic needle sensor for electrochemical detection of aflatoxin B1. *Anal Methods.* 2022;14:2063–71.
202. Adumitrăchioaie A, Tertiş M, Cernat A, Săndulescu R, Cristea C. Electrochemical methods based on molecularly imprinted polymers for drug detection. *Int J Electrochem Sci.* 2018;1:2556–76.

203. da Silva H, Pacheco J, Silva J, Viswanathan S, Delerue-Matos C. Molecularly imprinted sensor for voltammetric detection of norfloxacin. *Sens Actuators B*. 2015;219:301–7.
204. Shao Y, Duan J, Wang M, Cao J, She Y, et al. Application of molecularly imprinted electrochemical biomimetic sensors for detecting small molecule food contaminants. *Polymers*. 2023;15(1):187.
205. Ma C, Wu D, Xu E, Fan Y. Molecularly imprinted electrochemical sensor based on g-C3N4 for determination of endosulfan insecticide in agricultural food. *Int J Electrochem Sci*. 2022;17:1–12.
206. Li G, Qi X, Wu J, Wan X, Wang T, et al. Highly stable electrochemical sensing platform for the selective determination of pefloxacin in food samples based on a molecularly imprinted-polymer-coated gold nanoparticle/black phosphorus nanocomposite. *Food Chem*. 2024;436:137753.
207. Yang Y, Shi Z, Wang X, Bai B, Qin S, et al. Portable and on-site electrochemical sensor based on surface molecularly imprinted magnetic covalent organic framework for the rapid detection of tetracycline in food. *Food Chem*. 2022;395:133532.
208. Munawar H, Garcia-Cruz A, Majewska M, Karim K, Kutner W, et al. Electrochemical determination of fumonisin B1 using a chemosensor with a recognition unit comprising molecularly imprinted polymer nanoparticles. *Sensors Actuators B Chem*. 2020;321:128552.
209. Işık D, Şahin S, Caglayan MO, Üstündağ Z. Electrochemical impedimetric detection of kanamycin using molecular imprinting for food safety. *Microchem J*. 2021;160:105713.
210. Mugo SM, Lu W, Robertson SV. Molecularly imprinted polymer-modified microneedle sensor for the detection of imidacloprid pesticides in food samples. *Sensors*. 2022;22:8492.
211. Zeb S, Wong A, Khan S, Hussain S, Sotomayor MDPT. Using magnetic nanoparticles/MIP-based electrochemical sensor for quantification of tetracycline in milk samples. *J Electroanal Chem*. 2021;900:115713.
212. Lu Z, Wang Y, Li G. Covalent organic frameworks-based electrochemical sensors for food safety analysis. *Biosensors*. 2023;13:291.
213. Lin YX, Nie BB, Qu XY, Wang MH, Yang J, et al. Histostar-functionalized covalent organic framework for electrochemical detection of exosomes. *Biosensors*. 2022;12:704.
214. Chu HC, Sun X, Zha XQ, Khan SU, Wang Y. Ultrasensitive electrochemical detection of butylated hydroxy anisole via metalloporphyrin covalent organic frameworks possessing variable catalytic active sites. *Biosensors*. 2022;12:975.
215. Zhang T, Ma N, Ali A, Wei Q, Wu D, et al. Electrochemical ultrasensitive detection of cardiac troponin I using covalent organic frameworks for signal amplification. *Biosens Bioelectron*. 2018;119:176–81.
216. Zhu H, Li M, Cheng C, Han Y, Fu S, et al. Recent advances in and applications of electrochemical sensors based on covalent organic frameworks for food safety analysis. *Foods*. 2023;12:4274.

217. Chen S, Yuan B, Liu G, Zhang D. Electrochemical sensors based on covalent organic frameworks: A critical review. *Front Chem*. 2020;8:601044.
218. Li Y, Guo X, Li X, Zhang M, Jia Z, et al. Redox-active two-dimensional covalent organic frameworks (COFs) for selective reductive separation of valence-variable, redox-sensitive, and long-lived radionuclides. *Angew Chem Int Ed*. 2020;59:4168–75.
219. Yusran Y, Li H, Guan X, Fang Q, Qiu S. Covalent organic frameworks for catalysis. *EnergyChem*. 2020;2:100035.
220. Wang JY, Hu HY, Lu SL, Hu JD, Zhu H, et al. Conductive metal and covalent organic frameworks for electrocatalysis: Design principles, recent progress and perspective. *Nanoscale*. 2022;14:277–88.
221. Zhang X, Zhu JL, Wu Z, Wen W, Zhang, et al. Electrochemical sensor based on confined synthesis of gold nanoparticles@covalent organic frameworks for the detection of bisphenol A. *Anal Chim Acta*. 2023;1239:340743.
222. Pan F, Tong C, Wang Z, Han H, Liu P, et al. Nanocomposite based on graphene and intercalated covalent organic frameworks with hydrosulphonyl groups for electrochemical determination of heavy metal ions. *Mikrochim Acta*. 2021;188:295.
223. Wang J, Qu X, Zhao L, Yan B. Fabricating nanosheets and ratiometric detection of 5-fluorouracil by covalent organic framework hybrid material. *Anal Chem*. 2021;93:4308–16.
224. Wang X, Yang S, Bai X, Shan J. Bimetallic CoCu nanoparticles anchored on COF/SWCNT for electrochemical detection of carbendazim. *Sci Total Environ*. 2023;902:166530.
225. Yang S, Yang R, He J, Zhang Y, Yuan Y, et al. Au nanoparticles functionalized covalent-organic-framework-based electrochemical sensor for sensitive detection of ractopamine. *Foods*. 2023;12:842.
226. Luo Y, Yang Y, Wang L, Wang L, Chen S. An ultrafine ZnO/ZnNi2O4@ porous carbon@covalent-organic framework for electrochemical detection of paracetamol and tert-butyl hydroquinone. *J Alloys Compd*. 2022;906:164369.
227. Lu S, Wang S, Wu P, Wang D, Yi J, et al. A composite prepared from covalent organic framework and gold nanoparticles for the electrochemical determination of enrofloxacin. *Adv Powder Technol*. 2021;32(6):2106–15.
228. Yu L, Chen X, Sun L, Zhang Q, Yang B, et al. A covalent organic frameworks@gold nanoparticles@graphene nanocomposite based electrochemical sensor for simultaneous determination of trace Cd2+, Pb2+ and Cu2+. *React Funct Polym*. 2024;194:105770.

CHAPTER 6

Recent Advances in Nanomaterial-Based Electrochemical Sensors for Pharmaceutical and Biomedical Diagnosis

Sudenur Ozbey, Yusuf Yerli, Cem Erkmen, and Sevinc Kurbanoglu

6.1 DEFINITIONS OF SENSORS AND BIOSENSORS

Sensors are essential systems in modern technology, playing a critical role in the seamless integration of devices and systems across various industries. These devices are designed to detect and respond to different types of physical, chemical, or biological stimuli, converting these signals into readable data for analysis and action (Wan et al. 2023; Erkmen et al. 2022). In industrial automation, sensors monitor processes and machinery, ensuring efficiency and safety. Environmental sensors help in tracking climate changes, pollution levels, and natural disasters, providing crucial data for mitigating risks and preserving ecosystems. In healthcare, biosensors are used in diagnostics and patient monitoring, offering real-time insights into vital signs and disease markers (Lim et al. 2020; Asha

DOI: 10.1201/9781003512639-6

Sharma et al. 2022; Yarman, Kurbanoglu, and Scheller 2020; Kurbanoglu and Ozkan 2018).

Nowadays, consumer electronics, including smartphones and wearable devices, utilize sensors to offer personalized user experiences and fitness tracking. Moreover, the continuous advancements in sensor technology, such as miniaturization, improved sensitivity, and energy efficiency, offer opportunities for innovative applications (Purohit et al. 2020; Asha Sharma et al. 2022). Since the first biosensor construction, over the years, glucose sensors have undergone significant advancements, evolving from early enzyme-based electrodes to modern continuous glucose monitoring systems (Y. Wang et al. 2008; J. Wang 2005; Scheller and Schubert 1992). Early developments focused on improving accuracy and stability through enzyme immobilization techniques. The introduction of nanomaterials and wearable technologies has further enhanced sensor sensitivity, real-time monitoring, and user convenience. These innovations have revolutionized diabetes management, offering more precise and continuous glucose monitoring for better glycaemic control (Sumitha and Xavier 2023; Lin et al. 2017; Yarman, Kurbanoglu, and Scheller 2020).

Electrochemical sensors are specialized devices used to detect and quantify chemical substances by converting chemical information into an electrical signal. These sensors are widely employed in environmental monitoring, medical diagnostics, industrial process control, and food safety due to their high sensitivity, selectivity, and rapid response times. The fundamental components of an electrochemical sensor include the working electrode, reference electrode, counter electrode, electrolyte, and transducer (Privett, Shin, and Schoenfisch 2010; Baranwal et al. 2022; Soylemez, Kurbanoglu, and Kuralay 2024; Karadurmus et al. 2017). The working electrode is where the chemical reaction of interest occurs. It is typically made from inert materials like platinum, gold, or carbon, which provide a stable surface for electrochemical reactions. The reference electrode maintains a constant potential against which the working electrode's potential is measured. A common choice for this component is the silver/silver chloride (Ag/AgCl) electrode, known for its stable and reproducible potential (Erkmen et al. 2022; Privett, Shin, and Schoenfisch 2010). The counter electrode, also known as the auxiliary electrode, completes the electrical circuit by allowing the current to flow through the sensor. It is often made from inert materials similar to the working electrode to ensure durability and minimal interference in the measurements. The electrolyte

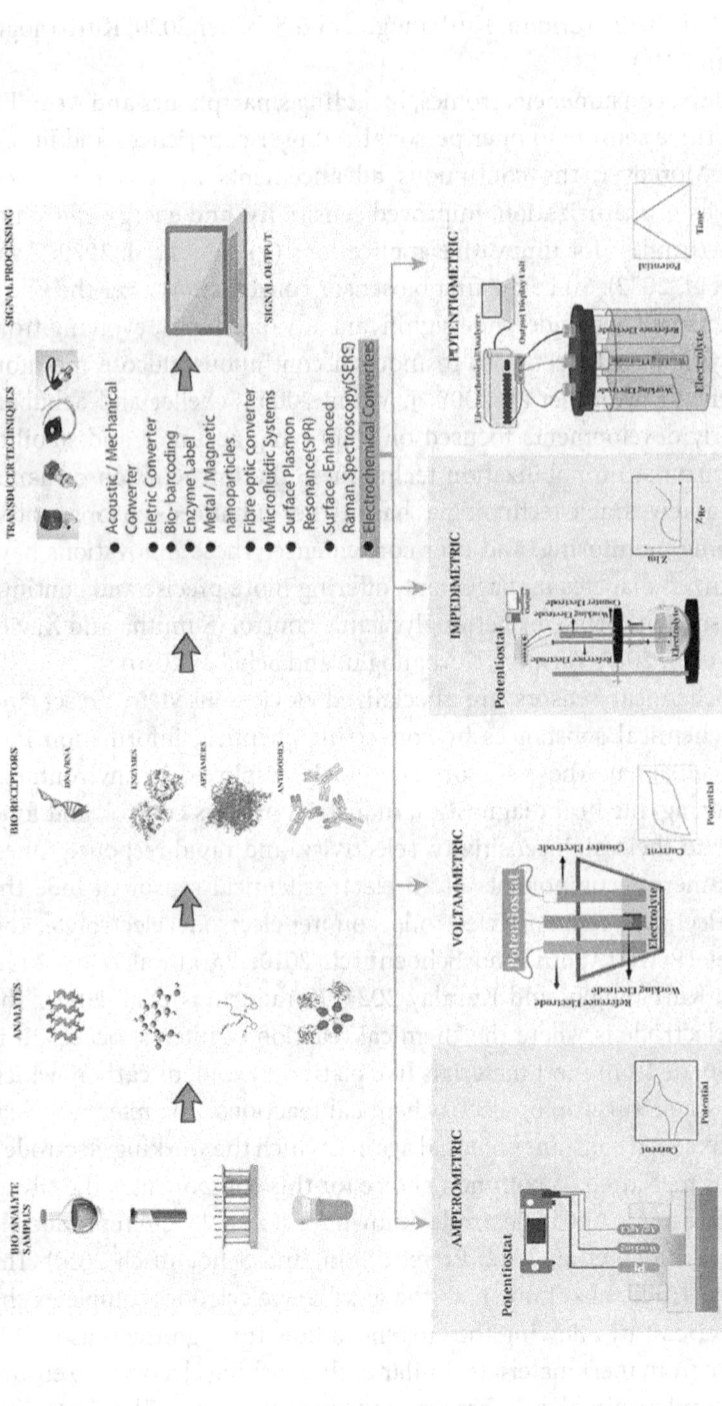

FIGURE 6.1 Basic components of electrochemical sensors/biosensors. (Reprinted from *Hybrid Advances*, 2, Sumitha, M.S. and Xavier, T.S., Recent Advances in Electrochemical Biosensors – A Brief Review, 100023, Copyright (2023), with permission from Elsevier.)

is a crucial component that facilitates the ionic conduction between the electrodes. It can be a liquid, gel, or solid substance containing ions that participate in the electrochemical reactions, providing a medium for the transfer of charge (Erkmen et al. 2022; Barhoum et al. 2023).

The transducer converts the electrochemical reaction into a measurable electrical signal. Depending on the type of electrochemical sensor, this could involve measuring changes in voltage (potentiometric sensors), current (amperometric sensors), or impedance (impedimetric sensors). The signal conditioning circuitry amplifies and processes the raw signal from the transducer, ensuring accurate and reliable data output. Overall, as shown in Figure 6.1, the interplay of these components enables electrochemical sensors to perform complex chemical analyses efficiently, making them indispensable in a wide range of applications (Sumitha and Xavier 2023; Kimmel et al. 2012; Baranwal et al. 2022).

6.2 PROPERTIES OF NANOMATERIALS AND THEIR USAGE IN ELECTROCHEMICAL SENSORS

Surface chemical functionalization is essential for enhancing the selectivity and sensitivity of electrochemical sensors, particularly in the presence of interfering substances and complex media (Bozal-Palabiyik et al. 2021; Erkmen, Kurbanoglu, and Uslu 2021; Baig et al. 2021; Meixiu Li et al. 2019). Nanomaterials have emerged as a particularly effective alternative due to their ability to achieve lower detection limits, robustness, and portability while also offering exceptional sensitivity and selectivity. Nanomaterials have been defined as substances engineered and utilized within the size range of 1 to 100 nanometres and benefit from significant advancements in synthetic methodologies. Their unique properties, such as high surface area-to-volume ratio, quantum effects, and exceptional electrical, mechanical, and thermal properties, ensure sensitivity, selectivity, and response time (Alhazimeh, Al-fandi, and Al-ebbini 2022; Szydłowska, Cai, and Hersam 2023). As a consequence of the recent advancements, producing nanomaterials with precisely controlled physicochemical properties, including surface charge, shape, and size, has been enabled. This precise monitoring allows the generation of sensitive contacts and electrodes, which, when functionalized on the electrode surface, facilitates the selective detection of analytes. Hence, nanomaterials provide numerous advantages when incorporated into electrochemical sensors due to their unique properties (Yang et al. 2015; A. Chen and Chatterjee 2013; Xie et al. 2019; Yun et al. 2023; C. Zhu et al. 2015).

One of the primary advantages of using nanomaterials in electrochemical sensors is the substantial increase in surface area. Nanomaterials such as nanoparticles, nanowires, and graphene sheets provide a larger active surface for the target analytes to interact with, significantly enhancing the sensor's sensitivity. This increased surface area provides more active sites for adsorption or catalytic reactions, leading to a stronger and more detectable signal even at low analyte concentrations (Suni 2008; C. Zhu et al. 2015; Xie et al. 2019; Yun et al. 2023). Nanomaterials also exhibit remarkable electrical properties that contribute to improved sensor performance. For instance, carbon nanotubes and graphene possess excellent conductivity, which facilitates efficient electron transfer between the analyte and the electrode. This enhances the electrochemical reaction rates and results in faster response times. Furthermore, the high electrical conductivity of these nanomaterials helps in reducing the noise level in the measurements, thereby improving the signal-to-noise ratio and enhancing the overall accuracy of the sensor (Liu, Huang, and Qian 2021; Burress et al. 2010; Wu et al. 2013; B. T. Zhang et al. 2013; Pyrzynska 2011). Selectivity is another crucial aspect where nanomaterials play a significant

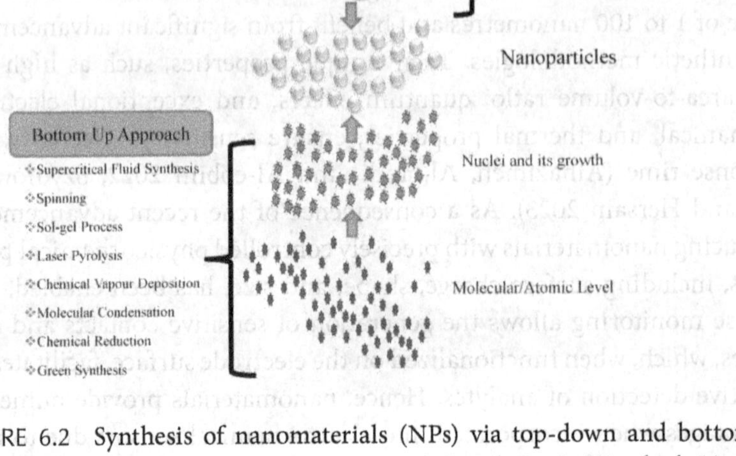

FIGURE 6.2 Synthesis of nanomaterials (NPs) via top-down and bottom-up approaches. (From Khanna, P., Kaur, A. and Goyal, D., *J. Microbiol. Methods*, 163, 105656, 2019.)

role. Functionalization of nanomaterials with specific chemical groups or biomolecules allows the selective binding of target analytes. For example, gold nanoparticles can be functionalized with antibodies, aptamers, or molecularly imprinted polymers (MIPs) to selectively detect specific proteins, DNA sequences, or small molecules. This functionalization capability enables the development of highly selective sensors that can discriminate between different analytes in complex sample matrices (Lahcen and Amine 2019; Xie et al. 2019; Murthy et al. 2022; Vadukumpully et al. 2011; Meixiu Li et al. 2019).

The synthesis of nanomaterials (NMs) is accomplished by creating materials with nanoscale dimensions. To achieve this, scientists have developed top-down and bottom-up approaches (Figure 6.2) (J. Lee et al. 2020; Kour et al. 2020; Yun et al. 2023; Khanna, Kaur, and Goyal 2019). Top-down methods involve starting with bulk materials and breaking them down into nanoscale structures, while bottom-up methods construct nanomaterials from smaller units, such as atoms or molecules, to form nanoscale structures. Top-down methods, such as mechanical milling, involve grinding bulk materials into nanoparticles, which are commonly applied to metal powders to produce metal nanopowders (Jamkhande et al. 2019; Anirudh Sharma and Das 2019; Baig et al. 2021). On the other hand, bottom-up approaches include techniques like chemical reduction, electrochemical methods, and chemical vapour deposition. In particular, bottom-up techniques are notably popular in the fabrication of electrochemical sensors due to their efficiency and precision. For example, through the utilization of a chemical reduction technique, metallic nanoparticles are obtained after the metallic nanoparticles undergo chemical reduction from their metal salts. In addition to that, electrochemical methods facilitate the synthesis of nanomaterials, such as graphene oxide, through various electrochemical reactions with the use of different techniques, such as electrodeposition and electrochemical exfoliation (Escorcia-d, Rend, and Ram 2023; Mekuye and Abera 2023; Shaheen et al. 2024; Ijaz et al. 2020; Corr 2016).

Nanomaterials can be systematically categorized based on a variety of criteria, including their dimensionality, structural attributes, and inherent properties. When classified by dimensionality, nanomaterials are segmented into four main categories: Zero-dimensional (0D), one-dimensional (1D), two-dimensional (2D), and three-dimensional (3D). Zero-dimensional nanomaterials, such as quantum dots and fullerenes, are defined by their extremely confined spatial dimensions, which typically result in unique electronic and optical properties (Tiwari, Tiwari, and Kim 2012; Z. Wang et al. 2020; Ding et al. 2020). One-dimensional nanomaterials, including nanotubes, nanowires, and nanorods, are characterized by their extended lengths relative to their width and height, which influence their mechanical and electrical behaviours (Gong and Cheng 2017; Hahm 2016; Kour et al. 2020). Two-dimensional nanomaterials, such as nanosheets, graphene, and other layered materials, feature extensive lateral dimensions with thicknesses on the nanometre scale. These materials exhibit remarkable properties, including high surface area and exceptional electrical and thermal conductivity (G. Wang, Ma, and Wang 2023; Ge, Ramachandran, and Wang 2020). In contrast, three-dimensional nanomaterials, which encompass nanostructured bulk materials and porous structures, possess a more complex configuration with significant surface area and volume attributes at the nanoscale, generally less than 100 nm. This dimensional complexity imparts distinctive mechanical, thermal, and chemical properties, enhancing their functionality in various applications such as catalysis, sensing, and energy storage. The diverse structural configurations of these nanomaterials enable a wide range of applications and functionalities, reflecting their versatility and the ongoing advancements in nanotechnology (Darwish, Abd-elaziem, and Elsheikh 2024; Kour et al. 2020; G. Wang, Ma, and Wang 2023).

On the other hand, to further understand the nature of NMs, they can also be classified based on their composition. ***Organic nanomaterials*** are organic molecules scaled down to the nanoscale, typically ranging from 1 to 100 nanometres. These materials can manifest in various forms, including organic nanoparticles, nanofibres, nanorods, nanowires, nanofilms, and other nanoscale configurations. They are primarily composed of carbon-based materials and include substances like carbon nanotubes, which are in high demand in electrochemical analysis (Saleh 2020; Kalia and Haldorai 2014; Joshi and Kim 2020; J. Lee et al. 2020). In contrast, ***inorganic nanomaterials*** are composed of non-carbon-based constituents, lacking carbon–hydrogen bonds. Examples of inorganic nanomaterials include metals, metal oxides, ceramics, and quantum dots. These materials find broad applications across various scientific disciplines. For instance, colloidal mesoporous silica nanoparticles (MSNPs) are particularly noteworthy for their enhanced utility in dispensing systems, such as those used in drug delivery (Modekwe et al. 2024; Malwal and Packirisamy 2018; X. Wang et al. 2021; J. Lee et al. 2020).

Conversely, ***composite nanomaterials*** (or nanocomposites) are solid materials composed of two or more distinct phases, with at least one of these phases existing at the nanoscale (1–100 nm), or featuring nanoscale gaps between phases. These composites can include combinations of nanoparticles with other nanoparticles, nanoparticles with larger-scale materials, or nanomaterials with bulk materials. The integration of diverse materials into a composite structure imparts a range of unique properties, such as improved bending strength, moisture absorption, optical characteristics, durability, and lustre maintenance (P. Kumar, Deep, and Kim 2015; Zhou and Kitagawa 2014; G. Wang, Ma, and Wang 2023). The creation of these composite structures consistently involves the utilization of nanometre-sized dimensions (Datta, Priyadarshini, and Deepak 2022). As an example, metal–organic frameworks (MOFs) represent a prominent subgroup of composite NMs, combining metal ions or clusters with organic components. MOFs are notable for their exceptional surface area, high porosity, ease of surface modification, and well-organized structure. Their high surface reactivity makes them widely used as support materials in various applications (Falcaro et al. 2016; Daniel et al. 2022; Zhou and Kitagawa 2014; G. Wang, Ma, and Wang 2023). In contrast, nanosheet-based composites are composed of materials such as graphene, MXenes, and other layered substances. These components impart distinctive structural and functional attributes to the composites (Jampasa et al. 2023; J. Zhang et al. 2021; Baig et al. 2021).

In the field of medical diagnostics, nanomaterial-based electrochemical sensors are used for the detection of biomarkers, pathogens, and metabolic compounds. These sensors offer rapid, accurate, and non-invasive diagnostic capabilities (Yonzon et al. 2005; Kaya, Kurbanoglu, and Ozkan 2019; Aftab and Kurbanoglu 2024). For example, glucose sensors using carbon nanotubes or gold nanoparticles provide real-time monitoring of blood glucose levels in diabetes management. Similarly, sensors functionalized with antibodies or DNA probes are used for the early detection of diseases such as cancer, cardiovascular diseases, and infectious diseases by detecting specific biomarkers in blood, urine, or saliva samples (Nemčeková and Labuda 2021; Sanko, Erkmen, and Kuralay 2024). In addition, nanomaterial-based sensors are increasingly used in the development of wearable and portable devices for personal health monitoring and environmental sensing. The integration of flexible nanomaterials like graphene and nanofibres into wearable sensors enables continuous monitoring of physiological parameters such as heart rate, sweat composition, and hydration levels, providing valuable data for personalized health management (Zheng et al. 2023; Liu, Huang, and Qian 2021; Kurbanoglu et al. 2016; Kaya, Kurbanoglu, and Ozkan 2019; Jampasa et al. 2023).

6.3 ELECTROCHEMICAL METHODS

In electrochemical sensors, transducer systems play a crucial role in converting chemical information into measurable electrical signals. These systems are essential for detecting and quantifying analytes with high sensitivity and accuracy. The main types of transducers used in electrochemical sensors include voltammetric, amperometric, potentiometric, and conductometric transducers. Each type of transducer offers unique advantages, enabling the development of versatile and efficient electrochemical sensors for various applications in medical diagnostics, environmental monitoring, and industrial processes (Y. Wang et al. 2008; Baranwal et al. 2022). ***Cyclic voltammetry*** (CV) is a widely used electrochemical technique that measures the current response of a redox-active substance to a linearly cycled potential. The theory behind CV involves applying a potential sweep to an electrochemical cell and recording the resulting current. The potential is varied linearly with time, typically in a triangular waveform, and the current is measured as the potential is swept back and forth. This creates a cyclic voltammogram, a plot of current versus potential, which provides detailed information about the electrochemical properties of the analyte, including its redox potentials, reaction kinetics, and diffusion coefficients (H.-W. Wang et al. 2021; Marken, Neudeck, and Bond 2010; Mahale et al. 2022). The primary advantage of cyclic voltammetry lies in its ability to provide a comprehensive electrochemical profile of a substance. It is particularly useful for identifying redox processes, studying reaction mechanisms, and determining the stability of intermediates. CV is also sensitive to changes in the electrode surface, making it ideal for investigating surface modifications and interactions at the electrode interface (H.-W. Wang et al. 2021; Marken, Neudeck, and Bond 2010). ***Differential pulse voltammetry*** (DPV) is an advanced electrochemical technique used to measure the current response of an analyte to a series of potential pulses superimposed on a linear potential sweep. In DPV, the potential is incrementally increased by small pulses, and the current is measured just before and after each pulse. The difference in current is recorded, resulting in a differential current plot versus potential. This technique enhances the resolution of the electrochemical signal by minimizing background currents and highlighting the faradaic current associated with the redox process of the analyte (Ángela Molina and Joaquín González 2016; Spinola Machado and Cincotto 2022; Deffo et al. 2024; Inesi 1986). The main advantage of DPV is its high sensitivity and selectivity, which allows the detection of low concentrations of analytes. The differential measurement technique significantly reduces the impact of capacitive currents and noise, leading to clearer and more distinct peak signals. This makes DPV particularly useful for trace analysis and detecting species with closely spaced redox potentials. Additionally, DPV provides detailed information about the redox behaviour, including peak potentials and current magnitudes, which can be used to identify and quantify specific analytes (Ángela Molina and Joaquín González 2016; Spinola Machado and Cincotto 2022). ***Square wave voltammetry*** (SWV) is an electrochemical technique that applies a square wave potential modulation on top of a staircase potential. In SWV, the current is measured at the end of each forward and reverse pulse, and the difference between these measurements is plotted against the applied potential. The technique offers high sensitivity and resolution by minimizing background currents and enhancing faradaic current responses. SWV is particularly effective for rapid and quantitative analysis of electroactive species, making it ideal for trace detection in complex samples (A. Chen and Shah 2013; Kokoskarova et al. 2023; Galeano-Dıaz et al. 2008; Deffo et al. 2024; Spinola Machado and Cincotto 2022).

As another electrochemical technique, ***Electrochemical impedance spectroscopy*** (EIS) is a powerful electrochemical technique that measures the impedance of a system over a range of frequencies. In EIS, a small alternating current (AC) signal is applied to an electrochemical cell, and the resulting current response is measured. The impedance, which is the ratio of the voltage to the current, is then analysed as a function of frequency. This technique provides detailed information about the electrochemical processes occurring at the electrode interface, including charge transfer resistance, double-layer capacitance, and diffusion processes (Srinivasan and Fasmin 2021; Lazanas and Prodromidis 2023; Prodromidis 2010; Bahadır et al. 2016; Aydın 2020). The primary advantage of EIS is its ability to decompose complex electrochemical reactions into individual components, offering insights into the kinetics and mechanisms of these reactions. EIS is highly sensitive to changes in the electrode surface and the properties of the electrolyte, making it ideal for studying corrosion, battery performance, and sensor interfaces. It can detect subtle changes in system properties, enabling the monitoring of surface modifications and the detection of low-concentration analytes (Srinivasan and Fasmin 2021; Lazanas and Prodromidis 2023; Prodromidis 2010; Bahadır et al. 2016; Aydın 2020). ***Amperometry*** is an electrochemical technique that measures the current resulting from the oxidation or reduction of an analyte at a fixed potential. In an amperometric sensor, a constant potential is applied to an electrode, and the resulting current, which is directly proportional to the concentration of the analyte, is measured. The main advantages of amperometry include its simplicity, high sensitivity, and rapid response time. The continuous measurement capability enables real-time monitoring, making it ideal for applications requiring quick and precise analysis (Bollella and Gorton 2018; Iwuoha et al. 2013; Amine and Mohammadi 2019; Patel 2020; Rafiee et al. 2024). ***Conductometry*** involves measuring the electrical conductivity of a solution, which is related to the concentration of ions present in the solution. This technique is based on the principle that the electrical conductivity of a solution changes with the ionic content (Safonova and Kolker 1992; Majeed et al. 2022). In a conductometric sensor, two electrodes are placed in the solution, and an alternating current is applied. The resulting conductance, which is inversely proportional to the resistance, is measured. It is a non-specific technique, meaning it does not require specific reagents for each analyte, making it versatile for various applications (Jaffrezic-Renault and Dzyadevych 2008; Majeed et al. 2022). ***Potentiometry*** is an electrochemical technique that measures the voltage of an electrochemical cell without drawing significant current. The potential difference between a reference electrode and an indicator electrode is measured, related to the concentration of a specific ion in the solution. Potentiometric sensors, such as ion-selective electrodes (ISEs), are widely used for measuring pH, ion concentrations (e.g., sodium, potassium, chloride), and other analytes (Zdrachek and Bakker 2019; Bratov, Abramova, and Ipatov 2010; Koncki 2007; Kahlert 2010).

6.4 RECENT APPLICATIONS OF NANOMATERIAL-BASED ELECTROCHEMICAL SENSORS FOR PHARMACEUTICAL AND BIOMEDICAL DIAGNOSIS

The United Nations' 2030 Sustainable Development Goals (SDGs) prioritize access to clean water, yet the presence of industrial wastewater containing hazardous synthetic dyes, such as Brilliant Green (BG), poses a significant challenge to achieving these objectives. Although widely used, BG is prohibited in the US and the EU due to its detrimental effects on both the environment and human health. It is imperative to detect and degrade BG; however, current techniques are intricate and not appropriate for on-site examination. Due to their ease of use, affordability, and high sensitivity, electrochemical sensors show promise in the detection and degradation (Rishpon 2002; Barhoum et al. 2023). Aslam et al. used Fe-doped ZnO as a photocatalyst in a recent study due to its greater surface area and efficiency. In order to achieve sensitive and efficient BG dye detection, the research presents a unique electrochemical approach using a glassy carbon electrode (GCE) modified with Fe–ZnO nanoparticles (NPs) and multi-walled carbon nanotubes (MWCNTs). By immediately adding the dye to the electrode, the newly designed sensor demonstrates great sensitivity. This provides an effective and environmentally acceptable way to identify and break down wastewater polluted with dye. Different techniques were used to produce and analyse ZnO and Fe–ZnO NPs, which were subsequently analysed using Fourier transform infrared spectroscopy (FTIR), X-ray diffraction (XRD), and scanning electron microscopy/energy-dispersive X-ray spectroscopy (SEM/EDX). For example, Fe content of 0.94% was detected by SEM and EDX in Fe–ZnO NPs. Electrochemical investigations were carried out by incorporating triple electrodes as platinum counter, Ag/AgCl reference, and GCE. Hydrothermal techniques were employed to synthesize NPs, and effective doping was confirmed by characterization. On modified GCEs, electrochemical detection of BG dye was carried out via pulse voltammetry, SWV, EIS, and CV. Furthermore, BG dye degradation investigations were carried out utilizing UV-visible spectrophotometry. Consequently, electrochemical experiments demonstrated that GCEs treated with MWCNT/Fe–ZnO had a six-fold higher electroactive surface area and a better charge transfer efficiency than bare electrodes. Better electrochemical performance was indicated by the noticeably greater exchange current density. Finally, square wave voltammetry was used to conduct the voltammetric investigation, which revealed that MWCNT/Fe–ZnO/GCE exhibited the strongest current response for 40 μM BG dye, outperforming bare GCE and other modifications. Phosphate-buffered saline (PBS) at pH 6, a deposition potential of −0.1 V, and an accumulation duration of 40 seconds provided the optimum sensitivity and dependability for optimal detection. In conclusion, with a limit of detection (LOD) of 0.40 nM and a limit of quantification (LOQ) of 1.34 nM, the MWCNT/Fe–ZnO/GCE sensor demonstrated good sensitivity for BG dye detectability. Besides, Fe–ZnO showed better photocatalytic activity than ZnO for the breakdown of BG dye, attaining 99% degradation in

30 minutes under sunlight as opposed to 69% with ZnO. Therefore, it can be claimed that the novel sensor demonstrated outstanding repeatability, reproducibility, and stability. For future investigations, additional doped semiconductors and carbon nanomaterials can be studied to improve dye degradation and detection (Aslam et al. 2023).

Pollutant discharges, particularly nitrites from industrial and agricultural sources, harm water quality and health. Therefore, it is essential to utilize analytical techniques such as ion chromatography, colorimetry, spectrophotometry, and high-performance liquid chromatography (HPLC) to detect trace nitrites in wastewater. Advances in screen printing have made electrodes more affordable, portable, and disposable. For example, MOFs, known for their high porosity and surface area, enhance electrochemical sensing due to their unique structure (Zhou and Kitagawa 2014; P. Kumar, Deep, and Kim 2015; G. Wang, Ma, and Wang 2023). In this respect, Cu–MOF materials, enhanced by Cu^{+}/Cu^{2+} redox reactions, improve electrocatalytic properties, and also adding carbon black (CB) boosts durability, conductivity, and affordability. In a very recent study, Ambaye et al. combined CB with Cu–MOF for better electrocatalysis. CB/Cu–MOF hybrid nanocomposites with screen-printed carbon electrodes (SPCEs) offer a cost-effective and portable solution for detecting nitrite in wastewater. The Cu–MOF was synthesized via solvothermal methods. Characterization using SEM, EDS, transmission electron microscopy (TEM), XRD, FTIR, and Raman spectroscopy confirmed the material's structure and composition. Electrochemical tests showed that CB/Cu–MOF/SPCEs effectively detect nitrite, with analyses revealing particle size, crystal structure, and functional groups. The CB/Cu–MOF modified electrode exhibited significantly higher current responsiveness through the utilization of the linear sweep voltammetry (LSV) technique at PBS pH 7.0 for nitrite detection. As a result, after the measurements with the LSV method, an anodic peak current increase is observed with nitrite concentration. The detection and quantification limits for nitrite were 0.084 μM and 0.28 μM, respectively, within a linear range of 1–200 μM. The performance of the CB/Cu–MOF/SPCE has outstanding lower LOD results compared with most of the other published studies. Hence, this newly designed nanocomposite shows strong potential for electrochemical sensing, offering improved performance and enhanced stability for nitrite detection in wastewater. Its scalability and cost-effectiveness, combined with simple production methods, make it a promising choice for future environmental monitoring and sensing applications (Ambaye et al. 2023).

Pesticides, especially organophosphorus (OP) chemicals like fenitrothion (FNT), are vital in agriculture due to their effectiveness and lower toxicity compared with organochlorines. Despite its efficacy against pests, FNT is highly hazardous, since it inhibits acetylcholinesterase, an enzyme that breaks down the neurotransmitter acetylcholine, causing toxic acetylcholine build-up in the nervous system. Traditional detection methods, such as gas or liquid chromatography with mass spectrometry, are effective but complex and costly. These drawbacks highlight the demand for simpler, cost-effective, enzyme-free electrochemical sensors for detecting low pesticide concentrations in real samples, offering a promising alternative for environmental monitoring (Huang et al. 2023; Periasamy, Umasankar, and Chen 2009). For this reason, Bolat et al. have developed a novel sensor by combining poly-arginine (*p*(Arg)) with reduced graphene oxide (GO) on disposable electrodes to enhance the sensitivity and performance of electrochemical biosensors for FNT detection. This sensor, created using electropolymerization and characterized by various techniques, demonstrated effective electrocatalytic performance. It successfully detected FNT in water and cucumber samples, proving its potential for practical use in environmental and food safety applications. During electropolymerization, arginine forms peptide bonds, anchoring to the electrode substrate via α-amino free radicals. As CV cycles increase, successful *p*(Arg) adhesion onto GO modified pencil graphite electrode (PGE); GO/PGE; is indicated by decreasing oxidation peak currents. The resulting *p*(Arg) structure enhances hydrogen bonding and chemisorption, aiding in FNT sensing. Electrochemical analysis with CV and EIS revealed that *p*(Arg)-GO/PGE exhibits higher peak currents and lower resistance than GO/PGE, indicating better electrocatalytic activity and conductivity. SEM and attenuated total reflectance (ATR)-FTIR confirmed the creation of a uniform, porous structure and successful electropolymerization of *p*(Arg). The electrode is suited for sensing applications due to these improvements. The *p*(Arg)-GO/PGE electrode at pH 4.0 showed a dramatic reduction peak for FNT, demonstrating high sensitivity and effective sensing. The SWV technique was used for FNT quantification, showing high sensitivity and a relatively short analysis time. The optimal reduction peak at −0.6 V provided LOD and LOQ values of 0.173 μM and 0.571 μM, suitable for real sample detection within legal limits. The *p*(Arg)-GO/PGE sensor demonstrated excellent anti-interference ability and reliable recovery rates (102.9–104.5% for tap water, 94.1–110.8% for cucumber samples). The sensor showed good reproducibility (3.27% relative standard deviation (RSD) and repeatability (4.2% RSD), with 96% stability over 20 days. This cost-effective and sensitive platform is practical for FNT detection and suitable for mass production (Bolat et al. 2023).

Atrazine (2-chloro-4-methylamino-6-isopropylamino-1, 3, 5-triazine) is a selective herbicide from the triazine family. Atrazine is known as a harmful chemical that contaminates water supplies and presents serious dangers to human health and the environment. It is still present in drinking water, with levels in US tap water exceeding 60 μg/L (Rhoades et al. 2013; Almberg et al. 2018). Exposure to atrazine, mostly through water, can create health problems by interfering with endocrine functioning. While efficient, traditional detection techniques such as high-performance liquid chromatography/mass spectrometry (HPLC-MS), ultrahigh-performance liquid chromatography (UHPLC), and gas chromatography (GC) are expensive, time-consuming, and not appropriate for real-time applications (Almberg et al. 2018; Rhoades et al. 2013). Thus, the necessity for effective point-of-care sensing platforms arises. High-sensitivity electrochemical sensors utilizing nanomaterials, including carbon nitride and graphitic materials, are perfect for environmental monitoring. In electrochemical sensing, naphthalene tetracarboxylic dianhydride (NTDA), an organic molecule conjugated with π, shows great promise due to its high sensitivity and low detection limits (K. Y. Chen and Chang 2014; G. Zhu et al. 2012). A study conducted by Singh et al. involved the synthesis and characterization of a cysteine-substituted naphthalene diimide (NDI-Cys) utilizing a variety of methods, including UV-Vis, FTIR, field emission scanning electron microscopy (FESEM), Raman, and nuclear magnetic resonance (NMR) spectroscopy. The NDI-Cys was fabricated into a modified carbon paste electrode (MCPE) to design a new nanosensor (NDI-Cys/MCPE) for the electrochemical detection of atrazine. In the meantime, pH, electrolyte type, concentration, and scan rates were all adjusted to find the ideal circumstances for atrazine detection. The sensor exhibited a low detection limit of 0.094 μM and a linear response to atrazine concentrations ranging from 10 to 100 μM after the analysis with the DPV method. The sensor's sensitivity to reduced levels of atrazine was assessed using the concentration variation. The NDI-Cys/MCPE electrode demonstrated remarkable atrazine selectivity, sustaining steady peak currents in spite of external interference. Over 50 cycles, stability testing revealed very little variation, demonstrating long-term dependability. Tests for repeatability, including six electrodes to detect atrazine (50 μM), yielded reliable findings, with peak current intensity variations of just around 5%. Specific detection of atrazine was ensured by the confirmation of selectivity against common ions and other pesticides. Testing in real-time samples of drinking water, milk, and tap water successfully confirmed the sensor's practical application (Singh et al. 2023). Another novel study conducted by Ismaeel et al. comprises metal nanoparticles, carbon nanotubes, fullerenes, and liquid ionic-modified electrodes for the electrochemical detection of atrazine. The study examines developments, difficulties, and possible breakthroughs in the future. The working electrode was a modified gold electrode containing nickel and cobalt nanoparticles, while the selected reference electrode was Ag/AgCl. The modified electrode was fabricated at room temperature and designed for use at an optimal pH of 9, which was then stabilized by Britton–Robinson buffer, along with other chemical and device parameters. Furthermore, the modified gold electrode was used to detect and measure atrazine in wastewater and water samples. To investigate the disruptive impact of symbiotic species with analyte species, a calibration curve was obtained by DPV. The limit of detection was determined to be as low as 0.009 nM for atrazine, while the linear range extended from 0.01 to 100 nM. Refinery wastewater samples were used for real sample analysis with an RSD of 1.45%, revealing the sensor's precision. The surface coating of cobalt, nickel, and pyrrole nanoparticles showed the electrode's analytical accuracy and precision. To conclude, Ismaeel et al. achieved a lower detection limit, in other words higher sensitivity, compared with the NDI-Cys/MCPE sensor. The results indicate that both sensors are applicable for use in the analysis of wastewater and potable water samples (Ismaeel et al. 2024).

Primary aromatic amines (PAAs), which are toxic and carcinogenic chemicals, can leach from polyurethane adhesives in food packaging. Hence, European regulations now require risk assessments for PAAs in these materials. In this respect, various analytical methods exist for detecting PAAs, including spectrophotometry and electrochemical techniques (Gheni et al. 2024; González et al. 2023; Campanella et al. 2015; Perez et al. 2021; Schubert et al. 2011). Among them, screen-printed electrodes (SPEs) are highly regarded for their ability to detect small amounts of chemicals. Recently, SPEs have gained popularity in the food industry; however, no studies have focused on SPEs designed explicitly for detecting PAAs, making this work novel (Beitollahi et al. 2020; Hayat and Marty 2014). For this reason, SPEs enhanced with nanocellulose (CNCs) and gold nanoparticles (AuNPs) were designed by Büyüktas et al. to improve selectivity and lower detection limits for 4,4′-methylene diphenyl diamine (4,4′-MDA), which

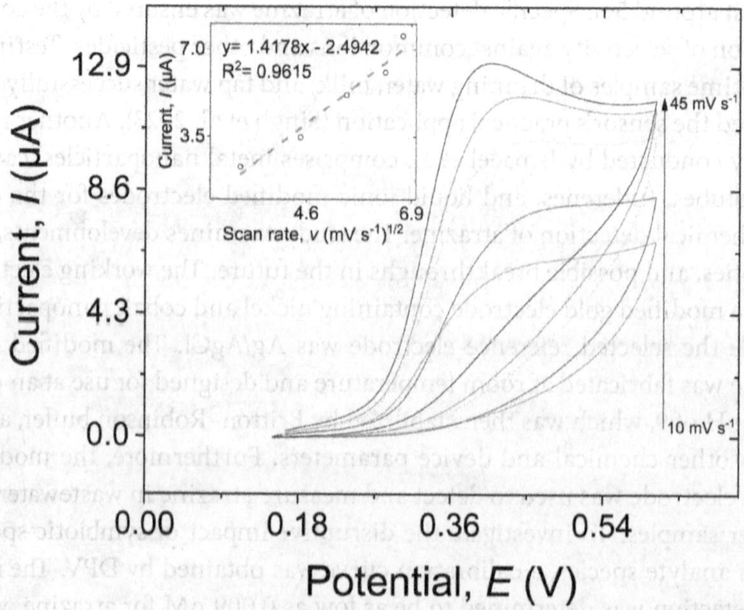

FIGURE 6.3 Cyclic voltammograms of AuNPs/CNCs/SPE. (Reprinted from *Heliyon*, 9 (4), Büyüktaş, D., Ghaani, M., Rovera, C., Carullo, D., Olsson, R. T., Korel, F., and Farris, S., A Screen-Printed Electrode Modified with Gold Nanoparticles/Cellulose Nanocrystals for Electrochemical Detection of 4,4′-Methylene Diphenyl Diamine, 1–10, Copyright (2023), with permission from Elsevier.)

is in the PAA group of toxic chemicals. SPEs with carbon and silver electrodes were modified with CNCs via drop casting. The surface of these modified SPEs was examined using a field emission scanning electron microscope. The modified electrodes' performance was assessed in real samples of laminate packaging materials with PU adhesives. The modified sensors were thoroughly characterized for their electrochemical properties. FESEM showed a rough surface for CNC-modified electrodes and small clusters of AuNPs, less visible on CNCs. CV analysis revealed that adding both CNCs and AuNPs achieved the highest peak current (9.81 μA), indicating improved electroactivity. Despite this, the modified SPE had lower peak currents than a modified GCE. With the application of the chronoamperometry method, a diffusion coefficient of 2.82×10^{-6} cm^2 s^{-1} was calculated, lower than that of modified GCEs. Furthermore, the sensor's performance is highly dependent on pH, with the highest oxidation peak at pH 7. As a result of the utilization of the DPV technique, a linear calibration for 4,4′-MDA with an LOD of 0.057 μM is observed. The sensor, tested with common additives, showed selectivity for 4,4′-MDA and excellent repeatability (3.65% RSD) and reproducibility (4.20% RSD). In real sample tests, it demonstrated high recovery (98.20–104.07%), proving its efficacy for detecting 4,4′-MDA migration from packaging. This innovation provides a promising alternative to traditional methods for PAA detection, offering potential improvements in food packaging quality control. Consequently, future investment could focus on applying this sensor to other contaminants, optimizing its performance, and validating its use in real-world scenarios (Büyüktaş et al. 2023) (Figure 6.3).

Even in trace amounts, heavy metals from industrial sources can crucially damage important organs. Therefore, different methods have been developed to detect these metals in environmental sources (Ming Li et al. 2013; Alhalili 2023; Huang et al. 2023). Electrochemical approaches are relatively simpler and less expensive than other traditional detection methods. Particularly, it is imperative to carry out research on electrochemical nanosensors for detection in water to achieve rapid, low-cost, and sensitive analysis. However, improving sensitivity and selectivity is crucial to obtain more reliable results. Carbon nanocomposites and metal nanoparticles are examples of advanced electrode materials that enhance detection. For example, MXene, a metallic nanomaterial, is perfect for electrochemical detection because of its exceptional conductivity (Pirzada and Altintas 2019; Xie et al. 2019). In a very recent study, Akhtar et al. blended graphitic carbon nitride nanorods, Fe_3O_4 nanoparticles, and MXene nanosheets into a composite (Ti_3C_2(HF)/Fe_3O_4/g-C_3N_4). Through the utilization of this novel nanocomposite, heavy metals such as Zn (II), Pb (II), Cd (II), Cu (II), and Hg (II) were successfully identified. The Ti_3C_2(HF)/Fe_3O_4/g-C_3N_4 composite was synthesized through the ultrasonication technique. Furthermore, by the application of the drop-casting method, the nanocomposite was modified on the surface of a GCE. Chronoamperometry, EIS, and CV were used to evaluate the electrochemical properties, while FTIR, SEM, and XRD were used for characterization. As a result of the data obtained by the SEM technique, it is observed that Fe_3O_4 and g-C_3N_4 nanorods were successfully incorporated into MXene sheets. After all the parameters were optimized for detection, NaAc–HAc buffer at pH 5.0 was determined to be the best-supporting electrolyte. Finally, the metal ions that were the subject of the experiment were analysed by the differential pulse anodic stripping voltammetry (DPASV) technique in potassium ferrocyanide solution. The metal ion simultaneous detection under these circumstances demonstrated a low LOD value of 0.26 nM for Zn (II), 0.21 nM for Cd (II), 0.10 nM for Pb (II), 0.11 nM for Cu (II), and 0.12 nM for Hg (II). Moreover, a relatively wide linear range was observed at the Ti_3C_2(HF)/Fe_3O_4/g-C_3N_4/GCE, ranging from 0.5 μM to 0.005 μM. The sensor showed minimal interference, stable performance (RSD 2.66%–4.22%), and reproducibility (RSD 1.05%–2.57%). Recovery experiments in real water samples yielded high recoveries, ranging between 96.7% and 104%, proving practical applicability. In conclusion, the newly synthesized nanocomposite material showed strong electroanalytical performance with high sensitivity and selectivity, proving its use for environmental monitoring. Future work could enhance electrode customization and broaden the sensor's application to more contaminants (Akhtar et al. 2023).

PCM (paracetamol) is widely used for pain relief and fever reduction, but excessive ingestion can damage the kidneys and liver, making accurate dosage crucial. Conventional detection methods are impractical for quick on-site analysis. On the other hand, electrochemical sensors offer a cost-effective, sensitive, and efficient solution for real-time PCM monitoring (Luhana and Mashazi 2023; Palmisano et al. 2000). Advances include modified electrodes like glassy carbon with zinc oxide nanoparticles, multi-walled carbon nanotubes, and carbon paste electrodes with La_2O_3. Nanotechnology, particularly spinel oxide–based nanostructures such as zinc oxide spinel and spinel ferrite, has improved electrochemical sensing by enhancing catalytic activity and durability. In this respect, Anh et al. synthesized spinel zinc ferrite ($ZnFe_2O_4$) nanomaterial via a hydrothermal method for PCM detection. The $ZnFe_2O_4$-modified electrode showed superior electron transfer, electrocatalytic activity, and adsorption capacity compared with standard electrodes. Deionized water and analytical-grade chemicals were used for solution preparation. $ZnFe_2O_4$ nanoparticles were hydrothermally synthesized and drop-casted onto SPEs. Moreover, the morphology of the samples was observed using field emission scanning electron microscopy, demonstrating that the newly synthesized $ZnFe_2O_4$ particles were spherical in shape and well-dispersed, with particle sizes ranging from 9 to 21 nm, with an average size of approximately 14.2 nm. CV and EIS experiments showed that the $ZnFe_2O_4$/SPE exhibited significantly higher peak currents (1.42 times for oxidation, 1.64 times for reduction) compared with bare SPE, indicating enhanced electron transfer and electrochemical performance. The $ZnFe_2O_4$-modified electrode achieved a limit of detection of 0.29 μM and demonstrated a wide linear range of 0.5 to 500 μM for PCM detection. The $ZnFe_2O_4$-modified electrode had a higher catalytic rate constant (kcat), indicating enhanced electrochemical activity. The DPV technique favoured the oxidation process, with the optimal pH of 5.0 providing the best peak current and reversible process characteristics. On top of that, the novel sensor revealed excellent reproducibility and effective quantification of PCM in tablets and urine samples, with recoveries of 96.3% to 110.9% and RSDs below 2.9%, maintaining a stable performance against potential interferents. Hence, it is reasonable to conclude that the $ZnFe_2O_4$-modified SPE sensor demonstrated excellent reproducibility, a wide dynamic range, low detection limits, and high sensitivity, making it ideal for affordable, compact, and on-site PCM analysis. It would be helpful to investigate other nanoparticles and composites in future research to improve sensitivity and selectivity further. In conjunction with this, creating sensors for the simultaneous detection of several analytes in complicated matrices and combining these sensors with portable devices for real-time monitoring would be potential avenues for the advancement of electrochemical sensing technology (Anh et al. 2023). On the other hand, the triple combination of PCM with propyphenazone (PRO) and caffeine (CA) is also widely used to enhance the anti-inflammatory effects. However, this combination can cause adverse effects when taken in excessive doses, including anaemia, respiratory issues, skin rashes, facial swelling, nausea, and liver damage (Burdan 2004). In clinical applications and diagnostic research, a sensitive and precise method is essential for the simultaneous detection of these active compounds (Karadurmus et al. 2018; Sato, Takahashi, and Toriyama 2011). Nevertheless, analysing multicomponent mixtures is challenging due to the varying conditions required for each analyte. In order to identify these drugs (PCM, PRO, CA), Urcuk et al. intended to create a sensitive, precise, inexpensive, and user-friendly voltammetric approach. To improve sensitivity and selectivity, conducting polymer sheets, metal nanoparticles, and carbon nanotubes were added to pencil graphite electrodes (PGEs). Because of their high catalytic activity, huge surface-to-volume ratio, active sites, and biological compatibility, AuNPs were used. Electrode performance was improved by polymerizing biological molecules on electrode surfaces. The drug ingredients were dissolved in methanol and water, and 0.01 M stock solutions of CA and PRO were used to make calibration solutions. AuNPs were electrodeposited, and L-cystine was electropolymerized. Using FESEM, the surface morphology of electrodes was analysed, and it was concluded that the bare PGE surfaces were normal, smooth, and homogeneous. By adjusting the quantities of L-cystine and $HAuCl_4$, the number of cycles, and the scan speeds utilized in L-cystine polymerization, the P(L-cyst)/AuNPs/PGE sensor was standardized. Peak currents of PCM, PRO, and CA increase by approximately 80% upon the fabrication of P(L-cyst)/AuNPs. Through the SWV method, PCM, PRO, and CA were measured concurrently using the newly modified electrode (P(L-cyst)/AuNPs/PGE). LOQ and LOD were calculated using linear curves. PCM, PRO, and CA had LOQ values of 0.023 μM, 0.009 μM, and 0.454 μM, and LODs of 0.007 μM, 0.003 μM, and 0.136 μM, respectively. The investigation into the influences of several interfering compounds on biological samples showed that these substances had no appreciable effect on the P(L-cyst)/AuNPs/PGE sensor's performance. Using traditional addition methods, the study assessed the recommended electrode's analytical applicability and dependability. Therefore, the sensor demonstrated exceptional sensitivity, precision, accuracy, and reproducibility for the electroanalysis of ternary mixtures. The recovery results proved its applicability in pharmaceutical formulations and human blood samples, offering an alternative approach to evaluate drug combination formulations and matrix mixtures (Urçuk et al. 2023). As one of the latest works regarding the detection of PCM, Foroughi et al. developed a unique nanocomposite, woollen ball–shaped lanthanum chloride (La^{3+}) and titanium butoxide (TiO^2) nanostructures (WB-S La^{3+}/TiO_2-NS), with the goal of detecting not only PCM but also tryptophan (Try), dopamine (DA), and ascorbic acid (AC). To quantify these compounds in body fluids, electrochemical sensors with high selectivity and sensitivity are required. Hence, the research used a three-cell electrochemical system, X-ray powder diffraction, EDX spectroscopy, and SEM to examine the structure of WB-S La^{3+}/TiO_2-NS. The designation process started with mixing TiO_2 and La^{3+} with glacial acetic acid and subsequently diluting in double-distilled water, which was then applied to a modified electrode to generate the nanostructure. Finally, the prepared nanocomposite was used to modify the surface of a screen-printed electrode. Using the DPV method, the electrochemical response of PCM, Try, DA, and AC was investigated. As a result, detection limits were obtained for PCM, DA, AC, and Try as 0.63, 0.51, 0,72, and 0.84 nM, respectively, while linear ranges were between 0.001 and 900.0 μM for all analytes. Furthermore, both specific pharmaceutical formulations and human urine samples were utilized for each analyte. The recovery results from the experiment ranged from 97.5% to 103.0%, confirming the novel sensor's ability to detect all analytes and its accuracy. To summarize, the WB-S La^{3+}/TiO_2-NS/SPE nanosensor showed strong potential for clinical and pharmacological assessments by effectively detecting PCM, Try, DA, and AC analytes while reducing the peak potential overlap and enhancing oxidation current responses. In summary, the WB-S La^{3+}/TiO_2-NS/SPE nanosensor demonstrated superior sensitivity, with the lowest detection limit for detecting PCM compared with the other two electrochemical sensors. Nevertheless, each sensor provides good selectivity, sensitivity, and accurate analysis performance. These are promising experiments with remarkable results for future studies (Mehdi Foroughi, Jahani, and Rashidi 2024).

Gentamicin sulfate (GNS) is a lipophilic aminoglycoside antibiotic produced through the fermentation of *Micromonospora purpurea* and is effective against both gram-positive and gram-negative bacteria. It is widely used in the treatment of respiratory and urinary infections, as well as in veterinary medicine, for its antibacterial properties. However, the presence of GNS in animal-derived food can pose serious health risks to humans, including ototoxicity, nephrotoxicity, neurological damage, bacterial resistance, and headaches, especially with overdose or prolonged use. To maintain public health, regulatory bodies such as those of the EU, the US, and China have established maximum residue limits (MRLs) for GNS in food. Therefore, accurate quantification of GNS is essential in the medical, chemical, and biological fields. Nanozeolites are structured, porous crystalline materials widely used due to their large surface areas and well-designed nanoporous channels, offering remarkable advantages over traditional support materials. Besides, the fabrication of zeolite modified electrodes (ZMEs) is rapid and inexpensive. Masihpour et al. proposed a nanosensor integrating ZMEs in a voltammetric detection analysis of GNS for the first time, according to their research. The designation was achieved by decorating the $Ni(OH)_2$ compound onto a nanozeolite NaX modified carbon paste electrode ($Ni(OH)_2$–NaX/CPE). The characteristics of the sensor were investigated with FESEM, revealing the presence of sphere-shaped nanozeolite NaX with diameters less than 100 nm. CV studies proved that $Ni(OH)^2$–NaX/CPE significantly enhanced the electrocatalytic oxidation of GNS compared with $Ni(OH)^2$/CPE, which was attributed to the mesoporous structure of nanozeolite NaX. DPV was used to determine the electrochemical performance of the new sensor, which resulted in a limit of detection of 0.82 μM and a linear range between 2.72 and 119.60 μM. Human serum samples were selected for real sample analysis and demonstrated good recoveries in the range of 97.04–104.14%. As a result of this study, the oxidation of GNS showed enhancement

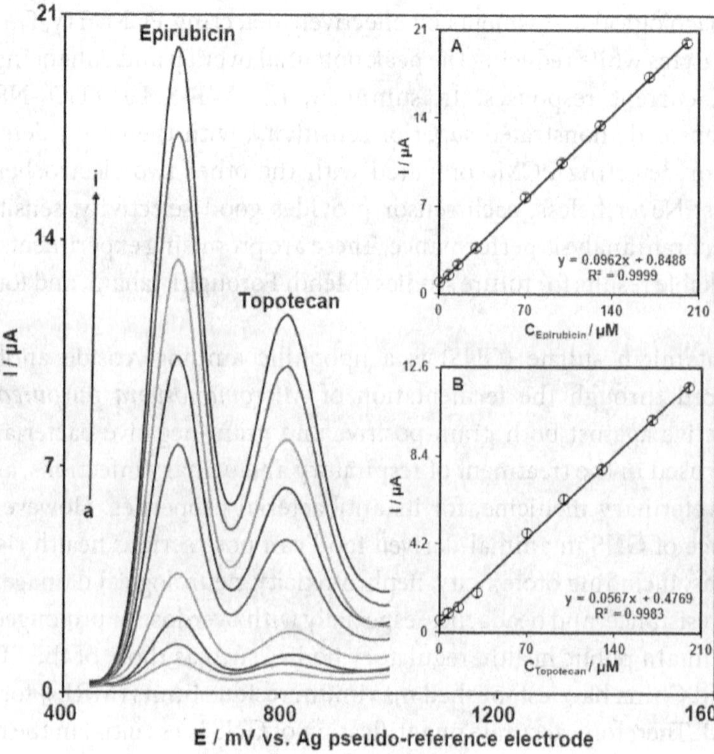

FIGURE 6.4 DPVs obtained at UiO-66-NH_2/GO/SPGE containing various concentrations of EP and TP. (Reprinted with permission from Tajik, S. et al., *Biosensors*, 14 (5), 229, 2024.)

with $Ni(OH)^2$–NaX/CPE sensor through a catalytic process and reduced over-potential compared with $Ni(OH)^2$/CPE. This process was explained by the porous structure of NaX facilitating the accommodation of Ni^{2+} ions, which transform into $Ni(OH)_2$ and NiOOH upon oxidation in basic conditions, thereby contributing to the electrocatalytic oxidation of GNS (Masihpour and Hassaninejad-Darzi 2023).

Anti-cancer medications like topotecan (TP) and epirubicin (EP) are commonly used in medicine, but their side effects can be fatal. Electrochemical techniques have gained popularity for detecting these substances, particularly with modified electrodes like carbon nanostructures and MOFs for energy storage and sensing. In a very recent study, Tajik et al. used a screen-printed graphite electrode (SPGE) and UiO-66-NH_2/GO nanocomposite to create an electrochemical sensing platform for the detection of EP and TP in their investigation. The UiO-66-NH_2/

GO nanocomposite was created by a solvothermal process. After modification of a SPGE with UiO-66-NH_2/GO nanomaterial, CV and chronoamperometry measurements were performed to analyse the modified electrode's electrochemical behaviour. The designed sensor's usefulness was demonstrated when it was employed to detect EP and TP injection samples simultaneously. The work reveals diffusion processes governing oxidation and reduction by examining the effect of scan rate on EP's redox reaction at UiO-66-NH_2/GO/SPGE with CV. With the DPV technique, a linear calibration curve for EP was obtained, which demonstrates a linear range between 0.008 and 200.0 M with an LOD of 0.003 μM (Figure 6.4). The enhanced mass transport, electron transfer, and electrocatalytic activity of the nanocomposite indicate that the new sensor is suitable for real-world investigation. Furthermore, subsequent studies may investigate the sensor's suitability for additional analytes and broaden its practical uses (Tajik et al. 2024).

Cancer comprises a range of diseases marked by uncontrolled cell proliferation. It preserves its place as one of the leading causes of death globally, largely due to the limitations of current treatments. Therefore, advancing cancer research is crucial for understanding its causes and devising effective strategies for prevention, diagnosis, treatment, and ultimately, a cure. 5-Fluorouracil is a chemotherapy drug that inhibits the enzyme thymidylate synthase and is used in the treatment of many cancer types. Nevertheless, overdosing can cause negative consequences, which necessitates the development of rapid and selective analytical devices to detect the concentration of this drug in human body fluids (Longley, Harkin, and Johnston 2003; J. J. Lee, Beumer, and Chu 2016). With high sensitivity, selectivity, reproducibility, low cost, and fast reaction times, electrochemical sensors have started to engage in this respect. Particularly, voltammetry could be the best choice, together with chemically modified electrodes, since they are able to improve the desired characteristics. Jahani et al. created a novel $CuFe_2O_4$ nanoparticle–modified SPEGs ($CuFe_2O_4$ NPs/SPGEs) sensor, using the drop-casting technique, to alter and enhance the ultimate performance of the working electrode. The study used DPV measurements of 50.0 μM 5-fluorouracil in the PBS range of 2.0 to 9.0 to examine the electrochemical behaviour of 5-fluorouracil at the surface of the electrodes. The findings demonstrated that the 5-fluorouracil oxidation peak current is pH dependent, rising with increasing pH until it reaches its maximum at pH 7.0. The greater peak current and optimum pH of 7.0 show that protons are involved in the 5-fluorouracil oxidation process. $CuFe_2O_4$ nanoparticles on the surface of $CuFe_2O_4$ NPs/SPGE dramatically increased the peak current and reduced the oxidation potential, suggesting that the $CuFe_2O_4$NPs/SPGE nanosensor is highly capable of catalysing the oxidation of 5-fluorouracil. The electrochemical performance of the sensor was tested using the DPV technique, which displayed a linear range from 0.1 to 270.0 μM with an LOD of 0.03 μM. The applicability of the sensor was evaluated using actual samples, comprising pharmaceutical injection and human urine samples, with recovery values ranging from 98.0% to 103.6%. The findings validate that $CuFe_2O_4$ NPs/SPGE have a superior capacity to detect 5-fluorouracil in actual samples with adequate sensitivity and selectivity. Even so, with more research and improved modifications, better results could be achieved in future studies (Jahani, Jafari, and Ravari 2023).

Due to its anticancer properties, antimutagenic effects, and antioxidant activities, gallic acid (GA) is an important natural polyphenolic compound. Therefore, the precise and effective analysis of GA is crucial for human health (Sun et al. 2020; Vilian et al. 2018). However, conventional detection methods such as thin-layer chromatography (TLC) and HPLC are costly and time-consuming. Using materials like Super-P carbon black (SPCB) for its conductivity and mesoporous silica nanoparticles (MSNs) for their vast surface area, electrochemical sensing provides a quicker and more effective substitute. In order to achieve sensitive GA detection, Zhao et al. designed an SPCB@MSNs/GCE sensor that combines these qualities. To develop the sensor, SPCB and MSNs were mixed together in N,N-dimethylformamide (DMF), ultrasonically dispersed, heated, stirred, and dried to make a nanocomposite that was then applied to a polished GCE. The materials were characterized using XRD, FTIR, SEM, and EDX. CV and DPV were used to evaluate the electrochemical performance. The SPCB@MSNs/GCE sensor produced the maximum GA oxidation peak current (47.7 μA) in 0.1 M PBS (pH 2.0). Using bare GCE, SPCB/GCE, and SPCB@MSNs/GCE sensors, the GA detection performance of the newly modified sensor was investigated. It was found that the peak responses of DPV increased with the increment in GA concentration (5.0–20.0 μM). The LOD of the bare GCE sensor was 0.151 μM, which responded linearly. The SPCB/GCE sensor performed better, with an LOD of 0.011 μM; however, with an LOD of 0.00491 μM, the SPCB@MSNs/GCE sensor demonstrated the best results because of the synergistic effects of MSNs' large surface area and SPCB's conductivity, which amplified surface accumulation and charge transfer. In DPV experiments for GA detection, the SPCB@MSNs/GCE sensor demonstrated high repeatability and reproducibility, with an

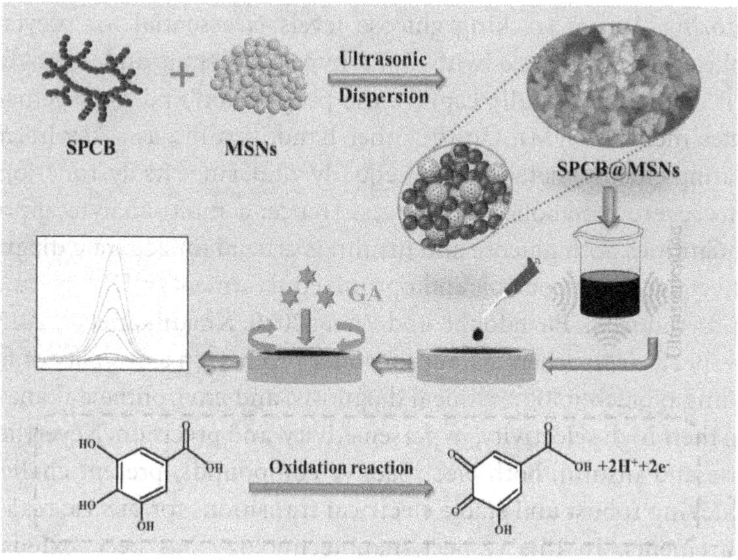

FIGURE 6.5 Fabrication of SPCB@MSNs/GCE sensor for GA detection. (Reprinted from *Journal of Materials Research and Technology*, 24, Zhao, H., Guo, M., Li, F., Zhou, Y., Zhu, G., Liu, Y., Ran, Q., Nie, F., and Dubovyk, V., Fabrication of Gallic Acid Electrochemical Sensor Based on Interconnected Super-P Carbon Black@mesoporous Silica Nanocomposite Modified Glassy Carbon Electrode, 2100–12, Copyright (2023), with permission from Elsevier.)

RSD of 2.53% for five distinct sensors and 2.11% for five measurements on a single sensor. The addition of other ions (Mg^{2+}, Ag^{+}, Na^{+}, Ni^{2+}, Ca^{2+}, Zn^{2+}, K^{+}) and organic molecules (ascorbic acid, quercetin, caffeic acid) as interferences did not significantly influence GA detection, further demonstrating the sensor's superior selectivity and resilience to interferents in the sample media. These outcomes demonstrate the sensor's dependability and resilience under various circumstances. The SPCB@MSNs/GCE sensor effectively detects GA in green and black tea samples. The sensor, made from a nanocomposite of SPCB and MSNs, offers a low detection limit of 4.911 nM for GA and excellent repeatability, reproducibility, and selectivity. Future research could focus on enhancing the sensor's performance, exploring applications for other analytes, and scaling for broader use. This study highlights the sensor's practicality and applicability in advanced electrochemical sensing across various fields (Zhao et al. 2023) (Figure 6.5).

The simultaneous quantification of diabetes-related biomarkers, particularly glucose and insulin, is increasingly emphasized for its critical role in medical diagnosis, patient health assessment, and treatment monitoring. While tracking glucose levels is essential for preventing complications associated with both hyperglycaemia and hypoglycaemia, it cannot alone fully capture the progression and management of diabetes mellitus (DM). On the other hand, insulin, as a key hormone regulating glucose metabolism, is equally vital, since its dysfunction can lead to severe metabolic imbalances. Hence, a multi-analyte approach that quantifies both glucose and insulin is crucial for accurate diagnosis, effective glycaemic control, and optimized treatment of DM (Sabu et al. 2019; Teymourian, Barfidokht, and Wang 2020; Khadilkar et al. 2013). In this respect, electrochemical sensors offer promise in a diversity of fields, including biotechnology, clinical diagnosis, and environmental analysis, due to their high selectivity, hypersensitivity, and precision. Nevertheless, glucose and insulin, both electroactive compounds, present challenges in achieving robust and stable electrical transitions for precise real-time measurements. In this respect, nanotechnology has been extensively employed to enhance the detection of these biomarkers, with CPEs being a versatile platform for sensor modification. Nanoparticles, CNTs, and MOFs are particularly effective due to their small size and unique characteristics, which improve electrocatalytic activity. Moreover, hybridizing nanoparticles with layered double hydroxides (LDHs) and incorporating CNTs as support materials can significantly enhance the sensor's ultimate performance by improving electrical conductivity, maintaining structural integrity, and reducing nanoparticle aggregation. Based on these new combinations, Khataee et al. designed a non-enzymatic electrochemical nanosensor by integrating copper aluminium layered double hydroxide (CuAl LDH)–based nanoparticles with two optimization steps. Merging nickel oxide/cobalt oxide (NiO/Co_3O_4) materials was the initial step, and subsequent assembly of these with multi-walled carbon nanotubes (MWCNTs) finalized the design process. The newly developed nanostructures were characterized by FTIR, XRD, and FESEM techniques. Lastly, the nanomaterials were employed as a CPE material for the concurrent dual-signal analysis of glucose and insulin. The performance of the modified electrode was evaluated using various electrochemical techniques including CV, EIS, DPV, and amperometric titration. The NiO/Co_3O_4 CuAl LDH@MWCNT electrode demonstrated outstanding capabilities, evidenced by two distinct and well-resolved peaks for detecting glucose and insulin. It exhibited a linear detection range of 0.01–26.5 mM for glucose and 0.02–70.0 nM for insulin. The limit of detection for glucose and insulin was determined to be as low as 2.8 μM and 17.3

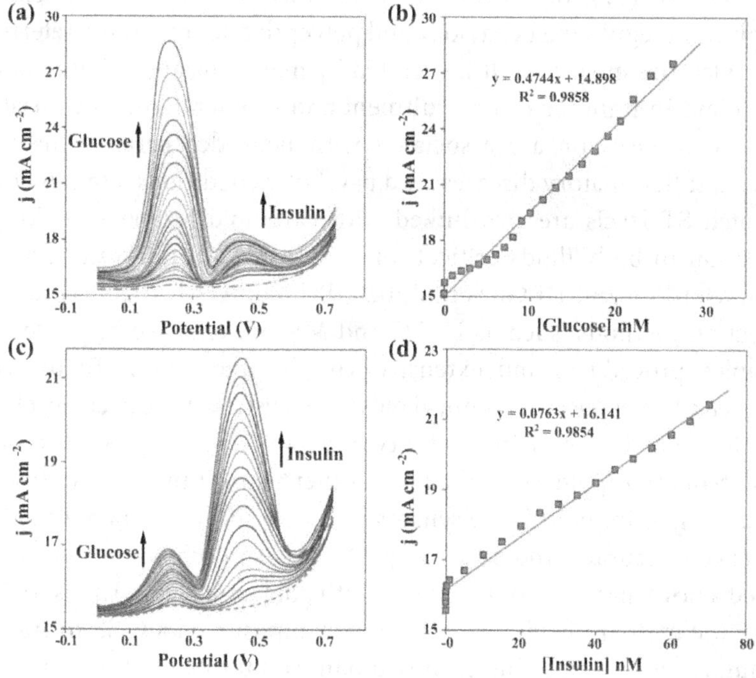

FIGURE 6.6 DPV curves and calibration curves for additional concentrations of glucose and insulin. (Reprinted from *Microchemical Journal*, 201 (March), Khataee, S., Dehghan, G., Shaghaghi, Z., Khataee, A., and Amini, M., A Novel Bifunctional Electrochemical Nanosensor for Simultaneous Detection of Glucose and Insulin Based on NiO/Co3O4@CuAl LDH-MWCNT Nanocomposite-Modified Carbon Paste Electrode, 110644, Copyright (2024), with permission from Elsevier.)

pM, respectively. Furthermore, the sensor was employed to measure glucose and insulin in human serum samples, achieving recovery rates of 101.0%–103.7% for glucose and 98.0%–106.0% for insulin, with all RSD values remaining below 3.0%. Besides, at the same concentration, glucose and insulin did not interfere with each other at the optimum potential. The stability tests confirmed that the new sensor could be stored in a dry place at a room temperature for 30 days without decomposition. To summarize, the objective of Khataee et al. was to investigate the potential of metal oxide nanomaterials, LDHs, and MWCNTs in fabricating a sensor through a flexible and adaptable approach, which resulted in high sensitivity, sufficient stability, and efficient selectivity (Khataee et al. 2024) (Figure 6.6).

Serotonin (ST), or 5-hydroxytryptamine, is a key neurotransmitter involved in regulating behaviour and perceptive actions, often referred to as a "pleasure molecule." It is secreted by mast cells and T lymphocytes, which aid in immune cell recruitment and proliferation, and variations in its concentration are associated with neurodegenerative disorders, autism, inflammatory diseases, and psychotic conditions. On top of that, elevated ST levels are also linked with carcinoid tumours, making its detection in body fluids critical for diagnosis (Ni and Watts 2006; De Deurwaerdère and Di Giovanni 2020; Pehrson et al. 2022). Traditional detection methods, such as HPLC and MS, are effective but necessitate complex procedures and extensive sample preparation. To overcome these challenges, electrochemical biosensor systems have been developed, yet they face issues of insensitivity and non-selectivity because of the close reduction potentials of ST and other neurotransmitters. Research is ongoing to improve these sensors using engineered nanomaterials for enhanced accuracy and selectivity. On this subject, Kumar et al. proposed a novel nanosensor fabricated with gold nanostars (Au-Nstars) and graphitic carbon nitride (g-C_3N_4) nanocomposite on a GCE to track ST in human body fluids. The proposed nanocomposite has been characterized thoroughly by several techniques, including UV-Vis, TEM, selected area electron diffraction (SAED), and EDX. CV and EIS were employed not only to establish the charge transfer kinetics but also to reveal the electrochemical performance at the electrode/electrolyte surface. After the experiment was conducted with different concentrations of ST by the SWV technique, a calibration curve for the Au-Nstar/g-C_3N_4/GCE sensor was generated. The results revealed a linear range from 5×10^{-7} to 1×10^{-3} M, with an LOD as low as 15.1 nM. Stability tests were performed to estimate the shelf-life of the new electrode, resulting in a lifetime of up to 8 weeks. The sensor was tested with both real samples (urine, serum) and in vitro analysis (HEK cells). The underlying reason for integrating the HEK cell line as an in vitro model was to evaluate its potential in clinical applications, given ST's role in promoting cell proliferation and inhibiting apoptosis, which demonstrates a relatively unconventional strategy to investigate the performance of a novel sensor. Furthermore, it is important to acknowledge that no pre-treatment of the urine sample, such as filtration or centrifugation, was performed during the analysis. Accordingly, 94.23–98.37% of ST could be detected in urine samples by the novel nanosensor. In summary, the Au-Nstar/g-C_3N_4/GCE sensor had several remarkable properties in combination with high sensitivity and simplicity, which could make it the choice for ST detection in clinical settings (A. Kumar et al. 2023).

6.5 CONCLUSION

Integrating nanomaterials into electrochemical sensors has particularly revolutionized the field of chemical detection and analysis, offering significant enhancements in sensitivity, selectivity, and response time. This distinctiveness could be attributed to their unique properties, such as high surface area-to-volume ratio, quantum effects, and exceptional electrical, mechanical, and thermal properties. The integration of nanomaterials into electrochemical sensors has ushered in a new era of precision and efficiency in pharmaceutical and biomedical diagnostics. These advanced sensors, leveraging the unique properties of nanomaterials, such as high surface area, enhanced conductivity, and tuneable chemical reactivity, have demonstrated remarkable improvements in sensitivity, selectivity, and detection limits. This progress has enabled the rapid and accurate detection of a wide range of pharmaceutical compounds and biomarkers, crucial for early diagnosis, therapeutic monitoring, and personalized medicine. As nanotechnology continues to evolve, the potential for further innovation in electrochemical sensing is vast. The development of multifunctional nanomaterials and the exploration of hybrid structures promise to push the boundaries of current diagnostic capabilities, offering more robust, portable, and cost-effective solutions. However, challenges remain, particularly in the scalability and reproducibility of nanomaterial synthesis, as well as in the integration of these sensors into practical diagnostic devices. In general, nanomaterial-based electrochemical sensors represent a significant leap forward in the field, with the potential to revolutionize how we detect and monitor diseases and pharmaceutical compounds. Continued research and interdisciplinary collaboration will be essential in overcoming existing challenges and unlocking the full potential of these innovative diagnostic tools, ultimately leading to more effective healthcare solutions worldwide.

ACKNOWLEDGEMENTS

S. Kurbanoglu acknowledges Turkish Academy of Sciences-The Young Scientists Award program (TUBA-GEBIP) and The Science Academy, Turkiye for Young Scientist Award Program (BAGEP).

REFERENCES

Aftab, Saima, and Sevinc Kurbanoglu. 2024. "Advanced Nanostructured Material-Based Biosensors in Clinical and Forensic Diagnosis." In *Novel Nanostructured Materials for Electrochemical Bio-Sensing Applications*, edited by Jamballi G. Manjunatha, 429–61. Elsevier.

Akhtar, Mehwish, Manzar Sohail, Muhammad Farooq Warsi, Mohamed M. Ibrahim, Mohammed A. Amin, and Muhammad Shahid. 2023. "Fe3O4 Nanochips Loaded MXenes/g-C3N4 Nanocomposite for Ultrasensitive Electrochemical Detection of Zinc (II), Cadmium (II), Lead (II), Copper (II) and Mercury (II) Metal Ions." *FlatChem* 41 (June): 100537. https://doi.org/10.1016/j.flatc.2023.100537.

Alhalili, Zahrah. 2023. "Metal Oxides Nanoparticles: General Structural Description, Chemical, Physical, and Biological Synthesis Methods, Role in Pesticides and Heavy Metal Removal through Wastewater Treatment." *Molecules* 28 (7): 3086. https://doi.org/10.3390/molecules28073086.

Alhazimeh, Hadeel, Mohamed Ghazi Al-fandi, and Lina M K Al-ebbini. 2022. "Development of an Inkjet-Printed Electrochemical Nanosensor for Ascorbic Acid Detection" 3 (February): 342–53. https://doi.org/10.1108/SR-11-2021-0434.

Almberg, Kirsten S., Mary E. Turyk, Rachael M. Jones, Kristin Rankin, Sally Freels, and Leslie T. Stayner. 2018. "Atrazine Contamination of Drinking Water and Adverse Birth Outcomes in Community Water Systems with Elevated Atrazine in Ohio, 2006–2008." *International Journal of Environmental Research and Public Health* 15 (9): 12–15. https://doi.org/10.3390/ijerph15091889.

Ambaye, Abera Demeke, Munkombwe Muchindu, Abongile Jijana, Shivani Mishra, and Edward Nxumalo. 2023. "Screen-Printed Electrode System Based on Carbon Black/Copper-Organic Framework Hybrid Nanocomposites for the Electrochemical Detection of Nitrite." *Materials Today Communications* 35 (February): 105567. https://doi.org/10.1016/j.mtcomm.2023.105567.

Amine, Aziz, and Hasna Mohammadi. 2019. "Amperometry." *Encyclopedia of Analytical Science* (January): 85–98. https://doi.org/10.1016/B978-0-12-409547-2.14204-0.

Ángela Molina, and Joaquín González. 2016. "Pulse Voltammetry in Physical Electrochemistry and Electroanalysis: Theory and Application." *Monographs in Electrochemistry* 671. https://doi.org/10.1007/978-3-319-21251-7.

Anh, Nguyen Tuan, Le Minh Tung, Le Khanh Vinh, Nguyen Van Quy, Ong Van Hoang, Ngo Xuan Dinh, and Anh Tuan Le. 2023. "An On-Site and Portable Electrochemical Sensing Platform Based on Spinel Zinc Ferrite Nanoparticles for the Quality Control of Paracetamol in Pharmaceutical Samples." *Nanoscale Advances* 6 (1): 256–67. https://doi.org/10.1039/d3na00749a.

Aslam, Faiza, Afzal Shah, Naimat Ullah, and Shamsa Munir. 2023. "Multiwalled Carbon Nanotube/Fe-Doped ZnO-Based Sensors for Droplet Electrochemical Detection and Degradation Monitoring of Brilliant Green." *ACS Applied Nano Materials* 6 (7): 6172–85. https://doi.org/10.1021/acsanm.3c00488.

Aydın, Elif Burcu. 2020. "Highly Sensitive Impedimetric Immunosensor for Determination of Interleukin 6 as a Cancer Biomarker by Using Conjugated Polymer Containing Epoxy Side Groups Modified Disposable ITO Electrode." *Talanta* 215 (120909): 1–11. https://doi.org/10.1016/j.talanta.2020.120909.

Bahadır, Elif Burcu, Mustafa Kemal Sezgintürk, Elif Burcu Bahadir, and Mustafa Kemal Sezgintürk. 2016. "A Review on Impedimetric Biosensors." *Artificial Cells, Nanomedicine and Biotechnology* 44 (1): 248–62. https://doi.org/10.3109/21691401.2014.942456.

Baig, Nadeem, Irshad Kammakakam, Wail Falath, and Irshad Kammakakam. 2021. "Nanomaterials: A Review of Synthesis Methods, Properties, Recent Progress, and Challenges." *Materials Advances* 2 (6): 1821–71. https://doi.org/10.1039/D0MA00807A.

Baranwal, Jaya, Brajesh Barse, Gianluca Gatto, Gabriela Broncova, and Amit Kumar. 2022. "Electrochemical Sensors and Their Applications: A Review." *Chemosensors* 10 (9): 363. https://doi.org/10.3390/chemosensors10090363.

Barhoum, Ahmed, Selma Hamimed, Hamda Slimi, Amina Othmani, Fatehy M. Abdel-Haleem, and Mikhael Bechelany. 2023. "Modern Designs of Electrochemical Sensor Platforms for Environmental Analyses: Principles, Nanofabrication Opportunities, and Challenges." *Trends in Environmental Analytical Chemistry* 38 (February): e00199. https://doi.org/10.1016/j.teac.2023.e00199.

Beitollahi, Hadi, Sayed Zia Mohammadi, Mohadeseh Safaei, and Somayeh Tajik. 2020. "Applications of Electrochemical Sensors and Biosensors Based on Modified Screen-Printed Electrodes: A Review." *Analytical Methods* 12 (12): 1547–60. https://doi.org/10.1039/c9ay02598g.

Bolat, G., Y. T. Yaman, S. Abacı, and S. Seyyar. 2023. "Poly-Arginine/Graphene Oxide Functionalized Disposable Sensor for Monitoring Fenitrothion Pesticide Residues in Water and Cucumber Samples." *Materials Today Chemistry* 30: 1–11. https://doi.org/10.1016/j.mtchem.2023.101517.

Bollella, Paolo, and Lo Gorton. 2018. "Enzyme Based Amperometric Biosensors." *Current Opinion in Electrochemistry* 10 (August): 157–73. https://doi.org/10.1016/j.coelec.2018.06.003.

Bozal-Palabiyik, Burcin, Sevinc Kurbanoglu, Cem Erkmen, and Bengi Uslu. 2021. "Future Prospects and Concluding Remarks for Electroanalytical Applications of Quantum Dots." In *Electroanalytical Applications of Quantum Dot-Based Biosensors*, 427–50.

Bratov, A, N Abramova, and A Ipatov. 2010. "Recent Trends in Potentiometric Sensor Arrays—A Review." *Analytica Chimica Acta* 678 (2): 149–59.

Burdan, Franciszek. 2004. "Developmental Effects of Propyphenazone in Analgesic and Antipyretic Combination with Caffeine or Paracetamol." *Human and Experimental Toxicology* 23 (5): 235–44. https://doi.org/10.1191/0960327104ht439oa.

Burress, Jacob W., Srinivas Gadipelli, Jamie Ford, Jason M. Simmons, Wei Zhou, and Taner Yildirim. 2010. "Graphene Oxide Framework Materials: Theoretical Predictions and Experimental Results." *Angewandte Chemie - International Edition* 49 (47): 8902–4. https://doi.org/10.1002/anie.201003328.

Büyüktaş, Duygu, Masoud Ghaani, Cesare Rovera, Daniele Carullo, Richard T. Olsson, Figen Korel, and Stefano Farris. 2023. "A Screen-Printed Electrode Modified with Gold Nanoparticles/Cellulose Nanocrystals for Electrochemical Detection of 4,4′-Methylene Diphenyl Diamine." *Heliyon* 9 (4): 1–10. https://doi.org/10.1016/j.heliyon.2023.e15327.

Campanella, Gaetano, Masoud Ghaani, Gianpiero Quetti, and Stefano Farris. 2015. "On the Origin of Primary Aromatic Amines in Food Packaging Materials." *Trends in Food Science and Technology* 46 (1): 137–43. https://doi.org/10.1016/j.tifs.2015.09.002.

Chen, Aicheng, and Sanghamitra Chatterjee. 2013. "Nanomaterials Based Electrochemical Sensors for Biomedical Applications." *Chemical Society Reviews* 42 (12): 5425–38. https://doi.org/10.1039/c3cs35518g.

Chen, Aicheng, and Badal Shah. 2013. "Electrochemical Sensing and Biosensing Based on Square Wave Voltammetry." *Analytical Methods* 5 (9): 2158–73. https://doi.org/10.1039/c3ay40155c.

Chen, Kew Yu, and Che Wei Chang. 2014. "Highly Soluble Monoamino-Substituted Perylene Tetracarboxylic Dianhydrides: Synthesis, Optical and Electrochemical Properties." *International Journal of Molecular Sciences* 15 (12): 22642–60. https://doi.org/10.3390/ijms151222642.

Corr, Serena A. 2016. "Metal Oxide Nanoparticles." *SPR Nanoscience* 3 (October): 31–56. https://doi.org/10.1039/9781782623717-00031.

Daniel, Miriam, Georgeena Mathew, Masakazu Anpo, and Bernaurdshaw Neppolian. 2022. "MOF Based Electrochemical Sensors for the Detection of Physiologically Relevant Biomolecules: An Overview." *Coordination Chemistry Reviews* 468: 214627. https://doi.org/https://doi.org/10.1016/j.ccr.2022.214627.

Darwish, Moustafa A, Walaa Abd-elaziem, and Ammar Elsheikh. 2024. "Nanoscale Advances Advancements in Nanomaterials for Nanosensors :" *Nanoscale Advances*. https://doi.org/10.1039/d4na00214h.

Datta, Deepshikha, Krishna Priyadarshini, and K. S. Deepak. 2022. *Candidates of Functionalized Nanomaterial-Based Membranes. Membranes with Functionalized Nanomaterials*. Elsevier Inc. https://doi.org/10.1016/B978-0-323-85946-2.00004-7.

Deffo, Gullit, Thierry Flavien Nde Tene, Liliane Medonbou Dongmo, Sherman Lesly Zambou Jiokeng, and Ranil Clément Tonleu Temgoua. 2024. "Differential Pulse and Square-Wave Voltammetry as Sensitive Methods for Electroanalysis Applications." *Encyclopedia of Solid-Liquid Interfaces* 1: 409–17. https://doi.org/10.1016/b978-0-323-85669-0.00040-4.

Deurwaerdère, Philippe De, and Giuseppe Di Giovanni. 2020. "Serotonin in Health and Disease." *International Journal of Molecular Sciences* 21 (10): 10–13. https://doi.org/10.3390/ijms21103500.

Ding, Xiaoteng, Yusheng Niu, Gong Zhang, Yuanhong Xu, and Jinghong Li. 2020. "Electrochemistry in Carbon-Based Quantum Dots." *Chemistry – An Asian Journal* 15 (8): 1214–24. https://doi.org/https://doi.org/10.1002/asia.202000097.

Erkmen, Cem, Sevinc Kurbanoglu, and Bengi Uslu. 2021. "Nanozymes in Electrochemical Analysis." *Nanozymes* (November): 159–98. https://doi.org/10.1201/9781003109228-9.

Erkmen, Cem, Didem N Unal, Sevinc Kurbanoglu, and Bengi Uslu. 2022. "Basics of Electrochemical Sensors BT - Organic Electrodes: Fundamental to Advanced Emerging Applications." In *Organic Electrodes*, edited by Ram K Gupta, 81–99. Cham: Springer International Publishing. https://doi.org/10.1007/978-3-030-98021-4_5.

Escorcia-d, Daniel, Leidy Rend, and Margarita Ram. 2023. "Advancements in Nanoparticle Deposition Techniques for Diverse Substrates : A Review," 1–25.

Falcaro, Paolo, Raffaele Ricco, Amirali Yazdi, Inhar Imaz, Shuhei Furukawa, Daniel Maspoch, Rob Ameloot, Jack D. Evans, and Christian J. Doonan. 2016. "Application of Metal and Metal Oxide Nanoparticles@MOFs." *Coordination Chemistry Reviews* 307 (January): 237–54. https://doi.org/10.1016/J.CCR.2015.08.002.

Galeano-Dıaz, T., A. Guiberteau-Cabanillas, A. Espinosa-Mansilla, and M. D. Lopez-Soto. 2008. "Adsorptive Stripping Square Wave Voltammetry (Ad-SSWV) Accomplished with Second-Order Multivariate Calibration Determination of Fenitrothion and Its Metabolites in River Water Samples." *Analytica Chimica Acta* 618 (2): 131–39. https://doi.org/10.1016/j.aca.2008.04.058.

Ge, Chengjie, Rajendran Ramachandran, and Fei Wang. 2020. "Metal-Organic Framework Derived CeO2 Based Two-Dimensional Layered Nanocomposites for Selective Electrochemical Dopamine Detection." *15th IEEE International Conference on Nano/Micro Engineered and Molecular System, NEMS 2020*, 512–17. https://doi.org/10.1109/NEMS50311.2020.9265548.

Gheni, Saba A., Mudheher M. Ali, Goh C. Ta, Hannah J. Harbin, and Saad A. Awad. 2024. "Toxicity, Hazards, and Safe Handling of Primary Aromatic Amines." *ACS Chemical Health and Safety* 31 (1): 8–21. https://doi.org/10.1021/acs.chas.3c00073.

Gong, Shu, and Wenlong Cheng. 2017. "One-Dimensional Nanomaterials for Soft Electronics." *Advanced Electronic Materials* 3 (3): 1600314. https://doi.org/https://doi.org/10.1002/aelm.201600314.

González, Natalia, Maite V. Aguinaga Martínez, Claudia E. Domini, and Carolina C. Acebal. 2023. "Current Trends in Sample Preparation for the Determination of Primary Aromatic Amines in Environmental Samples." *Trends in Environmental Analytical Chemistry* 37 (September 2022): e00197. https://doi.org/10.1016/j.teac.2023.e00197.

Hahm, Jong In. 2016. "Fundamental Properties of One-Dimensional Zinc Oxide Nanomaterials and Implementations in Various Detection Modes of Enhanced Biosensing." *Annual Review of Physical Chemistry* 67 (2016): 691–717. https://doi.org/10.1146/ANNUREV-PHYSCHEM-031215-010949/CITE/REFWORKS.

Hayat, Akhtar, and Jean Louis Marty. 2014. "Disposable Screen Printed Electrochemical Sensors: Tools for Environmental Monitoring." *Sensors* 14 (6): 10432–53. https://doi.org/10.3390/S140610432.

Huang, Chi Wei, Chitsan Lin, Minh Ky Nguyen, Adnan Hussain, Xuan Thanh Bui, and Huu Hao Ngo. 2023. "A Review of Biosensor for Environmental Monitoring: Principle, Application, and Corresponding Achievement of Sustainable Development Goals." *Bioengineered* 14 (1): 58–80. https://doi.org/10.1080/21655979.2022.2095089.

Ijaz, Irfan, Ezaz Gilani, Ammara Nazir, and Aysha Bukhari. 2020. "Detail Review on Chemical, Physical and Green Synthesis, Classification, Characterizations and Applications of Nanoparticles." *Green Chemistry Letters and Reviews* 13 (3): 223–45. https://doi.org/10.1080/17518253.2020.1802517.

Inesi, Achille. 1986. "Instrumental Methods in Electrochemistry." *Bioelectrochemistry and Bioenergetics* 15 (3): 531. https://doi.org/10.1016/0302-4598(86)85047-4.

Ismaeel, Ghufran Lutfi, Shaymaa Abed Hussein, Gulrux Daminova, Jameel Mohammed Ameen Sulaiman, Mohaned Mohammed Hani, Eftikhaar Hasan Kadhum, Shahad Abdulhadi Khuder, et al. 2024. "Fabrication and Investigating of a Nano-Structured Electrochemical Sensor to Measure the Amount of Atrazine Pollution Poison in Water and Wastewater." *Chemical Data Collections* 51 (March): 101135. https://doi.org/10.1016/j.cdc.2024.101135.

Iwuoha, Emmanuel Iheanyichukwu, Amir Al-Ahmed, Mantoa Sekota, Tesfaye Waryo, and Priscilla Baker. 2013. "Amperometric Sensors." In *Encyclopedia of Supramolecular Chemistry-Two-Volume Set (Print)*, 1–18. CRC Press.

Jaffrezic-Renault, Nicole, and Sergei V. Dzyadevych. 2008. "Conductometric Microbiosensors for Environmental Monitoring." *Sensors* 8 (4): 2569–88.

Jahani, Peyman Mohammadzadeh, Maedeh Jafari, and Farhad Nazari Ravari. 2023. "CuFe2O4 Nanoparticles-Based Electrochemical Sensor for Sensitive Determination of the Anticancer Drug 5-Fluorouracil." *ADMET and DMPK* 11 (2): 201–10. https://doi.org/10.5599/admet.1691.

Jamkhande, Prasad Govindrao, Namrata W. Ghule, Abdul Haque Bamer, and Mohan G. Kalaskar. 2019. "Metal Nanoparticles Synthesis: An Overview on Methods of Preparation, Advantages and Disadvantages, and Applications." *Journal of Drug Delivery Science and Technology* 53 (October): 101174. https://doi.org/10.1016/J.JDDST.2019.101174.

Jampasa, Sakda, Wisarut Khamcharoen, Surinya Traipop, Whitchuta Jesadabundit, Tugba Ozer, and Orawon Chailapakul. 2023. "Recent Advances on Nanomaterial-Modified Film-Electrode-Based Sensors: Approach to Clinical Purpose." *Current Opinion in Electrochemistry* 42 (November 2023): 101420. https://doi.org/10.1016/j.coelec.2023.101420.

Joshi, Anju, and Ki Hyun Kim. 2020. "Recent Advances in Nanomaterial-Based Electrochemical Detection of Antibiotics: Challenges and Future Perspectives." *Biosensors and Bioelectronics* 153 (December 2019). https://doi.org/10.1016/j.bios.2020.112046.

Kahlert, Heike. 2010. *Reference Electrodes. Electroanalytical Methods: Guide to Experiments and Applications*. Elsevier B.V. https://doi.org/10.1007/978-3-642-02915-8_15.

Kalia, Susheel, and Yuvaraj Haldorai. 2014. "*Organic-Inorganic Hybrid Nanomaterials*. Edited by Susheel Kalia and Yuvaraj Haldorai." *Advances in Polymer Science* 267 (Springer). https://doi.org/10.1007/978-3-319-13593-9.

Karadurmus, Leyla, I. Firat Sahin, Sevinc Kurbanoglu, Sibel A. S. A. Ozkan, I. Firat Sahin, Sevinc Kurbanoglu, and Sibel A. S. A. Ozkan. 2018. "Electrochemical Determination of Non-Steroidal Anti-Inflammatory Drugs." *Current Analytical Chemistry* 15 (4): 485–501. https://doi.org/10.2174/1573411014666180917113920.

Karadurmus, Leyla, Sevinc Kurbanoglu, Bengi Uslu, and Sibel A. Ozkan. 2017. "Differential Pulse Voltammetric Determination of Rosuvastatin Via Glassy Carbon Electrode." *Revue Roumaine de Chimie* 62: 581–88.

Kaya, S. Irem, Sevinc Kurbanoglu, and Sibel A. Ozkan. 2019. "Nanomaterials-Based Nanosensors for the Simultaneous Electrochemical Determination of Biologically Important Compounds: Ascorbic Acid, Uric Acid, and Dopamine." *Critical Reviews in Analytical Chemistry* 49 (2): 101–25. https://doi.org/10.1080/10408347.2018.1489217.

Khadilkar, KrantiShreesh, Tushar Bandgar, Vyankatesh Shivane, Anurag Lila, and Nalini Shah. 2013. "Current Concepts in Blood Glucose Monitoring." *Indian Journal of Endocrinology and Metabolism* 17 (9): 643. https://doi.org/10.4103/2230-8210.123556.

Khanna, Prerna, Amrit Kaur, and Dinesh Goyal. 2019. "Algae-Based Metallic Nanoparticles: Synthesis, Characterization and Applications." *Journal of Microbiological Methods* 163: 105656. https://doi.org/10.1016/J.MIMET.2019.105656.

Khataee, Simin, Gholamrez Dehghan, Zohreh Shaghaghi, Alireza Khataee, and Mojtaba Amini. 2024. "A Novel Bifunctional Electrochemical Nanosensor for Simultaneous Detection of Glucose and Insulin Based on NiO/Co3O4@CuAl LDH-MWCNT Nanocomposite-Modified Carbon Paste Electrode." *Microchemical Journal* 201 (March): 110644. https://doi.org/10.1016/j.microc.2024.110644.

Kimmel, Danielle W., Gabriel LeBlanc, Mika E. Meschievitz, and David E. Cliffel. 2012. "Electrochemical Sensors and Biosensors." *Analytical Chemistry* 84 (2): 685–707. https://doi.org/10.1021/ac202878q.

Kokoskarova, Pavlinka, Leon Stojanov, Kosta Najkov, Natasha Ristovska, Tatjana Ruskovska, Sławomira Skrzypek, and Valentin Mirceski. 2023. "Square - Wave Voltammetry of Human Blood Serum." *Scientific Reports* 13 (1): 8485. https://doi.org/10.1038/s41598-023-34350-1.

Koncki, Robert. 2007. "Recent Developments in Potentiometric Biosensors for Biomedical Analysis." *Analytica Chimica Acta* 599 (1): 7–15. https://doi.org/10.1016/j.aca.2007.08.003.

Kour, Ravinder, Sandeep Arya, Sheng-Joue Young, Vinay Gupta, Pankaj Bandhoria, and Ajit Khosla. 2020. "Review—Recent Advances in Carbon Nanomaterials as Electrochemical Biosensors." *Journal of The Electrochemical Society* 167 (3): 037555. https://doi.org/10.1149/1945-7111/ab6bc4.

Kumar, Ashutosh, Buddhadev Purohit, Uday Pratap Azad, Mohammad Tavakkoli Yaraki, Daphika S. Dkhar, Yen Nee Tan, Ananya Srivastava, and Pranjal Chandra. 2023. "Gold Nanostar and Graphitic Carbon Nitride Nanocomposite for Serotonin Detection in Biological Fluids and Human Embryonic Kidney Cell Microenvironment." *Microchimica Acta* 190 (1): 1–11. https://doi.org/10.1007/s00604-022-05612-y.

Kumar, Pawan, Akash Deep, and Ki Hyun Kim. 2015. "Metal Organic Frameworks for Sensing Applications." *TrAC Trends in Analytical Chemistry* 73 (November): 39–53. https://doi.org/10.1016/J.TRAC.2015.04.009.

Kurbanoglu, Sevinc, Burcu Dogan-Topal, Eva Plata Rodriguez, Burcin Bozal-Palabiyik, Sibel A. S.A. Ozkan, and Bengi Uslu. 2016. "Advances in Electrochemical DNA Biosensors and Their Interaction Mechanism with Pharmaceuticals." *Journal of Electroanalytical Chemistry* 775 (August): 8–26. https://doi.org/10.1016/j.jelechem.2016.05.022Review.

Kurbanoglu, Sevinc, and Sibel A. Ozkan. 2018. "Electrochemical Carbon Based Nanosensors: A Promising Tool in Pharmaceutical and Biomedical Analysis." *Journal of Pharmaceutical and Biomedical Analysis* 147 (January): 439–57. https://doi.org/10.1016/j.jpba.2017.06.062.

Lahcen, Abdellatif Ait, and Aziz Amine. 2019. "Recent Advances in Electrochemical Sensors Based on Molecularly Imprinted Polymers and Nanomaterials." *Electroanalysis* 31 (2): 188–201. https://doi.org/10.1002/elan.201800623.

Lazanas, Alexandros Ch, and Mamas I Prodromidis. 2023. "Electrochemical Impedance Spectroscopy— a Tutorial." *ACS Measurement Science Au* 3 (3): 162–93.

Lee, James J., Jan H. Beumer, and Edward Chu. 2016. "Therapeutic Drug Monitoring of 5-Fluorouracil." *Cancer Chemotherapy and Pharmacology* 78 (3): 447–64. https://doi.org/10.1007/s00280-016-3054-2.

Lee, Junsoo, Taeyeon Kim, Bong Geun Kim, and Hyon Bin Na. 2020. "Recent Progress in Inorganic Nanoparticles with Enzyme-Mimetic Activities and Their Applications to Diagnosis and Therapy." *Applied Chemistry for Engineering* 31 (4): 352–59. https://doi.org/10.14478/ace.2020.1044.

Li, Meixiu, Tao Chen, J. Justin Gooding, and Jingquan Liu. 2019. "Review of Carbon and Graphene Quantum Dots for Sensing." *ACS Sensors* 4 (7): 1732–48. https://doi.org/10.1021/acssensors.9b00514.

Li Ming, Honglei Gou, Israa Al-Ogaidi, and Nianqiang Wu. 2013. "Nanostructured Sensors for Detection of Heavy Metals: A Review." *ACS Sustainable Chemistry and Engineering* 1 (7): 713–23. https://doi.org/10.1021/sc400019a.

Lim, Hyo Ryoung, Hee Seok Kim, Raza Qazi, Young Tae Kwon, Jae Woong Jeong, and Woon Hong Yeo. 2020. "Advanced Soft Materials, Sensor Integrations, and Applications of Wearable Flexible Hybrid Electronics in Healthcare, Energy, and Environment." *Advanced Materials* 32 (15): 1–43. https://doi.org/10.1002/adma.201901924.

Lin, Tamar, Avner Gal, Yulia Mayzel, Keren Horman, and Karnit Bahartan. 2017. "Non-Invasive Glucose Monitoring: A Review of Challenges and Recent Advances." *Curr Trends Biomedical Eng & Biosci* 6: 1–8. https://doi.org/10.19080/CTBEB.2017.06.555696.

Liu, Xun, Lin Huang, and Kun Qian. 2021. "Nanomaterial-Based Electrochemical Sensors: Mechanism, Preparation, and Application in Biomedicine." *Advanced NanoBiomed Research* 1 (6): 1–12. https://doi.org/10.1002/anbr.202000104.

Longley, Daniel B., D. Paul Harkin, and Patrick G. Johnston. 2003. "5-Fluorouracil: Mechanisms of Action and Clinical Strategies." *Nature Reviews Cancer* 3 (5): 330–38. https://doi.org/10.1038/nrc1074.

Luhana, Charles, and Philani Mashazi. 2023. "Simultaneous Detection of Dopamine and Paracetamol on Electroreduced Graphene Oxide–Cobalt Phthalocyanine Polymer Nanocomposite Electrode." *Electrocatalysis* 14 (3): 406–17. https://doi.org/10.1007/s12678-022-00806-7.

Mahale, Rayappa Shrinivas, Vasantha Kumar Shamanth, Krishna Hemanth, Rajendrachari Shashanka, P. C. Sharath, and N. V. Sreekanth. 2022. "Electrochemical Sensor Applications of Nanoparticle Modified Carbon Paste Electrodes to Detect Various Neurotransmitters: A Review." *Applied Mechanics and Materials* 908 (August): 69–88. https://doi.org/10.4028/P-MIZM85.

Majeed, Saadat, Sayed Tayyab Raza Naqvi, Muhammad Najam ul Haq, and Muhammad Naeem Ashiq. 2022. "Electroanalytical Techniques in Biosciences: Conductometry, Coulometry, Voltammetry, and Electrochemical Sensors." In *Analytical Techniques in Biosciences*, 157–78. INC. https://doi.org/10.1016/b978-0-12-822654-4.00004-x.

Malwal, Deepika, and Gopinath Packirisamy. 2018. *Recent Advances in the Synthesis of Metal Oxide (MO) Nanostructures. Synthesis of Inorganic Nanomaterials: Advances and Key Technologies*. Elsevier. https://doi.org/10.1016/B978-0-08-101975-7.00010-5.

Marken, Frank, Andreas Neudeck, and Alan M Bond. 2010. "Cyclic Voltammetry." In *Electroanalytical Methods: Guide to Experiments and Applications*, edited by Fritz Scholz, A. M. Bond, R. G. Compton, D. A. Fiedler, G. Inzelt, H. Kahlert, Š. Komorsky-Lovrić, et al., 57–106. Berlin, Heidelberg: Springer Berlin Heidelberg. https://doi.org/10.1007/978-3-642-02915-8_4.

Masihpour, Nafiseh, and Seyed Karim Hassaninejad-Darzi. 2023. "Constructing NaX Nanozeolite Modified Carbon Paste Electrode for Electro-Catalytic Measurement of Gentamicin Sulfate in Pharmaceutical Samples." *Journal of Inorganic and Organometallic Polymers and Materials* 33 (5): 1317–30. https://doi.org/10.1007/s10904-023-02587-y.

Mehdi Foroughi, Mohammad, Shohreh Jahani, and Soroush Rashidi. 2024. "Simultaneous Detection of Ascorbic Acid, Dopamine, Acetaminophen and Tryptophan Using a Screen-Printed Electrode Modified with Woolen Ball-Shaped La3+/TiO2 Nanostructure as a Quadruplet Nanosensor." *Microchemical Journal* 198 (July 2023): 110156. https://doi.org/10.1016/j.microc.2024.110156.

Mekuye, Bawoke, and Birhanu Abera. 2023. "Nanomaterials: An Overview of Synthesis, Classification, Characterization, and Applications." *Nano Select* 4 (8): 486–501. https://doi.org/10.1002/nano.202300038.

Modekwe, Helen U., Michael O. Daramola, Messai A. Mamo, and Kapil Moothi. 2024. "Heliyon Recent Advancements in the Use of Plastics as a Carbon Source for Carbon Nanotubes Synthesis - A Review." *Heliyon* 10 (2): e24679. https://doi.org/10.1016/j.heliyon.2024.e24679.

Murthy, H. C. Ananda, Ararso Nagari Wagassa, C. R. Ravikumar, and H. P. Nagaswarupa. 2022. *Functionalized Metal and Metal Oxide Nanomaterial-Based Electrochemical Sensors. Functionalized Nanomaterial-Based Electrochemical Sensors: Principles, Fabrication Methods, and Applications*. Elsevier Ltd. https://doi.org/10.1016/B978-0-12-823788-5.00001-6.

Nemčeková, Katarína, and Ján Labuda. 2021. "Advanced Materials-Integrated Electrochemical Sensors as Promising Medical Diagnostics Tools: A Review." *Materials Science and Engineering C*. https://doi.org/10.1016/j.msec.2020.111751.

Ni, Wei, and Stephanie W. Watts. 2006. "5-Hydroxytryptamine in the Cardiovascular System: Focus on the Serotonin Transporter (SERT)." *Clinical and Experimental Pharmacology and Physiology* 33 (7): 575–83. https://doi.org/10.1111/j.1440-1681.2006.04410.x.

Palmisano, F., R. Rizzi, D. Centonze, and P. G. Zambonin. 2000. "Simultaneous Monitoring of Glucose and Lactate by an Interference and Cross-Talk Free Dual Electrode Amperometric Biosensor Based on Electropolymerized Thin Films." *Biosensors and Bioelectronics* 15 (9–10): 531–39. https://doi.org/10.1016/S0956-5663(00)00107-X.

Patel, Bhavik A. 2020. "Chapter 2. Amperometry and Potential Step Techniques." In *Electrochemistry for Bioanalysis*, 9–26. Elsevier Inc. https://doi.org/10.1016/c2019-0-03108-1.

Pehrson, A. L., D. Roberts, A. Khawaja, and R. McNair. 2022. "The Role of Serotonin Neurotransmission in Rapid Antidepressant Actions." *Psychopharmacology* 239 (6): 1823–38. https://doi.org/10.1007/s00213-022-06098-5.

Perez, Mary Ângela Favaro, Daniela Daniel, Marisa Padula, Claudimir Lucio do Lago, and Carla Beatriz Grespan Bottoli. 2021. "Determination of Primary Aromatic Amines from Cooking Utensils by Capillary Electrophoresis-Tandem Mass Spectrometry." *Food Chemistry* 362 (November 2020): 1–11. https://doi.org/10.1016/j.foodchem.2021.129902.

Periasamy, Arun Prakash, Yogeswaran Umasankar, and Shen Ming Chen. 2009. *Nanomaterials - Acetylcholinesterase Enzyme Matrices for Organophosphorus Pesticides Electrochemical Sensors: A Review.* Sensors. Molecular Diversity Preservation International. https://doi.org/10.3390/s90604034.

Pirzada, Muqsit, and Zeynep Altintas. 2019. "Nanomaterials for Healthcare Biosensing Applications." *Sensors (Switzerland)* 19 (23): 1–56. https://doi.org/10.3390/s19235311.

Privett, Benjamin J., Jae Ho Shin, and Mark H. Schoenfisch. 2010. "Electrochemical Sensors." *Analytical Chemistry* 82 (12): 4723–41. https://doi.org/10.1021/ac101075n.

Prodromidis, Mamas I. MI. 2010. "Impedimetric Immunosensors—A Review." *Electrochimica Acta* 55 (14): 4227–33. https://doi.org/10.1016/J.ELECTACTA.2009.01.081.

Purohit, Buddhadev, Ashutosh Kumar, Kuldeep Mahato, and Pranjal Chandra. 2020. "Smartphone-Assisted Personalized Diagnostic Devices and Wearable Sensors." *Current Opinion in Biomedical Engineering* 13: 42–50. https://doi.org/10.1016/j.cobme.2019.08.015.

Pyrzynska, Krystyna. 2011. "Carbon Nanotubes as Sorbents in the Analysis of Pesticides." *Chemosphere* 83: 1407–13. https://doi.org/10.1016/j.chemosphere.2011.01.057.

Rafiee, Mohammad, Dylan J. Abrams, Luana Cardinale, Zachary Goss, Antonio Romero-Arenas, and Shannon S. Stahl. 2024. "Cyclic Voltammetry and Chronoamperometry: Mechanistic Tools for Organic Electrosynthesis." *Chemical Society Reviews* 53 (2): 566–85. https://doi.org/10.1039/d2cs00706a.

Rhoades, Martha G., Jane L. Meza, Cheryl L. Beseler, Patrick J. Shea, Andy Kahle, Julie M. Vose, Kent M. Eskridge, and Roy F. Spalding. 2013. "Atrazine and Nitrate in Public Drinking Water Supplies and Non-Hodgkin Lymphoma in Nebraska, USA." *Environmental Health Insights* 7: 15–27. https://doi.org/10.4137/EHI.S10629.

Rishpon, Judith. 2002. "Electrochemical Biosensors for Environmental Monitoring." *Reviews on Environmental Health* 17 (3): 219–47. https://doi.org/10.1515/REVEH.2002.17.3.219.

Sabu, Chinnu, T. K. Henna, V. R. Raphey, K. P. Nivitha, and K. Pramod. 2019. "Advanced Biosensors for Glucose and Insulin." *Biosensors and Bioelectronics* 141 (January): 111201. https://doi.org/10.1016/j.bios.2019.03.034.

Safonova, Lyubov P., and Arkadii M. Kolker. 1992. "Conductometry of Electrolyte Solutions." *Russian Chemical Reviews* 61 (9): 959–73. https://doi.org/10.1070/RC1992V061N09ABEH001009/XML.

Saleh, Tawfik A. 2020. "Environmental Technology & Innovation Nanomaterials: Classification, Properties, and Environmental Toxicities." *Environmental Technology & Innovation* 20: 101067. https://doi.org/10.1016/j.eti.2020.101067.

Sanko, Vildan, Cem Erkmen, and Filiz Kuralay. 2024. "The Role of Nanoparticles in Non-Invasive Electrochemical Immunosensor Technology and Recent Developments." *Electroanalysis* 36 (6): 1–19. https://doi.org/10.1002/elan.202400004.

Sato, Kazuomi, Hideki Takahashi, and Masaru Toriyama. 2011. "Depigmenting Mechanism of NSAIDs on B16F1 Melanoma Cells." *Archives of Dermatological Research* 303 (3): 171–80. https://doi.org/10.1007/s00403-010-1094-8.

Scheller, Frieder, and Florian Schubert. 1992. *Biosensors. Elsevier.* https://doi.org/10.1016/S0021-9673(01)93769-0.

Schubert, Jens, Oliver Kappenstein, Andreas Luch, and Thomas G. Schulz. 2011. "Analysis of Primary Aromatic Amines in the Mainstream Waterpipe Smoke Using Liquid Chromatography-Electrospray Ionization Tandem Mass Spectrometry." *Journal of Chromatography A* 1218 (33): 5628–37. https://doi.org/10.1016/j.chroma.2011.06.072.

Shaheen, Syed, Falak Niaz, Muhammad Ali, Himadri Tanaya, Muhammad Younas, Amir Sohail, Haroon Ur, S M Abu Nayem, Munetaka Oyama, and Abdul Aziz. 2024. "Advanced Strategies in Electrode Engineering and Nanomaterial Modifications for Supercapacitor Performance Enhancement: A Comprehensive Review." *Journal of Energy Storage* 79 (December 2023): 110152. https://doi.org/10.1016/j.est.2023.110152.

Sharma, Anirudh, and Joydeep Das. 2019. "Small Molecules Derived Carbon Dots: Synthesis and Applications in Sensing, Catalysis, Imaging, and Biomedicine." *Journal of Nanobiotechnology* 17 (1): 1–24. https://doi.org/10.1186/s12951-019-0525-8.

Sharma, Asha, Anoop Singh, Vinay Gupta, and Sandeep Arya. 2022. "Advancements and Future Prospects of Wearable Sensing Technology for Healthcare Applications." *Sensors and Diagnostics* 1 (3): 387–404. https://doi.org/10.1039/d2sd00005a.

Singh, Simranjeet, Pavithra Narasimhappa, Nadeem A. Khan, Vishakha Chauhan, Nabila Shehata, S. K. Behera, Joginder Singh, and Praveen C. Ramamurthy. 2023. "Effective Voltammetric Tool for Nano-Detection of Triazine Herbicide (1-Chloro-3-Ethylamino-5-Isopropylamino-2,4,6-Triazine) by Naphthalene Derivative." *Environmental Research* 236 (P2): 116808. https://doi.org/10.1016/j.envres.2023.116808.

Soylemez, Saniye, Sevinc Kurbanoglu, and Filiz Kuralay. 2024. "Nanoscale Physics of Electrochemistry." *Biophysics at the Nanoscale: Applications of Functional Materials* (January): 43–71. https://doi.org/10.1016/B978-0-443-15359-4.00006-1.

Spinola Machado, Sergio A., and Fernando Henrique Cincotto. 2022. "Electrochemical Methods Applied for Bioanalysis: Differential Pulse Voltammetry and Square Wave Voltammetry." In *Tools and Trends in Bioanalytical Chemistry*, 273–82. https://doi.org/10.1007/978-3-030-82381-8_14.

Srinivasan, Ramanathan, and Fathima Fasmin. 2021. *An Introduction to Electrochemical Impedance Spectroscopy.* CRC Press.

Sumitha, M. S., and T. S. Xavier. 2023. "Recent Advances in Electrochemical Biosensors – A Brief Review." *Hybrid Advances* 2 (January): 100023. https://doi.org/10.1016/j.hybadv.2023.100023.

Sun, Chongde, Chao Zhao, Esra Capanoglu Guven, Paolo Paoli, Jesus Simal-Gandara, Kunka Mohanram Ramkumar, Shengpeng Wang, et al. 2020. "Dietary Polyphenols as Antidiabetic Agents: Advances and Opportunities." *Food Frontiers* 1 (1): 18–44. https://doi.org/10.1002/fft2.15.

Suni, Ian I. 2008. "Impedance Methods for Electrochemical Sensors Using Nanomaterials." *TrAC - Trends in Analytical Chemistry* 27 (7): 604–11. https://doi.org/10.1016/j.trac.2008.03.012.

Szydłowska, Beata M., Zizhen Cai, and Mark C. Hersam. 2023. "Printed Nanomaterial Sensor Platforms for COVID-19 and Future Pandemics." *Current Opinion in Solid State and Materials Science* 27 (6): 101121. https://doi.org/10.1016/j.cossms.2023.101121.

Tajik, Somayeh, Parisa Shams, Hadi Beitollahi, and Fariba Garkani Nejad. 2024. "Electrochemical Nanosensor for the Simultaneous Determination of Anticancer Drugs Epirubicin and Topotecan Using UiO-66-NH2/GO Nanocomposite Modified Electrode." *Biosensors* 14 (5): 229. https://doi.org/10.3390/bios14050229.

Teymourian, Hazhir, Abbas Barfidokht, and Joseph Wang. 2020. "Electrochemical Glucose Sensors in Diabetes Management: An Updated Review (2010–2020)." *Chemical Society Reviews* 49 (21): 7671–709. https://doi.org/10.1039/D0CS00304B.

Tiwari, Jitendra N., Rajanish N. Tiwari, and Kwang S. Kim. 2012. "Zero-Dimensional, One-Dimensional, Two-Dimensional and Three-Dimensional Nanostructured Materials for Advanced Electrochemical Energy Devices." *Progress in Materials Science* 57 (4): 724–803. https://doi.org/10.1016/j.pmatsci.2011.08.003.

Urçuk, Ayça, Ceren Yıldız, Dilek Eskiköy Bayraktepe, and Zehra Yazan. 2023. "Electrochemical Sensor for Simultaneous Determination of Paracetamol, Propyphenazone, and Caffeine – Electrodes Based on Poly(L-Cystine)/Gold Nanoparticles Plating on Pencil Graphite." *Microchemical Journal* 193 (July): 1–8. https://doi.org/10.1016/j.microc.2023.109079.

Vadukumpully, Sajini, Jhinuk Gupta, Yongping Zhang, Guo Qin Xu, and Suresh Valiyaveettil. 2011. "Functionalization of Surfactant Wrapped Graphene Nanosheets with Alkylazides for Enhanced Dispersibility." *Nanoscale* 3 (1): 303–8. https://doi.org/10.1039/c0nr00547a.

Vilian, A. T.Ezhil, Ji Yoon Song, Yun Sung Lee, Seung Kyu Hwang, Hae Jin Kim, Young Si Jun, Yun Suk Huh, and Young Kyu Han. 2018. "Salt-Templated Three-Dimensional Porous Carbon for Electrochemical Determination of Gallic Acid." *Biosensors and Bioelectronics* 117 (June): 597–604. https://doi.org/10.1016/j.bios.2018.06.064.

Wan, Tianqing, Bangjie Shao, Sijie Ma, Yue Zhou, Qiao Li, and Yang Chai. 2023. "In-Sensor Computing: Materials, Devices, and Integration Technologies." *Advanced Materials* 35 (37): 1–14. https://doi.org/10.1002/adma.202203830.

Wang, Guozhi, Yangbo Ma, and Juan Wang. 2023. "Metal Functionalization of Two-Dimensional Nanomaterials for Electrochemical Carbon Dioxide Reduction." *Nanoscale* 15 (14): 6456–75. https://doi.org/10.1039/d3nr00484h.

Wang, Hsiang-Wei, Cameron Bringans, Anthony J. R. Hickey, John A. Windsor, Paul A. Kilmartin, and Anthony R. J. Phillips. 2021. "Cyclic Voltammetry in Biological Samples: A Systematic Review of Methods and Techniques Applicable to Clinical Settings." *Signals* 2 (1): 138–58. https://doi.org/10.3390/signals2010012.

Wang, Joseph. 2005. "Carbon-Nanotube Based Electrochemical Biosensors: A Review." *Electroanalysis* 17 (1): 7–14. https://doi.org/10.1002/elan.200403113.

Wang, Xianwen, Xiaoyan Zhong, Jianxiang Li, Zhuang Liu, and Liang Cheng. 2021. "Inorganic Nanomaterials with Rapid Clearance for Biomedical Applications." *Chemical Society Reviews* 50 (15): 8669–742. https://doi.org/10.1039/D0CS00461H.

Wang, You, Hui Xu, Jianming Zhang, and Guang Li. 2008. "Electrochemical Sensors for Clinic Analysis." *Sensors* 8 (4): 2043–81. https://doi.org/10.3390/s8042043.

Wang, Zhengdi, Tingting Hu, Ruizheng Liang, and Min Wei. 2020. "Application of Zero-Dimensional Nanomaterials in Biosensing." *Frontiers in Chemistry* 8 (320): 1–19. https://doi.org/10.3389/fchem.2020.00320.

Wu, Shixin, Qiyuan He, Chaoliang Tan, Yadong Wang, and Hua Zhang. 2013. "Graphene-Based Electrochemical Sensors." *Small* 9 (8): 1160–72. https://doi.org/10.1002/smll.201202896.

Xie, Feng, Meng Yang, Min Jiang, Xing Jiu Huang, Wen Qing Liu, and Pin Hua Xie. 2019. "Carbon-Based Nanomaterials – A Promising Electrochemical Sensor toward Persistent Toxic Substance." *TrAC - Trends in Analytical Chemistry* 119: 115624. https://doi.org/10.1016/j.trac.2019.115624.

Yang, Cheng, Madelaine E. Denno, Poojan Pyakurel, and B. Jill Venton. 2015. "Recent Trends in Carbon Nanomaterial-Based Electrochemical Sensors for Biomolecules: A Review." *Analytica Chimica Acta* 887: 17–37. https://doi.org/10.1016/j.aca.2015.05.049.

Yarman, Aysu, Sevinc Kurbanoglu, and F. W. Frieder W. Scheller. 2020. "Noninvasive Biosensors for Diagnostic Biomarkers." In *Commercial Biosensors and Their Applications: Clinical, Food, and Beyond*, 167. Elsevier. https://doi.org/10.1016/B978-0-12-818592-6.00007-4.

Yonzon, Chanda Ranjit, Douglas A. Stuart, Xiaoyu Zhang, Adam D. McFarland, Christy L. Haynes, and Richard P. Van Duyne. 2005. "Towards Advanced Chemical and Biological Nanosensors - An Overview." *Talanta* 67 (3): 438–48. https://doi.org/10.1016/j.talanta.2005.06.039.

Yun, Qinbai, Yiyao Ge, Zhenyu Shi, Jiawei Liu, Xixi Wang, An Zhang, Biao Huang, et al. 2023. "Recent Progress on Phase Engineering of Nanomaterials." *Chemical Reviews* 123 (23): 13489–692. https://doi.org/10.1021/acs.chemrev.3c00459.

Zdrachek, Elena, and Eric Bakker. 2019. "Potentiometric Sensing." *Analytical Chemistry* 91 (1): 2–26. https://doi.org/10.1021/ACS.ANALCHEM.8B04681.

Zhang, Bo Tao, Xiaoxia Zheng, Hai Fang Li, and Jin Ming Lin. 2013. "Application of Carbon-Based Nanomaterials in Sample Preparation: A Review." *Analytica Chimica Acta*. Elsevier. https://doi.org/10.1016/j.aca.2013.03.054.

Zhang, Jian, Lei Liu, Yan Yang, Qingwu Huang, Delong Li, and Dawen Zeng. 2021. "A Review on Two-Dimensional Materials for Chemiresistive- and FET-Type Gas Sensors." *Physical Chemistry Chemical Physics* 23 (29): 15420–39. https://doi.org/10.1039/D1CP01890F.

Zhao, Hongyuan, Meimei Guo, Fang Li, Yu Zhou, Gan Zhu, Yunhang Liu, Qiwen Ran, Fuquan Nie, and Volodymyr Dubovyk. 2023. "Fabrication of Gallic Acid Electrochemical Sensor Based on Interconnected Super-P Carbon Black@mesoporous Silica Nanocomposite Modified Glassy Carbon Electrode." *Journal of Materials Research and Technology* 24: 2100–12. https://doi.org/10.1016/j.jmrt.2023.03.129.

Zheng, Long, Mengzhu Cao, Yan Du, Quanyi Liu, Mohammed Y. Emran, Ahmed Kotb, Mimi Sun, Chong Bo Ma, and Ming Zhou. 2023. "Artificial Enzyme Innovations in Electrochemical Devices: Advancing Wearable and Portable Sensing Technologies." *Nanoscale* 16 (1): 44–60. https://doi.org/10.1039/d3nr05728c.

Zhou, Hong Cai Joe, and Susumu Kitagawa. 2014. "Metal–Organic Frameworks (MOFs)." *Chemical Society Reviews* 43 (16): 5415–18. https://doi.org/10.1039/C4CS90059F.

Zhu, Chengzhou, Guohai Yang, He Li, Dan Du, and Yuehe Lin. 2015. "Electrochemical Sensors and Biosensors Based on Nanomaterials and Nanostructures." *Analytical Chemistry* 87 (1): 230–49. https://doi.org/10.1021/ac5039863.

Zhu, Gangbing, Xiaohua Zhang, Pengbo Gai, and Jinhua Chen. 2012. "Enhanced Electrochemical Sensing for Persistent Organic Pollutants by Nanohybrids of Graphene Nanosheets That Are Noncovalently Functionalized with Cyclodextrin." *ChemPlusChem* 77 (9): 844–49. https://doi.org/10.1002/cplu.201200144.

CHAPTER 7

Applications of Nano Electrochemical Sensors to Determine Vitamins and Hormones

Simi M, Lakshmi Krishnaa Suresh, Lekshmi B R, Moumita Gangopadhyay, Rejithamol R, and Santhy A

7.1 INTRODUCTION

Electroanalytical techniques are one among the oldest analytical techniques used for the determination of analytes in various areas associated with our day-to-day life, including clinical analysis, forensics, food safety, environmental monitoring, etc. [1–5]. Electrochemical sensors are a class of electrochemical device that works using the principle of electroanalytical techniques, mainly voltammetric techniques. Electrochemical sensors are good analytical tools for practical applications, as they offer excellent stability, reproducibility, selectivity, etc. These sensors are more reliable, and they can detect even picomolar level concentrations of analytes with simple instrumentation compared with other analytical techniques [6–8]. A wide range of electroactive molecules, including biomolecules, can be quantitatively analysed using the electrochemical sensing method.

DOI: 10.1201/9781003512639-7

Vitamins are among the most significant biomolecules that are needed to sustain a healthy body. Vitamins help to maintain proper metabolism and are essential for boosting the immune system in our body [9, 10]. Lower and higher levels of vitamins in the human body lead to various health issues [11]. Hence, the normal levels of vitamins should be present in our body fluids to avoid serious health disorders. Hormones are yet another class of complex biomolecules that play a crucial role in the stress response in the human body [12]. Hormones are usually secreted by the cells in living organisms that are responsible for signal processing and controlling various metabolic and physiological processes [13, 14]. The presence of hormones in elevated concentrations may cause several abnormalities in the body's functioning [15]. Hence, the analytical determination of hormones from biological samples is extremely important.

Numerous electrochemical sensors have been developed for the analytical detection of vitamins and hormones [16–19]. The experimental setup for electrochemical sensors consists of a three-electrode system connected with a potentiostat. The electrodes involved are the working electrode, the reference electrode, and the counter electrode. The nature of the working electrode plays a very important role in the performance of the sensor, as the electrochemical oxidation or reduction of electrodes takes place on the surface of the working electrodes. Hence, to improve the sensor's characteristics, the surface of the working electrodes should be chemically modified with a variety of materials, including polymers, ionic liquids, nanomaterials, etc. Among the electrode-modifying materials, nanomaterials as well as different types of nanocomposites have been widely used in the development of electrochemical sensors [20].

The continued development in electrochemical sensors is heavily reliant on nanocomposites, which are hybrid structures that are produced by combining nanomaterials with other types of materials. The useful properties of polymers, metals, or ceramics are combined with the exceptional features of nanoparticles in these hybrid materials. These nanoparticles have a large surface area, exceptional electrical conductivity, and chemical reactivity [21]. The combination of these two factors results in excellent sensor performance, which enables applications such as electrochemical sensing to benefit from high levels of sensitivity, selectivity, and stability.

The significance of nanocomposites in improving the active surface area accessible for analyte interaction is one of the main advantages of the construction of electrochemical sensors. Greater electron transfer rates are made possible by this higher surface area and help the sensors to become more sensitive and responsive. For the detection of low-concentration analytes, including vitamins and hormones, carbon nanotubes (CNTs) coupled with metal nanoparticles, including gold or platinum, produce nanocomposites that show improved electrocatalytic activity [22]. The tuneable surface chemistry of nanocomposites is another major benefit. By functionalizing the nanoparticles inside the composite to target particular analytes, the selectivity of the sensors can be increased. For instance, graphene oxide (GO) functionalized with specific biomolecules can be used to precisely and consistently identify hormone levels in intricate biological matrices [23].

The incorporation of nanocomposites into electrochemical sensors enhances the stability and durability of the sensors. Nanocomposites can maintain consistent sensor performance over extended periods through the addition of polymers or other stabilizing chemicals, which enable them to resist degradation under adverse environmental conditions [24]. This aspect is especially crucial for practical applications in which sensors frequently encounter fluctuating temperatures and pH levels, and other demanding circumstances. In addition, nanocomposites enable the advancement of flexible and wearable electrochemical sensors. Researchers have developed sensors that can be worn on the skin to continuously monitor health-related biomarkers, such as vitamins and hormones. These sensors are made by combining nanomaterials with flexible substrates, such as conductive polymers or fabrics, to ensure comfort [22–24]. Nanocomposites play a crucial role in creating enhanced electrochemical sensors because they can improve sensitivity, selectivity, stability, and adaptability. Nanocomposites possess synergistic qualities that offer significant potential for the advancement of next-generation sensors, which might have a wide range of applications in medical diagnostics, environmental monitoring, and industrial process control.

7.2 ROLE OF NANOMATERIALS IN ELECTROCHEMICAL SENSORS

Nanomaterials have completely transformed the field of electrochemical sensors by providing exceptional levels of sensitivity, selectivity, and adaptability. Nanomaterials possess distinct characteristics, including a large surface area, exceptional electrical conductivity, and adjustable chemical activity, which render them well suited for improving the efficiency of electrochemical sensors [25]. A plethora of nanomaterials, including nanoparticles, nanosheets, nanorods, polymer nanocomposites, etc., are exploited for the modification of electrodes in electrochemical sensors. The nanomaterials are mainly classified into zero-dimensional, one-dimensional, two-dimensional, and three-dimensional based on the size range of the particles [21, 26]. Zero-dimensional (0D) nanoparticles are particles with sizes in the nanoscale range in all three dimensions. Examples of zero-dimensional nanoparticles include nanoparticles, quantum dots (QDs), etc. However, one-dimensional (1D) nanomaterials possess nanoscale ranges only in two dimensions, and the size is beyond the nano range in the other direction. Some examples of one-dimensional nanomaterials are nanorods, nanofibers, nanotubes, etc. [27–31]. Nanomaterials with the nanoscale size only in one dimension are considered two-dimensional (2D) nanomaterials; examples are nanofilms, nanosheets, etc. Materials that are not limited to the nanoscale in any dimension or range of dimensions are known as three-dimensional (3D) nanomaterials or bulk materials. They consist of dispersed nanoparticles, nanowire and nanotube bundles, and multi-nano layers with interfaces formed by close interactions between 0D, 1D, and 2D structural elements [26].

7.2.1 Zero-Dimensional Nanomaterials

Nanoparticles, specifically metallic ones such as gold and silver, have been extensively utilized in electrochemical sensors because of their expansive surface area and exceptional conductivity [27]. These characteristics enable efficient electron transmission and enhance the sensitivity of sensors. Gold nanoparticles have been used in glucose sensors, demonstrating improved sensitivity and stability.

Quantum dots are nanoscale semiconductor particles that exhibit optical and electrical properties that vary depending on their size. Due to their high surface-to-volume ratio and adjustable electrical characteristics, they are well suited for use in electrochemical sensors [28]. QDs have been employed to detect heavy metals and biological molecules due to their exceptional sensitivity and selectivity. An example of this is the incorporation of cadmium sulfide QDs into sensors designed to detect mercury ions (Hg^{2+}) with a high level of sensitivity [29, 30].

7.2.2 One-Dimensional Nanomaterials

Nanorods, usually made of metals or metal oxides, possess anisotropic characteristics that are advantageous for sensing applications. Their elongated form offers an increased surface area and a greater number of active sites for interaction with the analyte [31]. Zinc oxide nanorods have been employed for detecting glucose and other biomolecules, exhibiting enhanced response times and detection limits [32]. Nanofibres, often derived from polymers or composites, possess a large surface area and porous structure, which augment the interaction between the sensor surface and analytes. Conductive nanofibres, namely, those composed of polyaniline, have demonstrated significant promise in gas sensors because of their exceptional sensitivity and quick reaction to gaseous analytes, such as ammonia [33]. Nanotubes are widely explored one-dimensional nanomaterials used in the development of electrochemical sensors. Due to their remarkable electrical conductivity, mechanical strength, and large surface area, they are well suited for a range of sensing applications. Among them, the CNT sensors have been employed to detect many substances, like glucose, dopamine, and environmental contaminants, with exceptional sensitivity and specificity [34].

7.2.3 Two-Dimensional Nanomaterials

Nanofilms, which are often made out of metals, metal oxides, or polymers, provide a surface that is seamless and consistent along with the ability to be tailored for more accurate sensing. To detect DNA and proteins, ultra-thin gold nanofilms have been utilized in biosensors. These nanofilms offer high sensitivity and small detection thresholds during the detection process. The electrical, thermal, and mechanical properties of nanosheets are extraordinary [35]. Examples of nanosheets include graphene and transition metal dichalcogenides (TMDs). In addition to having a significant surface area, graphene nanosheets have outstanding conductivity, since they are composed of carbon atoms and have a single-layer structure [36]. As a consequence of this, they are very well suited for use as electrochemical sensors. Because of their extraordinary sensitivity and their ability to transmit electrons in a short amount of time, they have been utilized in the process of identifying a variety of compounds, including glucose, heavy metals, and organic pollutants.

The utilization of 1D and 2D nanomaterials in electrochemical sensors is widespread in many domains, such as medical diagnostics, environmental monitoring, and food safety [37]. For example, sensors made from CNTs have been employed to continuously monitor glucose levels in individuals with diabetes, and sensors made from graphene have been created to detect heavy metals in water, thereby providing environmental protection. 1D and 2D nanomaterials are essential for the progress of electrochemical sensors [38]. Further exploration and advancement in this field are expected to result in the development of sensors that are exceptionally sensitive, versatile, and adaptable in medicinal, environmental, and various other domains.

7.2.4 Three-Dimensional Nanomaterials

Advancements in nanomaterials have led to a new era of possibilities in the realm of science and technology. Among these breakthroughs, 3D nanomaterials have emerged as exceptional due to their superior properties when compared with their 1D and 2D counterparts [39]. These materials possess unique characteristics that make them highly suitable for a wide range of advanced applications. They offer a high surface area, strong mechanical properties, and effective electron transport pathways. 3D nanomaterials possess intricate architectures contributing to their exceptional surface area and porosity [40]. These properties promote greater interaction with target molecules and improve catalytic and electrochemical reactions.

In addition, the enhanced 3D structure improves the mechanical stability of these materials, resulting in increased durability across a range of applications. Exploring the world of 3D printing opens up a realm of possibilities where custom-designed 3D nanostructures come into existence [41]. One of the fascinating applications of this technology is the creation of 3D-printed graphene aerogels, showcasing the incredible potential of 3D printing techniques. These aerogels have proven to be incredibly versatile, finding use in a wide range of applications, such as energy storage and environmental remediation [17]. Their remarkable combination of lightweight properties and impressive strength has garnered significant attention in recent years.

7.3 SIGNIFICANCE OF ELECTROCHEMICAL SENSORS TO DETERMINE VITAMINS AND HORMONES

Electrochemical sensors have become effective devices for detecting vitamins and hormones, providing numerous benefits compared with conventional analytical methods. These sensors possess notable attributes, such as increased sensitivity, quick response, and the capability to function in complicated biological matrices. Consequently, they are highly suitable for applications in clinical diagnostics, environmental monitoring, and food quality control. A schematic portrait of the significance of electrochemical sensors to determine vitamins and hormones is represented in Figure 7.1.

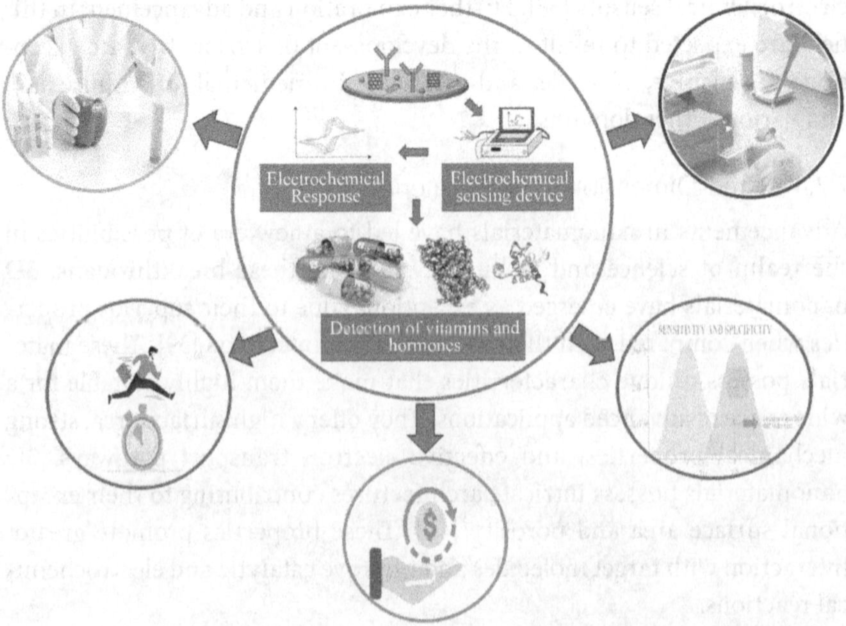

FIGURE 7.1 Schematic representation of the significance of electrochemical sensors to determine vitamins and hormones.

7.3.1 Clinical Diagnostics

Precise identification and measurement of vitamins and hormones are essential in healthcare facilities for the diagnosis and treatment of different health disorders. Vitamins such as B12, vitamin D, and folic acid are essential for metabolic functions, and not having enough of them can cause serious health problems [42]. Likewise, hormones such as insulin, cortisol, thyroid hormones, etc. play a crucial role in regulating physiological functioning. Electrochemical sensors facilitate point-of-care testing, enabling prompt and accurate monitoring of key biomarkers [43]. This instantaneous analysis enables prompt medical intervention, enhancing patient outcomes and minimizing healthcare expenses.

7.3.2 Sensitivity and Specificity

Electrochemical sensors demonstrate remarkable sensitivity and specificity, which are crucial for detecting minute amounts of vitamins and hormones [44]. Nanomaterials, including CNTs, graphene, and metal nanoparticles, are included in sensor designs to improve electron transport and signal amplification, resulting in enhanced detection limits [45]. Moreover, the process of modifying sensor surfaces with specialized bio-recognition elements, such as antibodies or aptamers, guarantees a high level of specificity, enabling the distinction between closely similar molecules.

7.3.3 Quick Response Analysis

An important benefit of electrochemical sensors is their capacity to offer prompt and instantaneous examination. Electrochemical sensors offer a faster alternative to traditional approaches, as they can provide data in a matter of minutes without the need for elaborate sample preparation or long processing durations [43]. The promptness of this response is especially beneficial in urgent medical scenarios, where speedy decision-making is crucial. In addition, the ability to monitor hormone levels in real time can offer valuable information about ongoing physiological changes, which can help diagnose and treat endocrine problems.

7.3.4 Portability and Cost-Effectiveness

When it comes to applications, portable electrochemical devices are suitable for various situations, including rural or resource-constrained areas, where access to advanced laboratory equipment is limited [46]. This level of accessibility ensures that essential diagnostic tests can be carried out in any location and at any time, which significantly broadens the range of services that are available in the medical field. Standard analytical procedures are often more expensive than electrochemical sensors [47]. There is a reduction in operational expenses as a result of the use of equipment that is less complicated, as well as the reduced necessity for chemicals and sample preparation. The cost-effectiveness of this technology is especially advantageous in programs that include intensive monitoring and regular screening.

7.3.5 Environmental and Food Quality Monitoring

To monitor both the environment and food quality, electrochemical sensors are necessary. When it comes to determining the nutritional worth and safety of dietary goods, vitamins and hormones are significant markers that play an important role. The detection of hormone residues in environmental samples, such as water sources, is of the utmost importance to assess the levels of pollution and the potential effects that it may have on ecosystems [48]. Overall, electrochemical sensors are not only rapid but also extremely responsive and economically efficient.

Electrochemical sensors offer several benefits that are particularly advantageous in the analysis of hormones and vitamins. As a result of their exceptional sensitivity, specificity, rapid reaction, mobility, and cost-effectiveness, these instruments are indispensable in the fields of clinical diagnostics, environmental monitoring, and food quality control [49]. It is expected that electrochemical sensors will gain improved adaptability and widespread application in environmental and medical fields as a result of continued breakthroughs in research and development and technological advancements.

7.4 ELECTROCHEMICAL SENSORS TO DETERMINE VITAMINS AND HORMONES

Precision and sensitivity are crucial when detecting the presence of certain substances that have a significant impact on various physiological systems. Vitamins and hormones have a vital role in the regulation of several physiological processes, such as metabolism, immunological response, and growth [50]. Inadequate levels can result in severe health complications, such as decreased immune function, metabolic abnormalities, and developmental impairments. It is crucial to diagnose these deficiencies in a timely way in order to avoid and manage illnesses such as scurvy, osteoporosis, or hormonal imbalances, which can have a negative influence on overall health [51]. Advancements in electrodes modified with nanostructured materials have led to significant progress in the electrochemical detection of vitamins and hormones [52]. These electrodes possess a high surface area, electrocatalytic activity, and exceptional sensitivity, making them ideal for detecting various analytes that have a significant impact on physiological functions. These electrodes can be created using various advanced materials, such as metal–organic frameworks (MOFs), zeolites, CNTs, metal nanoparticles, and 2D materials like TMDs, MXenes, graphene, conducting polymers, etc. [53]. These materials have the potential to revolutionize disease diagnosis and treatment, leading to better regulation of hormones and improved healthcare [54]. The focus of this chapter is to provide an overview of the recent developments in the electrochemical detection of hormones and vitamins using a variety of nanomaterials.

7.4.1 Metal and Metal Oxide–Based Sensors

Electrochemical sensors and biosensors that utilize metal nanoparticles are crucial components in diagnostic devices [55]. Metal nanoparticles (MNPs) possess exceptional conductivity and catalytic characteristics, rendering them ideal for serving as "electronic channels" to facilitate electron transfer between redox centres in target molecules and electrode surfaces [56]. Additionally, their nanoscale size and structure make them highly effective electrochemical catalysts. Metal oxide nanoparticles can function as semiconductor nanoparticles and are frequently employed for the immobilization of target molecules due to their biocompatibility [57]. Combinations of conducting polymers with MNPs are highly appropriate for achieving optimal sensitivity and stability. This is because the MNPs serve as redox mediators for biomolecules, while the polymer works as a selective adsorbate for biomolecules [58, 59]. Transition metals, including copper, zinc, nickel, silver, and cobalt, together with their oxides, have been synthesized and utilized in biosensors [60, 61].

This section of the chapter will give a comprehensive account of the application of metal and metal oxide–based nanoparticles in the electrochemical sensing of vitamins and hormones.

Recently, tin oxide nanoparticles coated on a glassy carbon electrode (GCE) were used for electrochemical detection of vitamin C and vitamin B6 by Baksh et al. [62]. The modified electrode exhibited excellent electrochemical performance for the oxidation of both analytes. It also exhibited enhanced sensitivity throughout a broad linear dynamic range of 10 to 140 μM, with a limit of detection of 0.067 μM for vitamin C and 0.048 μM for vitamin B6.

An electrochemical sensor utilizing a ZrO_2 and ZnO hybrid nanocomposites–modified GCE was fabricated by Wang et al. for simultaneous determination of epinephrine, uric acid, and folic acid [63]. The sensor exhibits exceptional selectivity and excellent sensitivity. The electrochemical studies showed that ZrO_2 and ZnO nanocomposite–modified electrodes exhibit a high degree of exposed catalytic sites, acceptable conductivity, and outstanding signal response. They also exhibit a low detection limit and a wide detection range. The real-time application of the sensor was conducted using a human blood serum sample, and the results showed a recovery rate of 97.3–103.8%. As a result, it is suggested that this technology will have widespread use in bioanalysis and disease diagnostics due to its simplicity and feasibility.

The electrochemical oxidation of vitamin D on bare electrodes is very difficult. To overcome this, Deepika and co-workers developed an effective electrochemical biosensor for the detection of vitamin D3 utilizing gadolinium oxide nanorods [64]. Gd nanorods functionalized by aspartic acid were used as the scaffold for electrochemical detection of vitamin D3.

Differential pulse voltammetry (DPV) tests demonstrated high sensitivity over a broad linear range and a low detection limit for vitamin D3. The sensor responded satisfactorily to a commercially available oral solution of Vitamin D3.

In other work, Jamil et al. modified a glassy carbon electrode with CuO nanostructures to develop a simple, sensitive, and reliable electrochemical sensor that enables the determination of ascorbic acid (AA) [65]. Electrochemical studies were done using DPV, and the sensor showed a wide linear range of 0.1–300 μM and a low detection limit of 0.01 μM. Moreover, the modified sensor was effectively used to determine AA in fruit juices and cosmetic samples. Furthermore, a mechanically alloyed material containing Y_2O_3 nanoparticles was synthesized and applied for the voltammetric detection of AA [66]. The synthesized composite was modified on a GCE to study the electrochemical oxidation of AA, and the sensor was able to detect AA at a high concentration range from 2 to 2.5 mM.

A novel electrode material, consisting of $NiCo_2O_4$, Zn, and Al layered double hydroxide nanocomposite, was developed by Roghayeh et al. for the precise and specific detection of pyridoxine (vitamin B6) [67]. The modified sensor demonstrated specific detection of pyridoxine even in the presence of other substances such as vitamins (B1, B2, B12, and ascorbic acid), inorganic ions, and other biomolecules. The fabricated sensor exhibited strong oxidation with pyridoxine across the concentration range of 0.2–200 μM, with an exceptionally low detection limit of 0.086 μM. Ultimately, the suggested approach was effectively utilized to determine the presence of pyridoxine in both commercial tablets and plasma samples, yielding satisfying outcomes.

MXenes are 2D nanomaterials that have high electrical conductivity, strong biocompatibility, and ease of functionalization, making them the materials of the future generation. Recently, Rajiv et al. fabricated a novel electrode of carbon fibre paper modified with manganese dioxide, inorganic phosphate, and MXene for the detection of vitamin B6 [68]. The synergistic effect of the materials used to modify the electrode enhanced the electrochemical performance and increased the surface area of the electrode. The fabricated electrochemical sensor demonstrated a broad linear dynamic range spanning from 0.06 to 650 μM, along with a low-level detection limit of 0.021 μM. The obtained results confirmed that the sensor exhibits improved selectivity, repeatability, and stability for the electrochemical detection of Vitamin B6 in real samples. The electrochemical

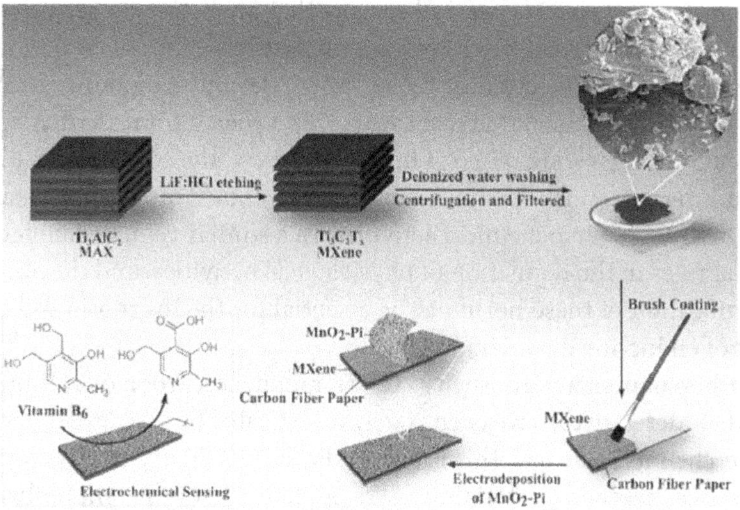

FIGURE 7.2 Graphical explanation of electrochemical detection of vitamin B6 using a nanocomposite of MXene and MnO_2/P. (Reprinted from *Materials Research Bulletin*, 169, Rajeev, R., Cherian, A. R., Thadathil, D. A., and Varghese, A., Electrochemical determination of Vitamin B_6 using coral-like MnO_2-Pi on $Ti_3C_2T_x$ MXene, 112523, Copyright (2024), with permission from Elsevier.)

detection of vitamin B6 using the modified electrode is portrayed in Figure 7.2.

A novel nanomaterial has been synthesized by Lokesh et al., consisting of manganese dioxide nanoparticles interconnected with iron and zinc metal ions through terephthalic acid for electrochemical application [69]. A GCE was coated with the synthesized nanomaterial, and the modified electrode was successful in recognition of vitamin D3 from other interfering species electrochemically and with a very low detection limit. The analytical utility of the sensor was confirmed in human plasma samples with a recovery rate ranging from 96% to 102%.

7.4.2 Carbon nanomaterial–based sensors

Electrochemical sensing is one of the many fields that have been greatly affected by the rapid growth of nanotechnology. Carbon-based nanomaterials have attracted a lot of attention among the many different types of nanomaterials because of their distinctive characteristics [70]. These characteristics include a high surface area, superior electrical conductivity, chemical stability, and the ease with which they may be functionalized. Because they possess these characteristics, they are excellent candidates for constructing electrochemical sensors that are both very sensitive and selective [71]. There has been growing interest in the use of carbon nanomaterials for the detection of biologically relevant compounds, such as vitamins and hormones, in recent years [72]. When it comes to diagnosing and monitoring a wide range of health disorders, the correct detection of vitamins is necessary. Vitamins are vital micronutrients that are required for a multitude of biochemical activities. In a similar vein, hormones play critical roles in the regulation of physiological activities, and the accurate quantification of these hormones is essential for the diagnosis and treatment of endocrine disorders [73].

Carbon nanomaterials, such as CNTs, graphene, carbon quantum dots, and their derivatives, have been widely investigated for the development of electrochemical sensors. This is due to the extraordinary electrochemical properties that these materials possess. A platform for the immobilization of bio-recognition components is provided by these nanomaterials, which also allow the quick transfer of electrons, increase the sensitivity of the sensors, and boost the sensors' overall performance. It is possible to further increase the selectivity and performance of the sensors through the functionalization of carbon nanomaterials with particular compounds [74].

Recent developments in carbon-based nanomaterial electrochemical sensors have greatly improved vitamin and hormone–detecting capabilities. Using the unique properties of graphene, CNTs, and carbon quantum dots, carbon nanomaterial sensors have shown amazing sensitivity and selectivity [75]. In particular, the great surface area, great conductivity, and biocompatibility of these nanomaterials have led to revelations in detecting hormones like cortisol and oestrogen as well as in detecting low concentrations of vital vitamins like B_{12} and D.

Extensive research has been conducted on carbon-based nanomaterials to sense a wide range of biomolecules. Recently, Hema Bora and co-workers modified a GCE using CNTs. In this work, the challenge posed by the hydrophobic nature of pure carbon was overcome by utilizing nitrogen-doped CNTs. The developed sensor had the remarkable ability to exhibit a greater selectivity towards vitamin D3 even in the presence of other interfering substances. The sensor also possessed long-term stability, reproducibility, a wide linear range, and exceptional ultralow detection capability [74]. In other work, Anusha et al. fabricated a GCE modified with fullerene-C60 and copper–nickel bimetallic nanoparticles to measure vitamin D levels in blood samples [76]. The nanocomposite film on GCE showed a

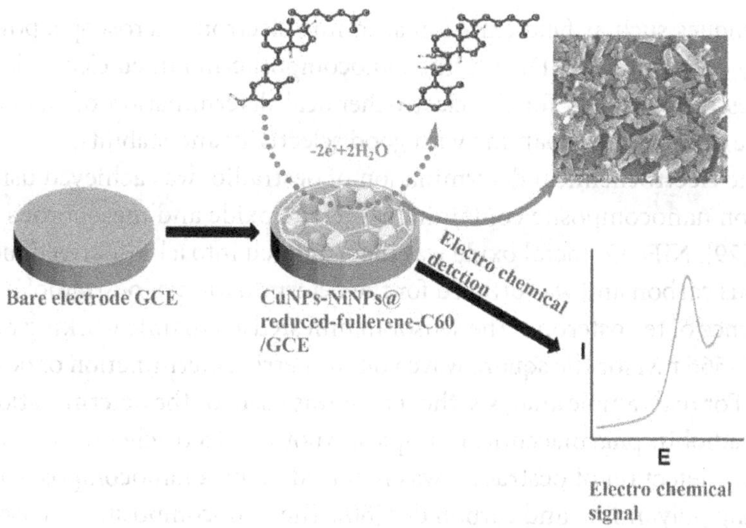

FIGURE 7.3 Schematic illustration of electrochemical sensing vitamin D using fullerene-C60 and copper-nickel bimetallic nanoparticles. (Reprinted from *Diamond and Related Materials*, 104, Anusha, T., Bhavani, K. S., Kumar, J. V. S., and Brahman, P. K., Designing and fabrication of electrochemical nanosensor employing fullerene-C60 and bimetallic nanoparticles composite film for the detection of vitamin D3 in blood samples, 107761, Copyright (2020), with permission from Elsevier.)

clear oxidation peak for vitamin D3, with a dynamic concentration range of 1.25–475 μM and a limit of detection of 0.0025 μM. A schematic representation of the electrochemical determination of vitamin D is shown in Figure 7.3.

The application of lanthanum nanoparticles (LaNPs) in electrochemical sensing was reported by Anusha and co-workers [77]. The nanocomposite was synthesized using LaNPs, graphene quantum dots, and a zeolitic imidazolate framework and used to modify a GCE for electrochemical detection of vitamin D3. The sensor showed an outstanding performance for vitamin D3 oxidation from 0.00625 to 1.25 μM, with a detection limit of 0.00610 μM. Due to its good selectivity, stability, and repeatability, the nanocomposite-modified sensor could be a valuable tool for clinical diagnostic purposes in the future.

A nanomaterial-based sensor for the determination of riboflavin was reported recently using a pencil graphite composite paste electrode modified using sodium lauryl sulfate and modified CNTs [78]. The modified electrode was thoroughly examined and analysed using advanced techniques such as field emission scanning electron microscope, powder X-ray diffraction (XRD), etc. The nanocomposite-modified electrode was successfully utilized for the electrochemical determination of riboflavin in the presence of dopamine with good selectivity and stability.

The electrochemical determination of oestradiol was achieved using a carbon nanocomposite containing Ni ferrite oxide and mesoporous carbon [79]. $NiFe_2O_4$ metal oxide was incorporated into a bio-derived mesoporous carbon and was utilized for the determination of oestradiol in the presence of testosterone. The sensor manifested a dynamic working range of 20–566 nM for the square wave voltammetric determination of oestradiol. For real sample analysis, the sensor was used for the determination of oestradiol in pharmaceutical samples. Another electrochemical method for the detection of oestradiol was reported using a nanocomposite containing polyaniline and carbon dot [80]. The nanocomposite was used to modify a GCE to study the electrochemical performance of oestradiol. The synergistic effect of all particles in the nanocomposite enabled the electrochemical detection of oestradiol with enhanced current response, and this analytical method was proved to be a good method for the detection of oestradiol in the presence of various interfering species.

7.4.3 Polymer Nanocomposite–Based Sensors

The unique properties of polymer nanocomposites (PNCs), such as their easy synthesis and variety of morphologies, make them superior transducers in electroanalytical sensors [81]. Low cost, improved electrical properties, tuneable chemical and electrochemical properties, fast electrode kinetics, biocompatibility, and environmental stability make PNCs one of the most versatile components in electrochemical sensors [82]. To improve sensor performance, nanocomposites have been explored in a variety of combinations, including conductive polymers containing graphene, CNTs, and metal nanoparticles, which provide high electrical conductivity, effective surface area, and a rapid electron transfer rate [83].

A novel electrochemical sensor was fabricated by Juang et al. for the detection of parathyroid hormone (PTH) [83]. The sensor was developed based on a nanocomposite of conducting polymer poly(3, 4-ethoxylenedioxythiophene), reduced graphene oxide (rGO), and 2D MoS_2 nanoparticles. The synergistic effect of the materials enables the sensor to detect PTH in the picomolar range with good selectivity. The high selectivity of the sensor was validated for PTH in the presence of interfering compounds such as p-cresol, urea, and creatinine.

Ghanbari et al. fabricated a three-dimensional porous nanocomposite of reduced graphene oxide embellished with polypyrrole nanofibres and zinc oxide–copper oxide p-n junction heterostructures on a GCE for the simultaneous voltammetric determination of ascorbic acid and dopamine [84]. The modified electrode showed a high level of selectivity and a lower detection limit. The real-time application of the sensor was done using commercially available human blood serum samples.

Another polymer nanocomposite material, polyaniline–polypyrrole doped with silver and cobalt, was utilized for the determination of vitamin D3 electrochemically in serum and urine samples [85]. The nanocomposite was used to modify both GCE and paper electrodes. The modified sensor showed a linear range in the oxidation of vitamin D3 from 0.0125 to 22.5 μM and from 0.025 to 0.125 μM for GCE and paper electrodes, respectively. The electrochemical sensing of vitamin D using polyaniline–polypyrrole doped with silver and cobalt nanoparticles is graphically illustrated in Figure 7.4.

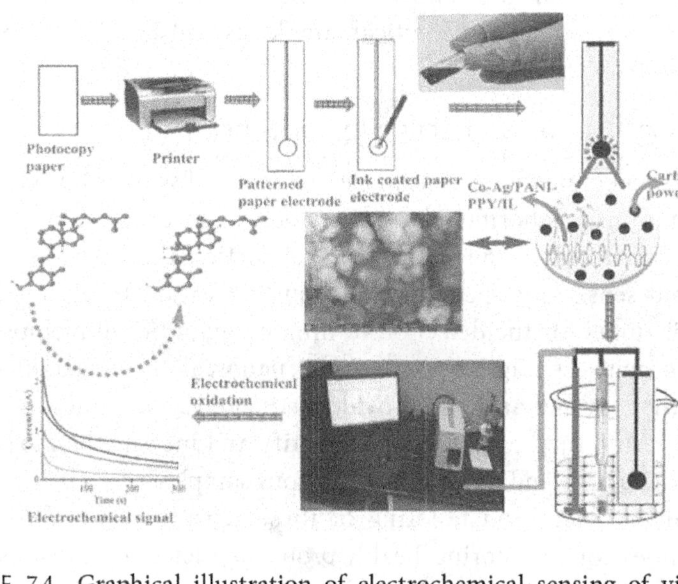

FIGURE 7.4 Graphical illustration of electrochemical sensing of vitamin D using polyaniline–polypyrrole doped with silver and cobalt nanoparticles[85]. (*Microchemical Journal*, 161, Anusha, T., Bhavani, K. Sai, Shanmukha Kumar, J. V., Bonanni, A., and Brahman, P. K., Fabrication of handmade paper sensor based on silver-cobalt doped copolymer-ionic liquid composite for monitoring of vitamin D3 level in real samples, 105789, Copyright (2021), with permission from Elsevier.)

In other work, a polymer nanocomposite of graphene oxide and poly(aniline-co-thionine) was employed for the determination of ascorbic acid [86]. DPV was used to assess the electrochemical performance of AA. A linear response of 0.5 to 5 mM was obtained for AA, and good selectivity and a low detection limit were also demonstrated.

The use of multi-walled CNTs and a PNC was reported by Filik et.a l. for the electrochemical detection of vitamin B12. A GCE was modified with poly(2,2′-(1,4-phenylenedivinylene) bis-8-hydroxyquinoline) and multi-walled CNTs [87]. As for vitamin B12, the electrode showed an excellent linear range of 0.1 to 10 μM and a low detection limit of 0.01 μM. The developed electrode was able to accurately and precisely measure vitamin B12 in pharmaceutical formulations. The use of another PNC was reported for the electrochemical sensing of vitamin B12 by Li et al. recently [88]. The nanocomposite was synthesized using gold, polypyrrole nanoparticles, and CNT, and the composite was characterized using scanning electron microscopy (SEM), XRD analysis, etc. The nanocomposite was then modified on a GCE for the successful determination of vitamin B12 with a detection limit of 0.9 nM. The analytical utility of the sensor was confirmed in pharmaceutical samples as well as in human plasma samples (Table 7.1).

7.5 CONCLUSION AND FUTURE PERSPECTIVES

The utilization of nanomaterials in electrochemical sensors for the detection of vitamins and hormones signifies noteworthy progress in the fields of analytical chemistry and biomedical diagnostics. These sensors possess exceptional sensitivity, specificity, and quick reaction times, rendering them well suited for the detection of minute quantities of biomolecules in diverse matrices. The sensors possess nanoscale characteristics that increase their surface area and provide distinct electrochemical capabilities. As a result, they can accurately identify and measure biomolecules, especially vitamins and hormones, in various samples.

The advancement and tailoring of these sensors have created new opportunities for monitoring health problems, detecting diseases, and evaluating nutritional status. The precise and accurate detection of vitamins and hormones has significance for personalized treatment since it allows more customized and timely interventions.

Despite these advances, reproducibility, stability, and large-scale manufacturing remain challenging. Further research should solve these difficulties, optimize sensor designs, and explore novel nanomaterials to

TABLE 7.1 Comparison of Some of the Reported Nanomaterial based Electrochemical Sensors to Determine Vitamins and Hormones

Nanomaterials used for electrode modification	Analyte	Electrochemical method	Linear range (μM)	Reference
SnO_2	Vitamin C, B6	Electrochemical impedance spectroscopy	10–140	[62]
ZrO_2/ZnO	Folic acid	DPV	2–480	[63]
Gd_2O_3	Vitamin D3	DPV	10–100	[64]
CuO	Ascorbic acid	DPV	0.1–300	[65]
$NiCo_2O_4$ / Zn and Al hydroxide	Vitamin B6	DPV	0.2–200	[67]
MnO_2, inorganic phosphate, and MXene	Vitamin B6	Electrochemical impedance spectroscopy	0.06–650	[68]
CNT	Vitamin D3	Chronoamperometry	0–0.600	[74]
Fullerene C60,Cu/Ni NP	Vitamin D	Square wave voltammetry	1.25–475	[76]
LaNPs, graphene quantum dots	Vitamin D	Chronoamperometry, square wave voltammetry	0.00625–1.25	[77]
Sodium lauryl sulfate, CNT	Riboflavin	DPV	0.2–0.8	[78]
Ni ferrite oxide and mesoporous carbon	Oestradiol	Square wave voltammetry	0.020–0.566	[79]
Polyaniline, carbon dot	Oestradiol	Linear sweep voltammetry	0.001–100	[80]
Poly(3,4-ethoxylenedioxythiophene), reduced graphene oxide, MoS_2	Parathyroid hormone	DPV	0.08–475	[89]
Polypyrrole, ZnO, CuO	Ascorbic acid	DPV	500–5000	[84]
Polyaniline, polypyrrole doped with Ag/Co	Vitamin D3	Chronoamperometry	0.0125–22.5	[85]
Poly(aniline-co-thionine), graphene oxide	Ascorbic acid	DPV	500–5000	[86]
Poly(2,2′-(1,4-phenylene vinylene) bis-8-hydroxyquinoline) and MWCNT	Vitamin B12	Square wave voltammetry	0.1–10	[87]
Gold, polypyrrole nanoparticles, CNT	Vitamin B12	DPV	0–85	[88]

maximize their potential. This field's ongoing progress will transform analytical methods and enable next-generation sensing devices, improving health and disease monitoring. In conclusion, nanomaterial-based electrochemical sensors are a frontier in analytical science, with their new applications and enhancements promising future technological advances.

REFERENCES

1. K. Lubert and K. Kalcher, "History of electroanalytical methods," *Electroanalysis*, vol. 22, no. 17–18, pp. 1937–1946, Sep. 2010, doi: 10.1002/elan.201000087.
2. M. Mazloum-Ardakani, A. Khoshroo, and L. Hosseinzadeh, "Simultaneous determination of hydrazine and hydroxylamine based on fullerene-functionalized carbon nanotubes/ionic liquid nanocomposite," *Sensors Actuators B Chem.*, vol. 214, pp. 132–137, Jul. 2015, doi: 10.1016/j.snb.2015.03.010.
3. R. G. Krishnan, R. Rejithamol, and B. Saraswathyamma, "Non-enzymatic electrochemical sensor for the simultaneous determination of adenosine, adenine and uric acid in whole blood and urine," *Microchem. J.*, vol. 155, p. 104745, Jun. 2020, doi: 10.1016/j.microc.2020.104745.
4. R. G. Krishnan and B. Saraswathyamma, "Disposable electrochemical sensor for coumarin induced milk toxicity in raw milk samples," *Measurement*, vol. 170, p. 108709, Jan. 2021, doi: 10.1016/j.measurement.2020.108709.
5. L. Shaw and L. Dennany, "Applications of electrochemical sensors: Forensic drug analysis," *Curr. Opin. Electrochem.*, vol. 3, no. 1, pp. 23–28, Jun. 2017, doi: 10.1016/j.coelec.2017.05.001.
6. S. Antherjanam, B. Saraswathyamma, and S. Murugesan Senthil Kumar, "Simultaneous electrochemical determination of the tumor biomarkers homovanillic acid and vanillylmandelic acid using a modified pencil graphite electrode," *Microchem. J.*, vol. 190, p. 108659, Jul. 2023, doi: 10.1016/j.microc.2023.108659.
7. A. Santhy, R. Rejithamol, R. Chandana, P. Sivanand, and S. Beena, "Electrochemical sensor applications of mechanically alloyed materials," 2024, pp. 33–53. doi: 10.4018/978-1-6684-9385-4.ch002.
8. J. Thomas, P. K. Anitha, T. Thomas, and N. Thomas, "BaZrO3 based non-enzymatic single component single step ceramic electrochemical sensor for the picomolar detection of dopamine," *Ceram. Int.*, vol. 48, no. 5, pp. 7168–7182, Mar. 2022, doi: 10.1016/j.ceramint.2021.11.278.
9. A. Carr and S. Maggini, "Vitamin C and immune function," *Nutrients*, vol. 9, no. 11, p. 1211, Nov. 2017, doi: 10.3390/nu9111211.
10. R. M. Khadim and F. S. Al-Fartusie, "Antioxidant vitamins and their effect on the immune system," *J. Phys. Conf. Ser.*, vol. 1853, no. 1, p. 012065, Mar. 2021, doi: 10.1088/1742-6596/1853/1/012065.
11. S. Antherjanam, B. Saraswathyamma, R. G. Krishnan, and G. M. Gopakumar, "Electrochemical sensors as a versatile tool for the quantitative analysis of Vitamin B12," *Chem. Pap.*, vol. 75, no. 7, pp. 2981–2995, Jul. 2021, doi: 10.1007/s11696-021-01574-2.

12. Disha, M. K. Nayak, P. Kumari, M. K. Patel, and P. Kumar, "Functional nanomaterials based opto-electrochemical sensors for the detection of gonadal steroid hormones," *TrAC Trends Anal. Chem.*, vol. 150, p. 116571, May 2022, doi: 10.1016/j.trac.2022.116571.
13. J. B. Raval, V. N. Mehta, S. Jha, R. K. Singhal, H. Basu, and S. K. Kailasa, "Functional nanostructures in analytical chemistry: new insights into the optical and electrochemical sensing of animal hormones in food, environmental and biological samples," *Sensors & Diagnostics*, vol. 2, no. 4, pp. 815–836, 2023, doi: 10.1039/D3SD00071K.
14. D. G. Hardie, "The First Messengers — Hormones, Neurotransmitters and Local Mediators," in *Biochemical Messengers*, Dordrecht: Springer Netherlands, 1991, pp. 21–43. doi: 10.1007/978-94-011-3108-7_2.
15. E. Möstl and R. Palme, "Hormones as indicators of stress," *Domest. Anim. Endocrinol.*, vol. 23, no. 1–2, pp. 67–74, Jul. 2002, doi: 10.1016/S0739-7240(02)00146-7.
16. T. A. Saleh, "Nanomaterials: Classification, properties, and environmental toxicities," *Environ. Technol. Innov.*, vol. 20, p. 101067, Nov. 2020, doi: 10.1016/j.eti.2020.101067.
17. S. Pareek, U. Jain, S. Balayan, and N. Chauhan, "Ultra-sensitive nano-molecular imprinting polymer-based electrochemical sensor for Follicle-Stimulating Hormone (FSH) detection," *Biochem. Eng. J.*, vol. 180, p. 108329, Mar. 2022, doi: 10.1016/j.bej.2021.108329.
18. S. Antherjanam and B. Saraswathyamma, "Electrochemical preparation and the characterizations of poly(3,5-diamino 1,2,4-triazole) film for the selective determination of pyridoxine in pharmaceutical formulations," *Chem. Pap.*, vol. 77, no. 8, pp. 4285–4296, Aug. 2023, doi: 10.1007/s11696-023-02777-5.
19. B. Koçak, "A selective and sensitive electrochemical sensor for pyridoxine hydrochloride (Vitamin B6) in pharmaceutic tablet," *ChemistrySelect*, vol. 8, no. 39, Oct. 2023, doi: 10.1002/slct.202303112.
20. S. Ren, J. Zeng, Z. Zheng, and H. Shi, "Perspective and application of modified electrode material technology in electrochemical voltammetric sensors for analysis and detection of illicit drugs," *Sensors Actuators A Phys.*, vol. 329, p. 112821, Oct. 2021, doi: 10.1016/j.sna.2021.112821.
21. L. V. Jodar, F. A. Santos, V. Zucolotto, and B. C. Janegitz, "Electrochemical sensor for estriol hormone detection in biological and environmental samples," *J. Solid State Electrochem.*, vol. 22, no. 5, pp. 1431–1438, May 2018, doi: 10.1007/s10008-017-3726-9.
22. S. E. Elugoke *et al.*, "Progress in electrochemical detection of neurotransmitters using carbon nanotubes/nanocomposite based materials: A chronological review," *Nano Sel.*, vol. 1, no. 6, pp. 561–611, Dec. 2020, doi: 10.1002/nano.202000082.
23. V. Georgakilas *et al.*, "Noncovalent functionalization of graphene and graphene oxide for energy materials, biosensing, catalytic, and biomedical applications," *Chem. Rev.*, vol. 116, no. 9, pp. 5464–5519, May 2016, doi: 10.1021/acs.chemrev.5b00620.

24. K. S. S. Devi, J. Prakash, and S. Tsujimura, "Graphene oxide-based nanomaterials for electrochemical bio/immune sensing and its advancements in health care applications: A review," *Hybrid Adv.*, vol. 5, p. 100123, Apr. 2024, doi: 10.1016/j.hybadv.2023.100123.
25. Z. Wang, T. Hu, R. Liang, and M. Wei, "Application of zero-dimensional nanomaterials in biosensing," *Front. Chem.*, vol. 8, Apr. 2020, doi: 10.3389/fchem.2020.00320.
26. B. Mekuye and B. Abera, "Nanomaterials: An overview of synthesis, classification, characterization, and applications," *Nano Sel.*, vol. 4, no. 8, pp. 486–501, Aug. 2023, doi: 10.1002/nano.202300038.
27. C. Chen, Y. Fan, J. Gu, L. Wu, S. Passerini, and L. Mai, "One-dimensional nanomaterials for energy storage," *J. Phys. D. Appl. Phys.*, vol. 51, no. 11, p. 113002, Mar. 2018, doi: 10.1088/1361-6463/aaa98d.
28. D. Huang, C. Niu, M. Ruan, X. Wang, G. Zeng, and C. Deng, "Highly sensitive strategy for Hg 2+ detection in environmental water samples using long lifetime fluorescence quantum dots and gold nanoparticles," *Environ. Sci. Technol.*, vol. 47, no. 9, pp. 4392–4398, May 2013, doi: 10.1021/es302967n.
29. P. V. Govardhana Reddy *et al.*, "A review on multicomponent reactions catalysed by zero-dimensional/one-dimensional titanium dioxide (TiO2) nanomaterials: Promising green methodologies in organic chemistry," *J. Environ. Manage.*, vol. 279, p. 111603, Feb. 2021, doi: 10.1016/j.jenvman.2020.111603.
30. J. Chen *et al.*, "Functionalized CdS quantum dots-based luminescence probe for detection of heavy and transition metal ions in aqueous solution," *Spectrochim. Acta Part A Mol. Biomol. Spectrosc.*, vol. 69, no. 3, pp. 1044–1052, Mar. 2008, doi: 10.1016/j.saa.2007.06.021.
31. Z. Wu and K. H. Khayat, "Design, mechanism, and performance of cement-based materials with zero-dimensional nanomaterials," in *Nanotechnology for Civil Infrastructure*, Elsevier, 2023, pp. 69–92. doi: 10.1016/B978-0-12-817832-4.00009-2.
32. M. Morais *et al.*, "Visible photoluminescent zinc oxide nanorods for label-free nonenzymatic glucose detection," *ACS Appl. Nano Mater.*, vol. 5, no. 3, pp. 4386–4396, Mar. 2022, doi: 10.1021/acsanm.2c00474.
33. T. Rasheed, A. A. Hassan, F. Kausar, F. Sher, M. Bilal, and H. M. N. Iqbal, "Carbon nanotubes assisted analytical detection – Sensing/delivery cues for environmental and biomedical monitoring," *TrAC Trends Anal. Chem.*, vol. 132, p. 116066, Nov. 2020, doi: 10.1016/j.trac.2020.116066.
34. M. Šetka, J. Drbohlavová, and J. Hubálek, "Nanostructured polypyrrole-based ammonia and volatile organic compound sensors," *Sensors*, vol. 17, no. 3, p. 562, Mar. 2017, doi: 10.3390/s17030562.
35. A. Murali, G. Lokhande, K. A. Deo, A. Brokesh, and A. K. Gaharwar, "Emerging 2D nanomaterials for biomedical applications," *Mater. Today*, vol. 50, pp. 276–302, Nov. 2021, doi: 10.1016/j.mattod.2021.04.020.
36. H. Wang, H. Feng, and J. Li, "Graphene and graphene-like layered transition metal dichalcogenides in energy conversion and storage," *Small*, vol. 10, no. 11, pp. 2165–2181, Jun. 2014, doi: 10.1002/smll.201303711.

37. S. S. Fiyadh *et al.*, "Review on heavy metal adsorption processes by carbon nanotubes," *J. Clean. Prod.*, vol. 230, pp. 783–793, Sep. 2019, doi: 10.1016/j.jclepro.2019.05.154.
38. H. Zhang, H.-M. Cheng, and P. Ye, "2D nanomaterials: beyond graphene and transition metal dichalcogenides," *Chem. Soc. Rev.*, vol. 47, no. 16, pp. 6009–6012, 2018, doi: 10.1039/C8CS90084A.
39. X. Zhou, B. Liu, Y. Chen, L. Guo, and G. Wei, "Carbon nanofiber-based three-dimensional nanomaterials for energy and environmental applications," *Mater. Adv.*, vol. 1, no. 7, pp. 2163–2181, 2020, doi: 10.1039/D0MA00492H.
40. D. Fattakhova-Rohlfing, A. Zaleska, and T. Bein, "Three-dimensional titanium dioxide nanomaterials," *Chem. Rev.*, vol. 114, no. 19, pp. 9487–9558, Oct. 2014, doi: 10.1021/cr500201c.
41. Y. Tian *et al.*, "Ordered three-dimensional nanomaterials using DNA-prescribed and valence-controlled material voxels," *Nat. Mater.*, vol. 19, no. 7, pp. 789–796, Jul. 2020, doi: 10.1038/s41563-019-0550-x.
42. S. I. Kaya, A. Cetinkaya, N. K. Bakirhan, and S. A. Ozkan, "Trends in sensitive electrochemical sensors for endocrine disruptive compounds," *Trends Environ. Anal. Chem.*, vol. 28, p. e00106, Dec. 2020, doi: 10.1016/j.teac.2020.e00106.
43. B. Brunetti, "Recent advances in electroanalysis of vitamins," *Electroanalysis*, vol. 28, no. 9, pp. 1930–1942, Sep. 2016, doi: 10.1002/elan.201600097.
44. N. Dhull, G. Kaur, V. Gupta, and M. Tomar, "Highly sensitive and non-invasive electrochemical immunosensor for salivary cortisol detection," *Sensors Actuators B Chem.*, vol. 293, pp. 281–288, Aug. 2019, doi: 10.1016/j.snb.2019.05.020.
45. M. D. Meti *et al.*, "Nanostructured Au-graphene modified electrode for electrosensing of chlorzoxazone and its biomedical applications," *Mater. Chem. Phys.*, vol. 266, p. 124538, Jul. 2021, doi: 10.1016/j.matchemphys.2021.124538.
46. S. Wang and D. Du, "Differential pulse voltammetry determination of ascorbic acid with ferrocene-l-cysteine self-assembled supramolecular film modified electrode," *Sensors Actuators B Chem.*, vol. 97, no. 2–3, pp. 373–378, Feb. 2004, doi: 10.1016/j.snb.2003.09.026.
47. Y. Y. Chan, Y. Yue, and R. D. Webster, "Voltammetric studies on vitamins D2 and D3 in organic solvents," *Electrochim. Acta*, vol. 138, pp. 400–409, Aug. 2014, doi: 10.1016/j.electacta.2014.06.133.
48. G.-H. Zhang *et al.*, "Self-healable, malleable, and flexible ionic polyimine as an environmental sensor for portable exogenous pollutant detection," *ACS Mater. Lett.*, vol. 4, no. 1, pp. 136–144, Jan. 2022, doi: 10.1021/acsmaterialslett.1c00687.
49. O. Kanoun *et al.*, "A review of nanocomposite-modified electrochemical sensors for water quality monitoring," *Sensors*, vol. 21, no. 12, p. 4131, Jun. 2021, doi: 10.3390/s21124131.
50. S. D'souza, P. Udavant, J. Kadam, S. Khairnar, E. D. Ahire, and R. Sable, "Role of vitamins in metabolic diseases," in *Vitamins as Nutraceuticals*, Wiley, 2023, pp. 205–233. doi: 10.1002/9781394175543.ch9.

51. A.-L. Tardy, E. Pouteau, D. Marquez, C. Yilmaz, and A. Scholey, "Vitamins and minerals for energy, fatigue and cognition: A narrative review of the biochemical and clinical evidence," *Nutrients*, vol. 12, no. 1, p. 228, Jan. 2020, doi: 10.3390/nu12010228.
52. T. B. V. da Silva *et al.*, "Analytical validation of an ultraviolet–visible procedure for determining vitamin D3 in vitamin D3-loaded microparticles and toxigenetic studies for incorporation into food," *Food Chem.*, vol. 360, p. 129979, Oct. 2021, doi: 10.1016/j.foodchem.2021.129979.
53. A. Paudics *et al.*, "A pillararene-based indicator displacement assay for the fluorescence detection of vitamin B1," *Sensors Actuators B Chem.*, vol. 369, p. 132364, Oct. 2022, doi: 10.1016/j.snb.2022.132364.
54. G. Tesfaye, N. Negash, and M. Tessema, "Sensitive and selective determination of vitamin B2 in non-alcoholic beverage and milk samples at poly (glutamic acid)/zinc oxide nanoparticles modified carbon paste electrode," *BMC Chem.*, vol. 16, no. 1, p. 69, Sep. 2022, doi: 10.1186/s13065-022-00863-5.
55. J. M. Pingarrón, P. Yáñez-Sedeño, and A. González-Cortés, "Gold nanoparticle-based electrochemical biosensors," *Electrochim. Acta*, vol. 53, no. 19, pp. 5848–5866, Aug. 2008, doi: 10.1016/j.electacta.2008.03.005.
56. S. Prakash and V. K. Shahi, "Enzyme free glucose sensor exploiting a poly (dimethylaniline) grafted sulfonated ionomer-copper nanocomposite," *Anal. Methods*, vol. 3, no. 6, p. 1331, 2011, doi: 10.1039/c1ay05112a.
57. A. I. Gopalan, K. P. Lee, D. Ragupathy, S. H. Lee, and J. W. Lee, "An electrochemical glucose biosensor exploiting a polyaniline grafted multiwalled carbon nanotube/perfluorosulfonate ionomer–silica nanocomposite," *Biomaterials*, vol. 30, no. 30, pp. 5999–6005, Oct. 2009, doi: 10.1016/j.biomaterials.2009.07.047.
58. T. Qian *et al.*, "Ultrasensitive dopamine sensor based on novel molecularly imprinted polypyrrole coated carbon nanotubes," *Biosens. Bioelectron.*, vol. 58, pp. 237–241, Aug. 2014, doi: 10.1016/j.bios.2014.02.081.
59. T. Wu and S. Lin, "Synthesis, characterization, and electrical properties of polypyrrole/multiwalled carbon nanotube composites," *J. Polym. Sci. Part A Polym. Chem.*, vol. 44, no. 21, pp. 6449–6457, Nov. 2006, doi: 10.1002/pola.21724.
60. S. Tajik, Z. Dourandish, F. Garkani Nejad, H. Beitollahi, P. M. Jahani, and A. Di Bartolomeo, "Transition metal dichalcogenides: Synthesis and use in the development of electrochemical sensors and biosensors," *Biosens. Bioelectron.*, vol. 216, p. 114674, Nov. 2022, doi: 10.1016/j.bios.2022.114674.
61. S. Kumar *et al.*, "Electrochemical sensors and biosensors based on graphene functionalized with metal oxide nanostructures for healthcare applications," *ChemistrySelect*, vol. 4, no. 18, pp. 5322–5337, May 2019, doi: 10.1002/slct.201803871.
62. H. Bakhsh *et al.*, "SnO2 nanostructure based electroanalytical approach for simultaneous monitoring of vitamin C and vitamin B6 in pharmaceuticals," *J. Electroanal. Chem.*, vol. 910, p. 116181, Apr. 2022, doi: 10.1016/j.jelechem.2022.116181.

63. Q. Wang, H. Si, L. Zhang, L. Li, X. Wang, and S. Wang, "A fast and facile electrochemical method for the simultaneous detection of epinephrine, uric acid and folic acid based on ZrO2/ZnO nanocomposites as sensing material," *Anal. Chim. Acta*, vol. 1104, pp. 69–77, Apr. 2020, doi: 10.1016/j.aca.2020.01.012.
64. D. Chauhan, R. Kumar, A. K. Panda, and P. R. Solanki, "An efficient electrochemical biosensor for Vitamin-D3 detection based on aspartic acid functionalized gadolinium oxide nanorods," *J. Mater. Res. Technol.*, vol. 8, no. 6, pp. 5490–5503, Nov. 2019, doi: 10.1016/j.jmrt.2019.09.017.
65. J. A. Buledi *et al.*, "CuO nanostructures based electrochemical sensor for simultaneous determination of hydroquinone and ascorbic acid," *Electroanalysis*, vol. 32, no. 7, pp. 1600–1607, Jul. 2020, doi: 10.1002/elan.202000083.
66. R. Shashanka, D. Chaira, and B. E. K. Swamy, "Electrocatalytic response of duplex and yittria dispersed duplex stainless steel modified carbon paste electrode in detecting folic acid using cyclic voltammetry," *Int. J. Electrochem. Sci.*, vol. 10, no. 7, pp. 5586–5598, Jul. 2015, doi: 10.1016/S1452-3981(23)17279-8.
67. R. Amini and K. Asadpour-Zeynali, "Cauliflower-like NiCo 2 O 4 –Zn/Al layered double hydroxide nanocomposite as an efficient electrochemical sensing platform for selective pyridoxine detection," *Electroanalysis*, vol. 32, no. 6, pp. 1160–1169, Jun. 2020, doi: 10.1002/elan.201900600.
68. R. Rajeev, A. R. Cherian, D. A. Thadathil, and A. Varghese, "Electrochemical determination of Vitamin B6 using coral-like MnO2-Pi on Ti3C2Tx MXene," *Mater. Res. Bull.*, vol. 169, p. 112523, Jan. 2024, doi: 10.1016/j.materresbull.2023.112523.
69. S. Lokesh Kumar, S. Kumar, and K. K. R. Tetala, "A manganese dioxide nanoparticle–bimetallic metal organic framework composite for selective and sensitive detection of vitamin D3 in human plasma," *Microchim. Acta*, vol. 190, no. 9, p. 345, Sep. 2023, doi: 10.1007/s00604-023-05904-x.
70. Y. Yang *et al.*, "A novel electrochemical sensor for sensitive detection of carbendazim based on zeolitic-imidazolate-framework-67/carbon nanohorns nanocomposite," *Mater. Res. Bull.*, vol. 179, p. 112975, Nov. 2024, doi: 10.1016/j.materresbull.2024.112975.
71. S. Kumar, A. Awasthi, M. D. Sharma, K. Singh, and D. Singh, "Functionalized multiwall carbon nanotube-molybdenum disulphide nanocomposite based electrochemical ultrasensitive detection of neurotransmitter epinephrine," *Mater. Chem. Phys.*, vol. 290, p. 126656, Oct. 2022, doi: 10.1016/j.matchemphys.2022.126656.
72. J. N. Tiwari, V. Vij, K. C. Kemp, and K. S. Kim, "Engineered carbon-nanomaterial-based electrochemical sensors for biomolecules," *ACS Nano*, vol. 10, no. 1, pp. 46–80, Jan. 2016, doi: 10.1021/acsnano.5b05690.
73. T. Anusha, K. S. Bhavani, J. V. Shanmukha Kumar, P. K. Brahman, and R. Y. A. Hassan, "Fabrication of electrochemical immunosensor based on GCN-β-CD/Au nanocomposite for the monitoring of vitamin D deficiency," *Bioelectrochemistry*, vol. 143, p. 107935, Feb. 2022, doi: 10.1016/j.bioelechem.2021.107935.

74. H. Bora, D. Mandal, and A. Chandra, "High-performance, nitrogen-doped, carbon-nanotube-based electrochemical sensor for vitamin D3 detection," *ACS Appl. Bio Mater.*, vol. 5, no. 4, pp. 1721–1730, Apr. 2022, doi: 10.1021/acsabm.2c00094.
75. F. C. Moraes, B. Rossi, M. C. Donatoni, K. T. de Oliveira, and E. C. Pereira, "Sensitive determination of 17β-estradiol in river water using a graphene based electrochemical sensor," *Anal. Chim. Acta*, vol. 881, pp. 37–43, Jun. 2015, doi: 10.1016/j.aca.2015.04.043.
76. T. Anusha, K. S. Bhavani, J. V. S. Kumar, and P. K. Brahman, "Designing and fabrication of electrochemical nanosensor employing fullerene-C60 and bimetallic nanoparticles composite film for the detection of vitamin D3 in blood samples," *Diam. Relat. Mater.*, vol. 104, p. 107761, Apr. 2020, doi: 10.1016/j.diamond.2020.107761.
77. T. Anusha, K. S. Bhavani, J. V. Shanmukha Kumar, and P. K. Brahman, "Synthesis and characterization of novel lanthanum nanoparticles-graphene quantum dots coupled with zeolitic imidazolate framework and its electrochemical sensing application towards vitamin D3 deficiency," *Colloids Surfaces A Physicochem. Eng. Asp.*, vol. 611, p. 125854, Feb. 2021, doi: 10.1016/j.colsurfa.2020.125854.
78. G. Tigari and J. G. Manjunatha, "A surfactant enhanced novel pencil graphite and carbon nanotube composite paste material as an effective electrochemical sensor for determination of riboflavin," *J. Sci. Adv. Mater. Devices*, vol. 5, no. 1, pp. 56–64, Mar. 2020, doi: 10.1016/j.jsamd.2019.11.001.
79. E. Tanrıkut, İ. Özcan, E. Sel, S. Köytepe, and E. K. Savan, "Simultaneous electrochemical detection of estradiol and testosterone using nickel ferrite oxide doped mesoporous carbon nanocomposite modified sensor," *J. Electrochem. Soc.*, vol. 167, no. 8, p. 087509, May 2020, doi: 10.1149/1945-7111/ab927f.
80. P. Supchocksoonthorn, M. C. Alvior Sinoy, M. D. G. de Luna, and P. Paoprasert, "Facile fabrication of 17 <math altimg="si1.svg"> <mi mathvariant="normal">β</mi> </math> -estradiol electrochemical sensor using polyaniline/carbon dot-coated glassy carbon electrode with synergistically enhanced electrochemical stability," *Talanta*, vol. 235, p. 122782, Dec. 2021, doi: 10.1016/j.talanta.2021.122782.
81. S. Tajik *et al.*, "Recent developments in polymer nanocomposite-based electrochemical sensors for detecting environmental pollutants," *Ind. Eng. Chem. Res.*, vol. 60, no. 3, pp. 1112–1136, Jan. 2021, doi: 10.1021/acs.iecr.0c04952.
82. S. Shrivastava, N. Jadon, and R. Jain, "Next-generation polymer nanocomposite-based electrochemical sensors and biosensors: A review," *TrAC Trends Anal. Chem.*, vol. 82, pp. 55–67, Sep. 2016, doi: 10.1016/j.trac.2016.04.005.
83. H. Zeng, Y. Xie, T. Liu, Z. Chu, E. Dempsey, and W. Jin, "Conductive polymer nanocomposites: recent advances in the construction of electrochemical biosensors," *Sensors & Diagnostics*, vol. 3, no. 2, pp. 165–180, 2024, doi: 10.1039/D3SD00160A.

84. K. Ghanbari and S. Bonyadi, "An electrochemical sensor based on reduced graphene oxide decorated with polypyrrole nanofibers and zinc oxide–copper oxide p–n junction heterostructures for the simultaneous voltammetric determination of ascorbic acid, dopamine, paracetamol, and tryptoph," *New J. Chem.*, vol. 42, no. 11, pp. 8512–8523, 2018, doi: 10.1039/C8NJ00857D.
85. T. Anusha, K. Sai Bhavani, J. V. Shanmukha Kumar, A. Bonanni, and P. K. Brahman, "Fabrication of handmade paper sensor based on silver-cobalt doped copolymer-ionic liquid composite for monitoring of vitamin D3 level in real samples," *Microchem. J.*, vol. 161, p. 105789, Feb. 2021, doi: 10.1016/j.microc.2020.105789.
86. N. Song *et al.*, "A novel electrochemical biosensor for the determination of dopamine and ascorbic acid based on graphene oxide /poly(aniline-co-thionine) nanocomposite," *J. Electroanal. Chem.*, vol. 873, p. 114352, Sep. 2020, doi: 10.1016/j.jelechem.2020.114352.
87. H. Filik, A. A. Avan, and S. Aydar, "Electrochemical determination of vitamin B-12 in food samples by Poly(2,2′-(1,4-phenylenedivinylene) Bis-8 hydroxyquinaldine)/multi-walled carbon nanotube-modified glassy carbon electrode," *Food Anal. Methods*, vol. 9, no. 8, pp. 2251–2260, Aug. 2016, doi: 10.1007/s12161-016-0420-y.
88. L. Guo and Z. Yang, "Electrochemical sensors based on Au-PPy NPs@f-CNTs nanocomposite modified glassy carbon electrode for determination of Vitamin B12 as a key agent in human motion coordination," *Int. J. Electrochem. Sci.*, vol. 16, no. 10, p. 211029, Oct. 2021, doi: 10.20964/2021.10.31.
89. R.-S. Juang, C.-T. Hsieh, C.-M. Liu, T.-A. Lin, and Y. A. Gandomi, "Electrochemical detection of parathyroid hormone using graphene-incorporated conductive polymer decorated with MoS2 nanoparticles," *Colloids Surfaces A Physicochem. Eng. Asp.*, vol. 666, p. 131336, Jun. 2023, doi: 10.1016/j.colsurfa.2023.131336.

CHAPTER 8

Determination of Medical Wastes by Nanomaterial-Based Electrodes

Raghavendra P. Bakale, Sheela A. Sangam, S.N. Mathad, Vinayak Adimule, Gangadhar Bagihalli, Rajeev S. Joshi, and Santosh Nandi

8.1 INTRODUCTION

Medical wastes, encompassing a broad spectrum of hazardous materials like sharps, pharmaceuticals, laboratory waste, and infectious substances, present substantial risks to human health and the environment if not managed and disposed of properly [1]. Improper management can lead to contamination of soil and water, the spread of infectious diseases, and harmful exposure to healthcare workers and the public [2–3]. For instance, the World Health Organization (WHO) estimates that improper handling of healthcare waste contributes to over 5.2 million deaths annually due to diseases like hepatitis and HIV [4–6].

Conventional methods for medical waste detection often lack sensitivity, portability, and cost-effectiveness [7–9]. Traditional detection techniques may fail to accurately identify and quantify low concentrations of

 DOI: 10.1201/9781003512639-8

hazardous substances, which is essential for ensuring safe handling and disposal [10–12]. Many of these methods rely on bulky and stationary equipment, making it difficult to perform on-site assessments in diverse healthcare settings. Additionally, the high costs associated with these technologies can be prohibitive, especially for resource-limited facilities, leading to inconsistent waste management practices and increased health risks [13–15].

Traditional methods for medical waste detection rely on labour-intensive techniques such as chromatography, spectroscopy, and immunoassays. Chromatography, which separates and analyses complex mixtures, requires extensive sample preparation and long analysis times [16–18]. Spectroscopy, used to identify substances based on their interaction with electromagnetic radiation, necessitates sophisticated and often expensive instruments. Immunoassays, which detect specific molecules through antigen—antibody interactions, also demand specialized reagents and protocols [19–21].

These conventional techniques, while highly accurate and reliable, are often time-consuming and expensive. They typically require specialized equipment and trained personnel to operate, which limits their accessibility and practicality, especially in resource-constrained environments. The high costs associated with these methods can be prohibitive for smaller healthcare facilities, leading to potential lapses in effective waste management.

Electrochemical sensing offers a promising alternative for medical waste detection [22–24]. This approach leverages the inherent advantages of electrochemical techniques, such as high sensitivity and rapid response times. Electrochemical sensors can detect minute changes in the chemical composition of waste samples, making them highly effective for identifying hazardous substances even at low concentrations [25–27]. Additionally, these sensors are generally more portable and cost-effective compared with traditional methods [28–30].

The simplicity and affordability of electrochemical sensing make it particularly suitable for on-site applications. Portable electrochemical sensors can be easily deployed in various healthcare settings, allowing real-time monitoring and immediate decision making [31]. This not only enhances the efficiency of medical waste management but also reduces the risk of environmental contamination and exposure to hazardous materials. The field of electrochemical sensing has witnessed significant advancements with the integration of nanomaterials, particularly for the detection and management of medical wastes [32]. These innovations offer enhanced sensitivity, selectivity, and portability, addressing the limitations of traditional detection methods.

In summary, while traditional methods for medical waste detection are valuable, they are often constrained by their complexity, cost, and the need for specialized resources. Electrochemical sensing presents a viable alternative, combining high sensitivity, speed, and affordability to improve the detection and management of medical waste. Nanomaterial-based electrodes represent a significant advancement in the field of medical waste detection. Their superior sensitivity, selectivity, and potential for portability address the limitations of traditional methods, offering a promising solution for effective waste management. Ongoing research and development in this field have the potential to significantly improve the safety and sustainability of healthcare practices.

8.2 NANOMATERIAL-BASED ELECTRODES

Nanomaterials exhibit unique physicochemical properties (Figure 8.1), including high surface area [33–35], enhanced conductivity [36–38], and catalytic activity [39–41], making those ideal candidates for electrode modification in electrochemical sensors. Carbon-based materials such as graphene [42–44], carbon nanotubes [45–47], and carbon nanofibres [48–50] have attracted considerable interest and acclaim due to their outstanding electrical conductivity and exceptional chemical stability. Metal nanoparticles, including gold [51–53], silver [54–56], and platinum [57–59],offer high catalytic activity and biocompatibility, enabling the detection of various medical waste components with high sensitivity and selectivity. Hybrid nanocomposites, comprising combinations of different nanomaterials, further enhance the electrochemical performance of electrodes by synergistically combining their unique properties.

8.2.1 Nanomaterials in Electrochemical Sensing

Nanomaterials have transformed the field of electrochemical sensing due to their distinctive physicochemical properties. Characterized by high surface areas, exceptional conductivity, and adaptable surface chemistry, they are perfectly suited for electrode modification. The most commonly employed nanomaterials include carbon-based variants like graphene and carbon nanotubes, metal and metal oxide nanoparticles, and composite

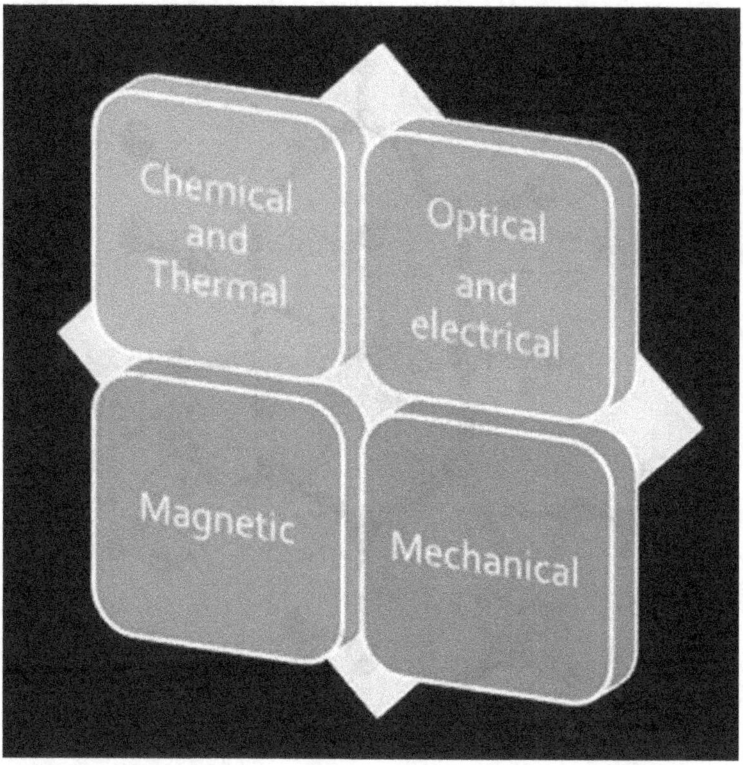

FIGURE 8.1 Physicochemical properties of nanomaterials.

nanomaterials, which collectively offer unparalleled opportunities for sensor development and optimization.

8.2.2 Carbon-Based Nanomaterials

Graphene and Graphene Oxide: Graphene-based electrodes have demonstrated outstanding performance in detecting various medical waste constituents, such as pharmaceuticals and pathogens, leveraging their exceptional electrical conductivity and expansive surface area. By functionalizing graphene with targeted chemical groups, its selectivity and sensitivity can be significantly enhanced [60–62].

Carbon Nanotubes (CNTs): CNTs offer high conductivity and mechanical strength, making them suitable for robust sensor applications(Figure 8.2). Their ability to be functionalized with various molecules improves detection capabilities for specific medical waste components [63–65].

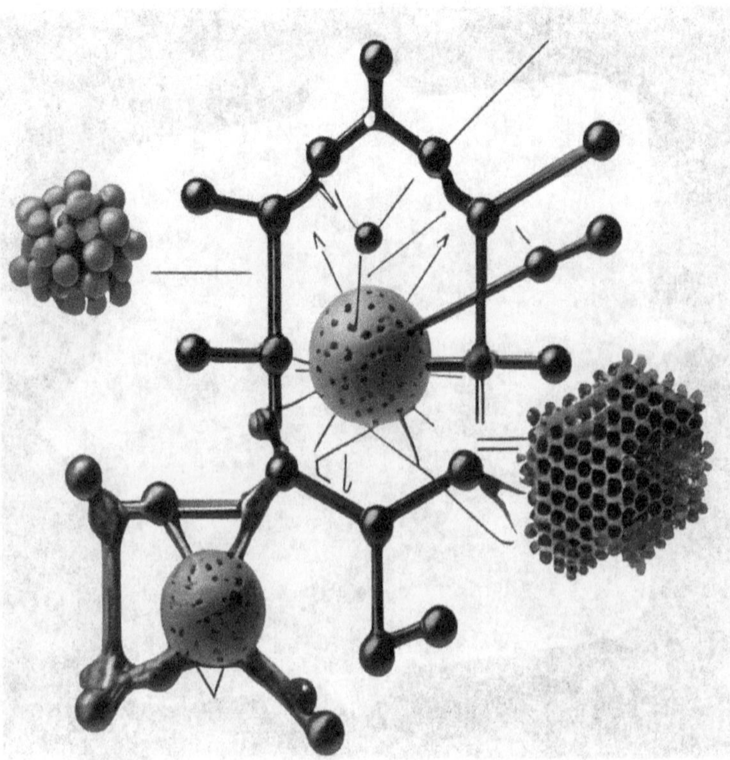

FIGURE 8.2 Carbon-based materials.

8.2.3 Metal and Metal Oxide Nanoparticles

Gold Nanoparticles (AuNPs): AuNPs are widely used due to their excellent conductivity and ease of surface modification [66]. They have been employed in the detection of medical waste biomarkers through various electrochemical techniques, such as voltammetry and impedance spectroscopy [67].

Silver and Platinum Nanoparticles: These nanoparticles enhance electron transfer rates and provide high catalytic activity, essential for the sensitive detection of waste constituents like antibiotics and heavy metals [68].

8.2.4 Composite Nanomaterials

Hybrid Nanocomposites: Combining different nanomaterials can synergistically enhance sensor performance. For instance, composites of graphene and metal nanoparticles combine the high surface area of graphene

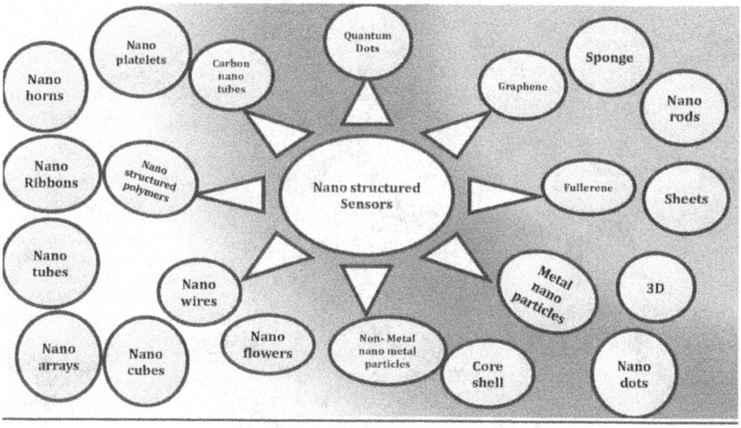

FIGURE 8.3 Applications of nano sensors.

with the catalytic properties of metal nanoparticles(Figure 8.4), resulting in highly sensitive and selective electrodes [69–70].

PolymerNanoparticle Composites: Incorporating polymers with nanoparticles provides flexibility and mechanical strength, making them

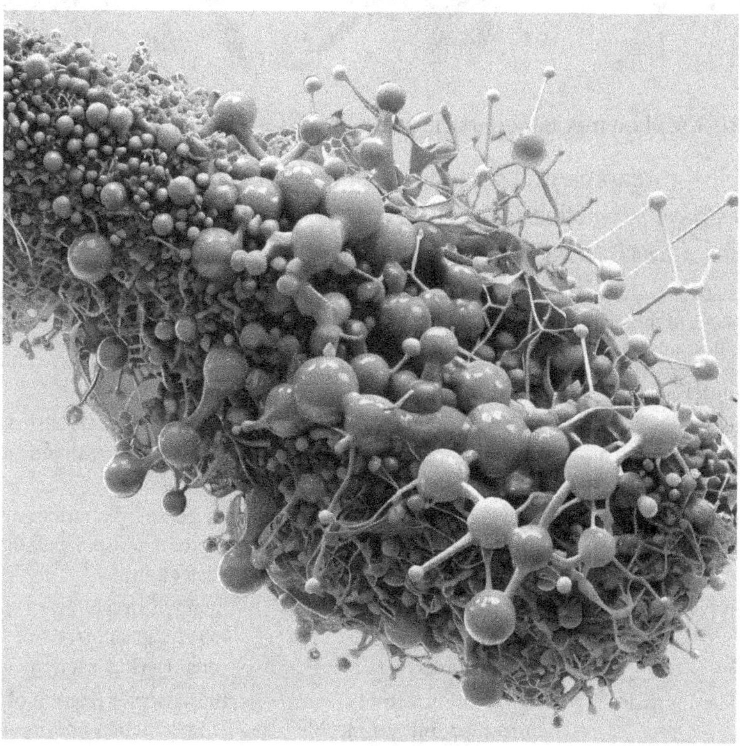

FIGURE 8.4 Hybrid nanocomposites.

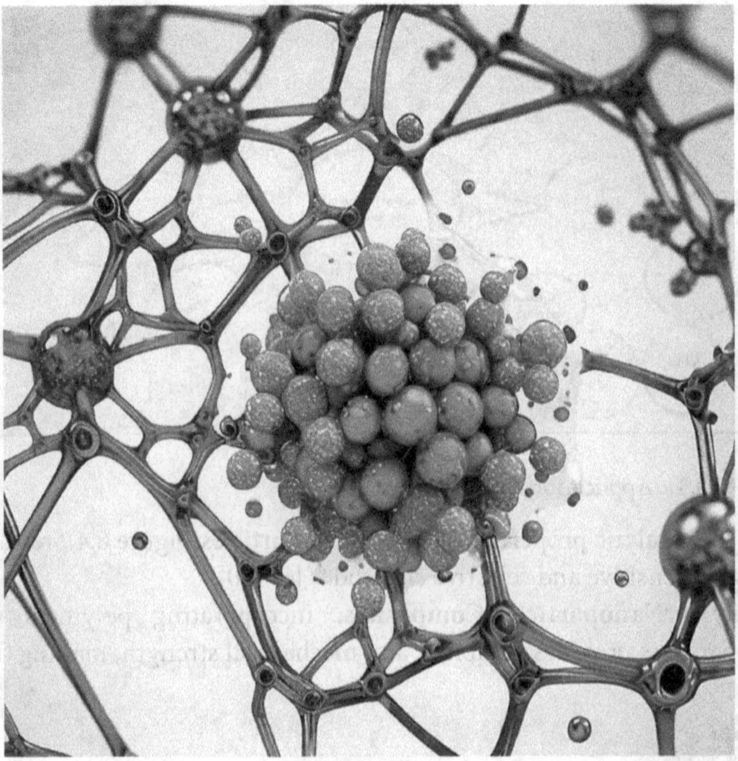

FIGURE 8.5 Polymer nanoparticle composites.

suitable for wearable and portable sensing devices (Figure 8.5). These composites have been effectively used to detect various biological and chemical substances in medical waste [71–72].

List of some of the frequently used nanomaterial-based electrodes along with their properties and applications (Figure 8.3)

Nanomaterial	Properties	Applications
Graphene	High electrical conductivity, large surface area, chemical stability	Electrochemical sensors, DNA sensors, chemical sensors
Carbon Nanotubes	Excellent electrical conductivity, high aspect ratio, mechanical strength	Detection of pharmaceuticals, glucose sensors, pathogen detection
Carbon Nanofibres	High electrical conductivity, mechanical flexibility, chemical inertness	Energy storage, supercapacitors, tissue engineering, catalysis
Gold Nanoparticles	High catalytic activity, biocompatibility, tuneable surface plasmon resonance	Biosensors, detection of heavy metals, enzyme-based sensors

(*Continued*)

Silver Nanoparticles	Strong antibacterial properties, high electrical conductivity, catalytic activity	Pathogen detection, environmental monitoring, chemical sensors
Platinum Nanoparticles	Excellent catalytic activity, chemical stability, resistance to corrosion	Chemical sensors, biosensors, gas sensors
Hybrid Nanocomposites	Synergistic combination of properties from different nanomaterials, enhanced electrochemical performance	Electronics and sensors, biomedical applications, environmental remediation
Zinc Oxide (ZnO) Nanoparticles	High electron mobility, good chemical stability, piezoelectric properties	Gas sensors, biosensors, UV detection
Titanium Dioxide (TiO_2) Nanoparticles	Good stability, photocatalytic properties, high surface area	Detection of organic compounds, photocatalytic degradation of contaminants

8.3 RECENT APPLICATIONS IN MEDICAL WASTE DETERMINATION USING NANOMATERIAL-BASED ELECTRODES

The rise of nanotechnology has paved the way for significant advancements in medical waste management. Nanomaterial-based electrodes have become invaluable tools for detecting various contaminants in medical waste with unprecedented sensitivity and specificity. This detailed examination highlights recent applications of these innovative electrodes in detecting pharmaceutical residues, pathogens, heavy metals, chemical contaminants, and organic waste.

8.3.1 Pharmaceutical Detection

Nanomaterial-based electrodes have revolutionized the detection of pharmaceutical residues in medical waste. These advanced sensors provide highly sensitive detection capabilities, essential for monitoring and mitigating pharmaceutical contamination in the environment.Nanomaterial-based electrodes have been developed to detect pharmaceutical residues in medical waste with high sensitivity(Figure 8.6). For example, graphene-based sensors can identify trace amounts of drugs in wastewater, aiding in the assessment and mitigation of pharmaceutical contamination [73].

8.3.1.1 Graphene-based sensors

Graphene's exceptional electrical conductivity and large surface area make it an ideal material for sensor development. Graphene-based sensors can detect trace amounts of pharmaceuticals in wastewater, such as

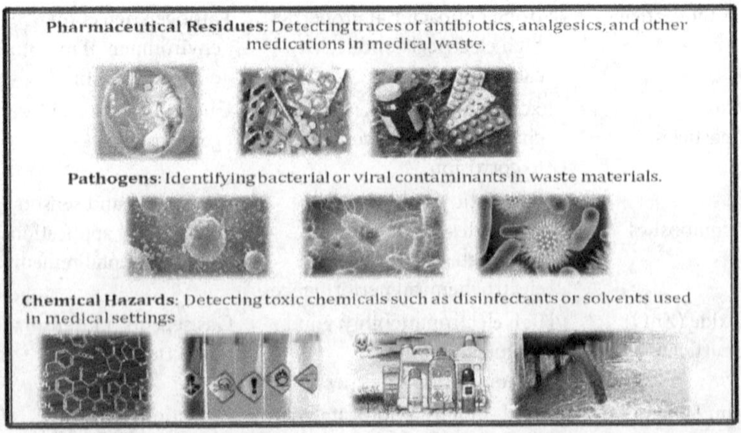

FIGURE 8.6 Medical waste detection.

antibiotics, analgesics, and hormones. These sensors function by adsorbing the pharmaceutical molecules onto the graphene surface, which alters the electrical properties of the graphene, providing a measurable signal that correlates with the concentration of the pharmaceuticals. This capability is crucial for assessing the extent of pharmaceutical contamination and implementing measures to mitigate its impact on ecosystems and human health.

8.3.1.2 Carbon Nanotubes (CNTs)

CNTs are another class of nanomaterials used in electrodes for pharmaceutical detection. Their unique cylindrical structure and excellent conductivity enhance the sensitivity and detection limits of these sensors. CNT-based electrodes can be functionalized with specific chemical groups to selectively interact with various pharmaceutical compounds, enabling precise detection in complex medical waste matrices.

8.3.2 Pathogen Identification

The rapid and accurate identification of pathogenic microorganisms in medical waste is vital for preventing the spread of infections and ensuring safe disposal practices. Electrochemical sensors with nanomaterial modified electrodes can rapidly and accurately detect pathogenic microorganisms in medical waste [74]. This capability is crucial for preventing the spread of infections and ensuring safe waste disposal. Nanomaterial-based electrochemical sensors have demonstrated significant potential in this area.

8.3.2.1 Metal Oxide Nanoparticles

Sensors incorporating metal oxide nanoparticles, such as ZnO, TiO_2, and Fe_3O_4, have been developed to detect bacteria, viruses, and other pathogens in medical waste. These nanoparticles enhance the sensitivity and selectivity of the sensors by providing a high surface area for pathogen adsorption and facilitating electron transfer processes. The presence of pathogens leads to changes in the electrochemical properties of the electrodes, producing detectable signals that indicate contamination.

8.3.2.2 Gold and Silver Nanoparticles

Electrodes modified with gold and silver nanoparticles offer excellent performance in pathogen detection. These noble metals possess unique optical and electronic properties, allowing the development of highly sensitive and specific sensors. Gold and silver nanoparticles can be functionalized with biomolecules, such as antibodies or aptamers, to target specific pathogens, enabling rapid and accurate identification.

8.3.3 Heavy Metal Monitoring

Heavy metals, such as lead, mercury, and cadmium, are toxic contaminants commonly found in medical waste. Sensors incorporating metal nanoparticles have shown exceptional performance in detecting heavy metals like lead and mercury in medical waste [75]. These sensors provide rapid on-site analysis, crucial for environmental monitoring and regulatory compliance.

8.3.3.1 Metal Nanoparticles

Metal nanoparticle–based sensors, featuring Au, Ag, and Pt, have shown exceptional sensitivity and selectivity for detecting heavy metals. By enhancing the electrochemical properties of electrodes, these nanoparticles enable the detection of even trace amounts of heavy metals. The interaction between the metal nanoparticles and heavy metal ions generates distinct changes in the sensor's electrical signal, allowing accurate and precise quantification of heavy metal concentrations.

8.3.3.2 Carbon-Based Nanomaterials

Carbon-based nanomaterials, including graphene and CNTs, are also used in heavy metal sensors. Their high conductivity and large surface area improve the sensitivity and detection limits of the sensors. Functionalizing these nanomaterials with specific ligands or chemical groups enhances their selectivity for different heavy metals, allowing for simultaneous detection and analysis.

8.3.4 Detection of Chemical Contaminants

Chemical contaminants, such as disinfectants, solvents, and other hazardous chemicals, pose significant risks in medical waste. Nanomaterial-based electrodes offer advanced detection capabilities for these contaminants, ensuring safe and effective waste management.

8.3.4.1 Conducting Polymer Nanocomposites

Conducting polymer nanocomposites combine the advantages of conducting polymers and nanomaterials, resulting in sensors with improved electrochemical response and detection accuracy. These materials can detect various chemical contaminants by undergoing redox reactions or interacting with the target molecules. The changes in the electrochemical properties of the sensors provide measurable signals, indicating the presence and concentration of the contaminants.

8.3.4.2 Quantum Dots

Quantum dot–based sensors harness the distinctive optical and electronic properties of quantum dots to detect chemical contaminants with high sensitivity and rapid response times. As semiconductor nanocrystals, quantum dots display size-dependent optical properties, enabling the detection of a broad spectrum of chemical contaminants, including volatile organic compounds (VOCs) and toxic industrial chemicals. This technology provides critical data for effective waste management and regulatory compliance, offering a valuable tool for monitoring and mitigating environmental pollution.

8.3.5 Biosensors for Organic Waste

The detection of organic waste products, such as biomolecules and genetic material, is crucial for managing medical waste effectively. Nanomaterial-based biosensors offer innovative solutions for identifying and quantifying these organic contaminants.

8.3.5.1 Enzyme-Modified Nanomaterial Electrodes

Enzyme-modified electrodes utilize enzymes immobilized on nanomaterials to detect specific biomolecules in medical waste. Enzymes, such as glucose oxidase, urease, and protease, catalyse reactions with their respective substrates, producing measurable electrochemical signals. These biosensors can detect glucose, urea, proteins, and other organic waste products, providing insights into the composition and contamination levels of medical waste.

8.3.5.2 DNA-Based Sensors

Nanomaterial-based electrodes incorporating DNA sequences are designed to detect specific genetic material in medical waste. These sensors use DNA hybridization or aptamer binding to identify and quantify genetic contaminants, such as DNA or RNA from infectious agents. The interaction between the target genetic material and the DNA-functionalized nanomaterials alters the electrochemical properties of the electrodes, producing signals that indicate contamination.

8.3.6 Future Directions

Integration with Portable Devices: The development of handheld and portable electrochemical sensors using nanomaterial-based electrodes is a promising area [76]. These devices can provide real-time monitoring and data logging, enhancing waste management practices in diverse healthcare settings.

Enhanced Selectivity: Future research is focused on improving the selectivity of nanomaterial-based sensors through advanced functionalization techniques and the development of new composite materials [77].

Sustainability and Scalability: Addressing the scalability and environmental impact of nanomaterial production is essential. Efforts are being made to develop ecofriendly synthesis methods and to utilize sustainable materials [78].

8.4 ELECTROCHEMICAL DETECTION OF MEDICAL WASTES AND APPLICATIONS OF NANO SENSORS

The integration of nanomaterial-based electrodes with electrochemical techniques enables the selective detection of specific medical waste components (Figure 8.7). Voltammetric methods, such as cyclic voltammetry [79–81] and differential pulse voltammetry [82–84], allow the quantitative analysis of electroactive species, including heavy metals, pharmaceuticals, and organic pollutants. Amperometric sensors offer real-time monitoring of microbial contamination by measuring the current produced by enzymatic reactions [85–87]. Electrochemical impedance spectroscopy (EIS) offers valuable insights into the interfacial properties of

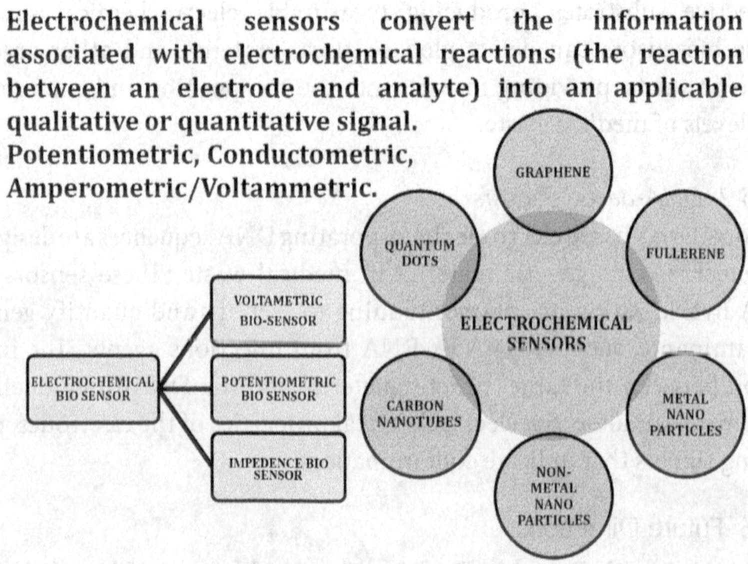

FIGURE 8.7 Integration of nanomaterial-based electrodes with electrochemical techniques.

nanomaterial-modified electrodes, facilitating the highly sensitive detection of pathogens and biochemical markers [88–90].

8.4.1 Voltammetric Methods

Voltammetric techniques, including cyclic voltammetry (CV) and differential pulse voltammetry (DPV), are potent tools for the quantitative analysis of electroactive species in medical waste. CV provides valuable insights into the redox reactions of analytes by measuring the current response as a function of the applied potential. DPV, meanwhile, offers enhanced sensitivity and resolution by applying a series of potential pulses and measuring the resulting current, making it particularly effective for detecting trace amounts of heavy metals, pharmaceuticals, and organic pollutants in complex matrices.

8.4.2 Amperometric Sensors

Amperometric sensors measure the current produced by redox reactions at the electrode surface. They are particularly useful for real-time monitoring of microbial contamination in medical waste. For example, enzymes immobilized on nanomaterial-modified electrodes can catalyse reactions with specific pathogens, producing measurable currents proportional to the pathogen concentration.

8.4.3 Impedance Spectroscopy

Impedance spectroscopy measures the impedance of an electrode–electrolyte interface over a range of frequencies. This technique provides valuable insights into the properties of nanomaterial-modified electrodes, such as charge transfer resistance and double-layer capacitance. It enables the sensitive detection of pathogens and biochemical markers in medical waste by monitoring changes in impedance caused by specific interactions with target analytes.

Electrochemical Technique	Principle	Medical Waste Application
Voltammetry	Measures current as a function of applied potential	Detection of electroactive species (e.g. metals, drugs)
Amperometry	Measures current produced by redox reactions	Real-time monitoring of microbial contamination
Impedance Spectroscopy	Measures impedance changes at electrode interfaces	Detection of pathogens and biochemical markers

8.4.4 Case Studies

a. Detection of Antibiotics in Wastewater: A case study demonstrated the use of graphene-based electrodes for the voltammetric detection of antibiotics in hospital wastewater [91–93]. The sensors exhibited high sensitivity and selectivity, allowing the quantification of trace levels of antibiotics, which is crucial for preventing the spread of antibiotic resistance [94–96].

b. Pathogen Detection in Medical Waste: Another study utilized Au nanoparticle electrodes modified for the amperometric detection of bacteria in medical waste [97–99]. The sensors achieved rapid and accurate detection, highlighting their potential for real-time monitoring of microbial contamination in healthcare settings.

8.4.5 Applications

Nanomaterial-based electrodes have been used in various settings for medical waste detection. Case studies highlight their effectiveness in real-world applications, demonstrating their potential for widespread adoption in healthcare facilities and environmental monitoring systems.

Table 8.1 provides examples of nanomaterial-based electrodes combined with various electrochemical techniques for the selective detection of specific medical waste components in different sample matrices.

TABLE 8.1 Nanomaterial based electrochemical detection of medical waste components

Medical Waste Component	Nanomaterial-Based Electrode	Electrochemical Technique	Target Analyte	Sample Matrix
Pathogenic bacteria	Graphene oxide/carbon nanotubes	Impedance spectroscopy	*Escherichia coli*	Water, clinical samples
Pharmaceutical residues	Gold nanoparticles	Voltammetry	Ibuprofen	Wastewater, soil
Heavy Metal Ions [100–102]	Silver nanoparticles	Anodic stripping voltammetry	Lead (Pb^{2+})	Industrial effluents
Organic pollutants	Carbon nanofibres	Amperometry	Phenol	Soil, industrial wastewater

8.5 CHALLENGES AND FUTURE PERSPECTIVES

Despite the promising applications, several challenges need to be addressed for the widespread adoption of nanomaterial-based electrodes in medical waste detection. Stability, reproducibility, and selectivity of nanomaterials are critical factors that influence sensor performance. The potential toxicity and environmental impact of nanomaterials also raise concerns that require careful consideration.

8.5.1 Challenges

Stability and Reproducibility: Ensuring the long-term stability and reproducibility of nanomaterial-based electrodes is essential for reliable sensor performance. Variability in nanomaterial synthesis and electrode fabrication(Figure 8.8) can lead to inconsistent results.

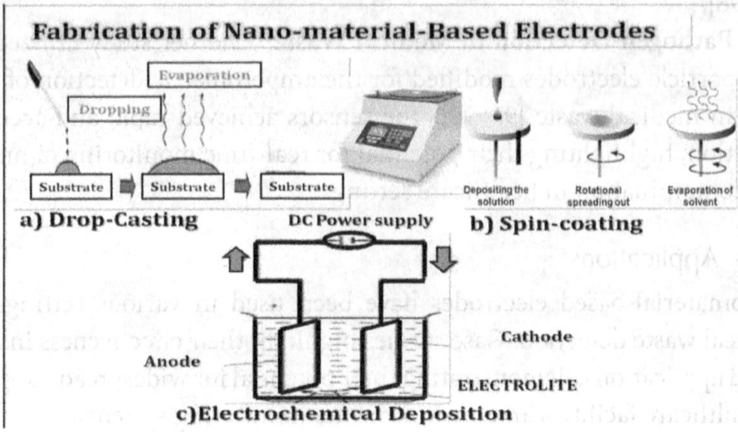

FIGURE 8.8 Fabrication of nanomaterial-based electrodes.

Selectivity: Achieving high selectivity for specific analytes in complex medical waste matrices remains a significant challenge. Interference from other substances can affect sensor accuracy and reliability.

Toxicity and Environmental Impact: The potential toxicity of nanomaterials to human health and the environment must be carefully evaluated. Sustainable synthesis methods and safe disposal practices are necessary to mitigate these risks.

8.5.2 Future Perspectives

Enhanced Functionalization Techniques: Advances in nanomaterial functionalization techniques can improve the selectivity and sensitivity of sensors. Developing novel functional groups and surface modification strategies will enable the detection of a broader range of analytes with higher precision.

Integration with Portable Devices: The development of portable and hand-held electrochemical sensors incorporating nanomaterial-based electrodes is a promising direction [103–105]. These devices can provide real-time monitoring and data logging, enhancing waste management practices in diverse healthcare settings.

Sustainable and Scalable Production: Addressing the scalability and environmental impact of nanomaterial production is essential for widespread adoption. Efforts are being made to develop ecofriendly synthesis methods and utilize sustainable materials to minimizeenvironmental impact [106–108].

8.5.2.1Future Research Directions

Ecofriendly Nanomaterials: Research into ecofriendly and biodegradable nanomaterials is gaining momentum. These materials can reduce the environmental impact of sensor production and disposal, aligning with sustainability goals [109–111].

Digital Integration: Integrating nanomaterial-based sensors with digital health technologies, such as Internet of Things (IoT) and data analytics, can enhance monitoring capabilities and enable remote management of medical waste. Real-time data collection and analysis can provide valuable insights for improving waste management practices[112–115].

Regulatory Compliance: Collaboration with regulatory bodies to establish standards and guidelines for the use of nanomaterial-based sensors in medical waste detection is crucial. Ensuring compliance with safety and environmental regulations will facilitate the adoption of these technologies in healthcare settings.

8.6 Conclusion

Nanomaterial-based electrodes represent a significant advancement in the field of medical waste detection. Their superior sensitivity, selectivity, and potential for portability address the limitations of traditional methods, offering a promising solution for effective waste management. The integration of these sensors with electrochemical techniques enables the selective detection of specific medical waste components, enhancing the accuracy and efficiency of waste management practices. Continued research and development in this area hold the potential to further enhance the safety and sustainability of healthcare practices.

8.7 CALL TO ACTION

Researchers and industry professionals are encouraged to collaborate and explore innovative solutions to the challenges faced in the development and application of nanomaterial-based electrodes. By advancing this technology and promoting its adoption in healthcare facilities and environmental monitoring systems, we can improve the management of medical wastes and protect public health and the environment.

REFERENCES

1. Bolan, Shiv, Lokesh P. Padhye, et al. "Review on distribution, fate, and management of potentially toxic elements in incinerated medical wastes." *Environmental Pollution* 321 (2023): 121080.
2. Endale, Habtamu, Mesfin Mathewos, et al. "Potential causes of spread of antimicrobial resistance and preventive measures in one health perspective—a review." *Infection and Drug Resistance* (2023): 7515–7545.
3. Adimule, V., B. C. Yallur, R. Keri, C. Bathula, and S. Batakurki. "Microstructure, photoluminescence and electrical properties of SmxGd(1–x):SrO hybrid nanomaterials synthesized via facile coprecipitation method." *Electronic Materials Letters* 19, no. 3 (2023): 278–297.
4. Cook, Ed, Anne Woolridge, et al. "Medical and healthcare waste generation, storage, treatment and disposal: A systematic scoping review of risks to occupational and public health." *Critical Reviews in Environmental Science and Technology* 53, no. 15 (2023): 1452–1477.
5. Adimule, V., R. Keri, S. Rajendrachari, P. Kendrekar, and C. V. Kulkarni. "Highly photoluminescence and wide band gap insulating metal hybrid nanoparticles array of samarium-doped SrO:CoO: Synthesis, characterizations and sensor characteristics." *Journal of Materials Science: Materials in Electronics* 34, no. 5 (2023): 442.

6. Challa, M., B. C. Yallur, M. R. Ambika, and V. Adimule. "Influence of nanoparticles on optical properties of Cu-MOFs." *Advanced Materials Research* 1173 (2022): 23–33.
7. Rypar, Tomas, Jaroslava Bezdekova, Kristyna Pavelicova, et al. "Low-tech vs. high-tech approaches in μPADs as a result of contrasting needs and capabilities of developed and developing countries focusing on diagnostics and point-of-care testing." *Talanta* 266 (2024): 124911.
8. Nandi, S. S., V. Adimule, S. A. Kadapure, and S. S. Kerur. "Rare earth based nanocomposite materials for prominent performance supercapacitor: A review." *Applied Mechanics and Materials* 908 (2022): 3–18.
9. Manhas, N., V. Adimule, P. Lakshminarayana, R. Keri, L. S. Kumar, and C. Bathula. "Utilization of Co2+ doped Eu2O3 nanostructures and their effect on the dielectric and photoisomerization of liquid crystals." *Physics of Fluids* 35, no. 10 (2023).
10. Mandal, Swagata, Rajlakshmi Poi, et al. "Review of extraction and detection techniques for the analysis of pesticide residues in fruits to evaluate food safety and make legislative decisions: Challenges and anticipations." *Journal of Chromatography B* 1215 (2023): 123587.
11. Nagalakshmi, G., I. M. Nandeesh, B. C. Yallur, V. Adimule, and S. Batakurki. "Synthesis and optical properties of copper terephthalate metal organic frameworks." *Engineering Chemistry* 2 (2023): 3–11.
12. Adimule, V., B. C. Yallur, and A. H. Gowda. "Advanced sensors based on carbon nanomaterials." In *Carbon Nanomaterials-Based Sensors*, 259–268. Elsevier, 2022.
13. Chen, Yang, Qinzhong Feng, and Zhongkui Zhang. *Innovative Disposal Technology and Management Practice for Medical Waste.* Springer, 2023.
14. Shashanka, R., Esgin, H., Yilmaz, V. M., & Caglar, Y. (2020). Fabrication and characterization of green synthesized ZnO nanoparticle based dye-sensitized solar cells. Journal of Science: Advanced Materials and Devices, 5(2), 185-191.
15. Shashanka, R., & Kumara Swamy, B. E. (2020). Simultaneous electro-generation and electro-deposition of copper oxide nanoparticles on glassy carbon electrode and its sensor application. SN Applied Sciences, 2, 1-10.
16. Kataoka, Hiroyuki. "Sample preparation for liquid chromatography." In *Liquid Chromatography*, 1–48. Elsevier, 2023.
17. Adimule, V., V. Sharma, P. Kumar, R. Keri, and K. Sharma. "Perylene-3,4,9,10-tetracarboxylic dianhydride dye intercalated liquid crystal molecules: Synthesis, spectroscopic characterizations and photoluminescence dynamics." *Journal of Molecular Liquids* (2024): 125990.
18. Rajendrachari, S., Swamy, B. K., Reddy, S., & Chaira, D. (2013). Synthesis of silver nanoparticles and their applications. Anal. Bioanal. Electrochem, 5(4), 455-466.
19. Schlotheuber, Luca Johannes, Ines Lüchtefeld, et al. "Antibodies, repertoires and microdevices in antibody discovery and characterization." *Lab on a Chip* (2024).

20. Reddy, S., Swamy, B. K., Aruna, S., Kumar, M., Shashanka, R., & Jayadevappa, H. (2012). Preparation of NiO/ZnO hybrid nanoparticles for electrochemical sensing of dopamine and uric acid. Chem Sens, 2(1), 1-8.
21. Shashanka, R., Jayaprakash, G. K., BG, P., Kumar, M., & Kumara Swamy, B. E. (2022). Electrocatalytic determination of ascorbic acid using a green synthesised magnetite nano-flake modified carbon paste electrode by cyclic voltammetric method. Materials Research Innovations, 26(4), 229-239.
22. Melo, Jaqueline F., Joadir H. S. Junior, Thais B. de M. Freire, et al. "Industrial waste reuse: An alternative source to reduced graphene oxide for preparing electrochemical sensors." *Electrochimica Acta* 454 (2023): 142382.
23. Adımule, V., S. Nandi, K. Sharma, and R. Keri. "Microstructure, zeta potential and enhanced photocatalytic dye degradation performances of mesoporous flower-like Nb_2O_5@CeO_2 bimetallic oxide nanomaterials decorated with reduced graphene oxide." *Inorganic Chemistry Communications* (2024): 112592.
24. Rajendrachari, S., Kamacı, Y., Taş, R., Ceylan, Y., Bülbül, A. S., Uzun, O., & Karaoğlanlı, A. C. (2019). Antimicrobial investigation of CuO and ZnO nanoparticles prepared by a rapid combustion method. Physical chemistry research, 7(4), 799-812.
25. Barhoum, Ahmed, Selma Hamimed, Hamda Slimi, et al. "Modern designs of electrochemical sensor platforms for environmental analyses: Principles, nanofabrication opportunities, and challenges." *Trends in Environmental Analytical Chemistry* 38 (2023): e00199.
26. Adimule, V., C. Bathula, K. Sharma, and R. Keri. "Study of structural, optical, dielectric properties poly-3-butyl thiophene reinforced with Co_3O_4 nanoparticles." *Nano-Structures & Nano-Objects* 38 (2024): 101150.
27. Adimule, V., C. Bathula, K. Sharma, and R. Keri. "Study of structural, optical, dielectric properties poly-3-butyl thiophene reinforced with Co_3O_4 nanoparticles." *Nano-Structures & Nano-Objects* 38 (2024): 101150.
28. He, Quanguo, Bing Wang, Jing Liang, et al. "Research on the construction of portable electrochemical sensors for environmental compounds quality monitoring." *Materials Today Advances* 17 (2023): 100340.
29. Rajendrachari, S., Kudur Jayaprakash, G., Pandith, A., Karaoglanli, A. C., & Uzun, O. (2022). Electrocatalytic investigation by improving the charge kinetics between carbon electrodes and dopamine using bio-synthesized CuO nanoparticles. Catalysts, 12(9), 994.
30. Sharma, V. and P. Kumar. "Electro-optically oriented Kerr and orientational phase study of normal mode polymer dispersed liquid crystals–Effect of dispersion of nanoparticles." *Journal of Molecular Liquids* 348 (2022): 118030.
31. Li Xin, Fenni Zhang, and Qingjun Liu. "Portable and wearable sensing technologies for biochemical detection." In *Portable and Wearable Sensing Systems: Techniques, Fabrication, and Biochemical Detection*, 1–17, 2024.
32. Thakur, Abhinay, and Ashish Kumar. "Recent trends in nanostructured carbon-based electrochemical sensors for the detection and remediation of persistent toxic substances in real-time analysis."*Materials Research Express*10, no. 3 (2023): 034001.

33. Bai, Hua, Chun Li, Xiaolin Wang, et al. "A pH-sensitive graphene oxide composite hydrogel."*Chemical Communications*46, no. 14 (2010): 2376–2378.
34. Chen, Da, Hongbin Feng, and Jinghong Li. "Graphene oxide: Preparation, functionalization, and electrochemical applications."*Chemical Reviews*112, no. 11 (2012): 6027–6053.
35. Geim, Andre Konstantin. "Graphene: Status and prospects."*Science*324, no. 5934 (2009): 1530–1534.
36. Sun, Xiaoming, Zhuang Liu, Kevin Welsher, et al. "Nano-graphene oxide for cellular imaging and drug delivery."*Nano Research*1 (2008): 203–212.
37. Zhu, Yanwu, Shanthi Murali, Weiwei Cai, et al. "Graphene and graphene oxide: Synthesis, properties, and applications."*Advanced Materials*22, no. 35 (2010): 3906–3924.
38. Allen, Matthew J., Vincent C. Tung, and Richard B. Kaner. "Honeycomb carbon: A review of graphene."*Chemical Reviews*110, no. 1 (2010): 132–145.
39. Liu, Yong, Dingshan Yu, Chao Zeng, et al. "Biocompatible graphene oxide-based glucose biosensors."*Langmuir*26, no. 9 (2010): 6158–6160.
40. Novoselov, Kostya S., Andre K. Geim, Sergei V. et al. "Electric field effect in atomically thin carbon films."*Science*306, no. 5696 (2004): 666–669.
41. He, Qiyuan, Shixin Wu, ZongyouYin,et al. "Graphene-based electronic sensors."*Chemical Science*3, no. 6 (2012): 1764–1772.
42. Stankovich, Sasha, Dmitriy A. Dikin, Geoffrey HB Dommett, et al. "Graphene-based composite materials."*Nature*442, no. 7100 (2006): 282–286.
43. Pumera, Martin. "Graphene-based nanomaterials and their electrochemistry."*Chemical Society Reviews*39, no. 11 (2010): 4146–4157.
44. Bonaccorso, Francesco, Zhipei Sun, Tawfique Hasan, et al. "Graphene photonics and optoelectronics."*Nature Photonics*4, no. 9 (2010): 611–622.
45. Balasubramanian, Kannan, and Marko Burghard. "Chemically functionalized carbon nanotubes."*Small*1, no. 2 (2005): 180–192.
46. Dai, Hongjie. "Carbon nanotubes: Synthesis, integration, and properties."*Accounts of Chemical Research*35, no. 12 (2002): 1035–1044.
47. Baughman, Ray H., Anvar A. Zakhidov, and Walt A. De Heer. "Carbon nanotubes--the route toward applications."*Science*297, no. 5582 (2002): 787–792.
48. Wei, B. Q., Robert Vajtai, and P. M. Ajayan. "Reliability and current carrying capacity of carbon nanotubes."*Applied Physics Letters*79, no. 8 (2001): 1172–1174.
49. Endo, Morinobu, Michael S. Strano, and Pulickel M. Ajayan. "Potential applications of carbon nanotubes." In*Carbon Nanotubes: Advanced Topics in the Synthesis, Structure, Properties and Applications*, 13–62. Berlin, Heidelberg: Springer Berlin Heidelberg, 2007.
50. Geim, Andre K., and Konstantin S. Novoselov. "The rise of graphene."*Nature Materials*6, no. 3 (2007): 183–191.
51. Daniel, Marie-Christine and Didier Astruc. "Gold nanoparticles: Assembly, supramolecular chemistry, quantum-size-related properties, and applications toward biology, catalysis, and nanotechnology."*Chemical Reviews*104, no. 1 (2004): 293–346.

52. Huang, Xiaohua, Prashant K. Jain, Ivan H. El-Sayed, et al. "Gold nanoparticles: Interesting optical properties and recent applications in cancer diagnostics and therapy."*Nanomedicine*2, no. 5 (2007): 681–693.
53. Murphy, Catherine J., Tapan K. Sau, Anand M. Gole,et al. "Anisotropic metal nanoparticles: Synthesis, assembly, and optical applications."*The Journal of Physical Chemistry B*109, no. 29 (2005): 13857–13870.
54. Sharma, Virender K., Ria A. Yngard, and Yekaterina Lin. "Silver nanoparticles: Green synthesis and their antimicrobial activities."*Advances in Colloid and Interface Science*145, no. 1–2 (2009): 83–96.
55. Sondi, Ivan, and Branka Salopek-Sondi. "Silver nanoparticles as antimicrobial agent: A case study on E. coli as a model for Gram-negative bacteria."*Journal of Colloid and Interface Science*275, no. 1 (2004): 177–182.
56. Rai, Mahendra, Alka Yadav, and Aniket Gade. "Silver nanoparticles as a new generation of antimicrobials."*Biotechnology Advances*27, no. 1 (2009): 76–83.
57. Narayanan, Radha, and Mostafa A. El-Sayed. "Shape-dependent catalytic activity of platinum nanoparticles in colloidal solution."*Nano Letters*4, no. 7 (2004): 1343–1348.
58. Wei, Ming, Yanxia Qiao, Haitao Zhao, et al. "Electrochemical non-enzymatic glucose sensors: Recent progress and perspectives."*Chemical Communications*56, no. 93 (2020): 14553–14569.
59. Sakamoto, Masanori, Mamoru Fujistuka, and Tetsuro Majima. "Light as a construction tool of metal nanoparticles: Synthesis and mechanism."*Journal of Photochemistry and Photobiology C: Photochemistry Reviews*10, no. 1 (2009): 33–56.
60. Wei, Liangqin, Hongdi Sun, TiantianYang,et al. "Iron carbide encapsulated by porous carbon nitride as bifunctional electrocatalysts for oxygen reduction and evolution reactions."*Applied Surface Science*439 (2018): 439–446.
61. Zhang, Hao, Xiaojun Lv, YuemingLi,et al. "P25-graphene composite as a high performance photocatalyst."*ACS Nano*4, no. 1 (2010): 380–386.
62. Xiang, Quanjun, Jiaguo Yu, and Mietek Jaroniec. "Graphene-based semiconductor photocatalysts."*Chemical Society Reviews*41, no. 2 (2012): 782–796.
63. Cao, L. I., Mohammed J. Meziani, Sushant Sahu,et al. "Photoluminescence properties of graphene versus other carbon nanomaterials."*Accounts of Chemical Research*46, no. 1 (2013): 171–180.
64. Peng, Sheng, and Shouheng Sun. "Synthesis and characterization of monodisperse hollow Fe3O4 nanoparticles."*AngewandteChemie International Edition*46, no. 22 (2007): 4155–4158.
65. Zhu, Qingpeng, Jiaojiao Song, Zhaofei Liu, et al. "Photothermal catalytic degradation of textile dyes by laccase immobilized on Fe3O4@ SiO2 nanoparticles."*Journal of Colloid and Interface Science*623 (2022): 992–1001.
66. Mahato, Kuldeep, Sahil Nagpal, Mahero Ayesha Shah, et al. "Gold nanoparticle surface engineering strategies and their applications in biomedicine and diagnostics."*3 Biotech*9 (2019): 1–19.

67. Mansuriya, Bhargav D., and Zeynep Altintas. "Enzyme-free electrochemical nano-immunosensor based on graphene quantum dots and gold nanoparticles for cardiac biomarker determination."*Nanomaterials*11, no. 3 (2021): 578.
68. Yadav, Neelam, Vinod Kumar Garg, Anil Kumar Chhillar, et al. "Detection and remediation of pollutants to maintain ecosustainability employing nanotechnology: A review."*Chemosphere*280 (2021): 130792.
69. Yin, Perry T., Tae-Hyung Kim, Jeong-Woo Choi, et al. "Prospects for graphene–nanoparticle-based hybrid sensors."*Physical Chemistry Chemical Physics*15, no. 31 (2013): 12785–12799.
70. Gupta, Sanju, and Romney Meek. "Metal nanoparticles-grafted functionalized graphene coated with nanostructured polyaniline 'hybrid'nanocomposites as high-performance biosensors."*Sensors and Actuators B: Chemical*274 (2018): 85–101.
71. Biswas, Manik C., Amit Chowdhury, Md Milon Hossain, et al. "Applications, drawbacks, and future scope of nanoparticle-based polymer composites." In*Nanoparticle-based Polymer Composites*, 243–275. Woodhead Publishing, 2022.
72. Alam, Mir Waqas, Shahidul Islam Bhat, Hassan S. Al Qahtani, et al. "Recent progress, challenges, and trends in polymer-based sensors: A review."*Polymers*14, no. 11 (2022): 2164.
73. Nabgan Walid, Aishah Abdul Jalil, Bahador Nabgan, et al. "A state of the art overview of carbon-based composites applications for detecting and eliminating pharmaceuticals containing wastewater."*Chemosphere*288 (2022): 132535.
74. Pourakbari, Ramin, Nasrin Shadjou, Hadi Yousefi, et al. "Recent progress in nanomaterial-based electrochemical biosensors for pathogenic bacteria."*Microchimica Acta*186 (2019): 1–13.
75. Liaquat, Hina, Muhammad Imran, Shoomaila Latif, et al. "Multifunctional nanomaterials and nanocomposites for sensing and monitoring of environmentally hazardous heavy metal contaminants."*Environmental Research*214 (2022): 113795.
76. Umapathi, Reddicherla, Seyed Majid Ghoreishian, Sonam Sonwal, et al. "Portable electrochemical sensing methodologies for on-site detection of pesticide residues in fruits and vegetables."*Coordination Chemistry Reviews*453 (2022): 214305.
77. Abdel-Karim, Randa. "Nanotechnology-enabled biosensors: A review of fundamentals, materials, applications, challenges, and future scope."*Biomedical Materials & Devices*2, no. 2 (2024): 759–777.
78. Sher, Farooq, Imane Ziani, Mariam Hameed, et al. "Advanced nanomaterials design and synthesis for accelerating sustainable biofuels production–A review."*Current Opinion in Green and Sustainable Chemistry*(2024): 100925.
79. Wang, J. "Electrochemical sensors."*Analytical Electrochemistry*3 (2006): 201–243.

80. Bard, Allen J., and Larry R. Faulkner. "Electrochemical methods: Fundamentals and applications."*Surf Technologies*20, no. 1 (1983): 91–92.
81. Kissinger, Peter T., and William R. Heineman. "Cyclic voltammetry."*Journal of Chemical Education*60, no. 9 (1983): 702.
82. Kissinger, Peter T., and William R. Heineman. "Cyclic voltammetry."*Journal of Chemical Education*60, no. 9 (1983): 702.
83. Wang, Joseph. *Analytical Electrochemistry*. New York: A John Willey and Sons.Inc.,2000, 81–84.
84. Bartle, Keith D., Christopher Gibson, Derek Mills, et al. "Differential-pulse voltammetry at the hanging-mercury-drop electrode for identification of aromatic structures in coal extracts."*Analytical Chemistry*54, no. 11 (1982): 1730–1733.
85. Wang, Joseph. "Stripping analysis: Principles."*Instrumentation and Applications, VCH, Deerfield Beach, FL*(1985): 30.
86. Fan, Fu-Ren F., Jose Fernandez, Biao Liu, et al. "Platinum and gold inlaid disks≥ 5 μm diameter." In*Handbook of Electrochemistry*, pp. 189–197. Elsevier, 2007.
87. Wang, Baozhen, Uichi Akiba, and Jun-ichi Anzai. "Recent progress in nanomaterial-based electrochemical biosensors for cancer biomarkers: A review."*Molecules*22, no. 7 (2017): 1048.
88. Justino, Celine IL, Armando C. Duarte, et al. "Recent progress in biosensors for environmental monitoring: A review."*Sensors*17, no. 12 (2017): 2918.
89. Daniels, Jonathan S., and Nader Pourmand. "Label-free impedance biosensors: Opportunities and challenges."*Electroanalysis: An International Journal Devoted to Fundamental and Practical Aspects of Electroanalysis*19, no. 12 (2007): 1239–1257.
90. Cui, Yi, Qingqiao Wei, Hongkun Park, et al. "Nanowire nanosensors for highly sensitive and selective detection of biological and chemical species."*Science*293, no. 5533 (2001): 1289–1292.
91. Tiwari, Jitendra N., Varun Vij, et al. "Engineered carbon-nanomaterial-based electrochemical sensors for biomolecules."*ACS Nano*10, no. 1 (2016): 46–80.
92. Shao, Yuyan, Jun Wang, Hong Wu,et al. "Graphene based electrochemical sensors and biosensors: A review."*Electroanalysis: An International Journal Devoted to Fundamental and Practical Aspects of Electroanalysis*22, no. 10 (2010): 1027–1036.
93. Liu, Zhuang, Joshua T. Robinson, Xiaoming Sun, et al. "PEGylated nanographene oxide for delivery of water-insoluble cancer drugs."*Journal of the American Chemical Society*130, no. 33 (2008): 10876–10877.
94. Wang, Joseph. "Carbon-nanotube based electrochemical biosensors: A review."*Electroanalysis: An International Journal Devoted to Fundamental and Practical Aspects of Electroanalysis*17, no. 1 (2005): 7–14.
95. Cai, Chenxin, and Jing Chen. "Direct electron transfer of glucose oxidase promoted by carbon nanotubes."*Analytical Biochemistry*332, no. 1 (2004): 75–83.

96. Ali, Mohamamd, and Mahmoudzadeh Ahmadi Nejad. "Carbon nanotubes for biomolecular sensing and photovoltaics." PhD diss., University of British Columbia, 2008.
97. Li, Yongxin, Rouzbeh Afrasiabi, Farkhondeh Fathi, et al. "Impedance based detection of pathogenic E. coli O157: H7 using a ferrocene-antimicrobial peptide modified biosensor."*Biosensors and Bioelectronics*58 (2014): 193–199.
98. Singh, Renu, Maumita Das Mukherjee, Gajjala Sumana, et al. "Biosensors for pathogen detection: A smart approach towards clinical diagnosis."*Sensors and Actuators B: Chemical*197 (2014): 385–404.
99. Zhao, Weian, Michael A. Brook, and Yingfu Li. "Design of gold nanoparticle-based colorimetric biosensing assays."*ChemBioChem*9, no. 15 (2008): 2363–2371.
100. Rashid, JahwarharIzuan Bin Abdul. "Development of dna electrochemical sensor based on silicon nanowires/gold nanoparticles-modifed electrode for early detection of dengue virus." (2016).
101. Tang, Libin, Xueming Li, Rongbin Ji, Kar Seng Teng, et al. "Bottom-up synthesis of large-scale graphene oxide nanosheets."*Journal of Materials Chemistry*22, no. 12 (2012): 5676–5683.
102. Liu, Wei, and Giorgio Speranza. "Functionalization of carbon nanomaterials for biomedical applications."*C - Journal of Carbon Research*5, no. 4 (2019): 72.
103. He, Qiyuan, Shixin Wu, Zongyou Yin, et al. "Graphene-based electronic sensors."*Chemical Science*3, no. 6 (2012): 1764–1772.
104. Tang, Yongchao, Zongbin Zhao, Han Hu, et al. "Highly stretchable and ultrasensitive strain sensor based on reduced graphene oxide microtubes–elastomer composite."*ACS Applied Materials & Interfaces*7, no. 49 (2015): 27432–27439.
105. Wang, Chuan, David Hwang, Zhibin Yu, et al. "User-interactive electronic skin for instantaneous pressure visualization."*Nature Materials*12, no. 10 (2013): 899–904.
106. Cui, Yi, Qingqiao Wei, Hongkun Park, and Charles M. Lieber. "Nanowire nanosensors for highly sensitive and selective detection of biological and chemical species."*Science*293, no. 5533 (2001): 1289–1292.
107. Song, Qi, Xindi Sun, Ziyi Dai, et al. "Point-of-care testing detection methods for COVID-19."*Lab on a Chip*21, no. 9 (2021): 1634–1660.
108. Huang, Xian, Yuhao Liu, et al. "Materials and designs for wireless epidermal sensors of hydration and strain."*Advanced Functional Materials*24, no. 25 (2014): 3846–3854.
109. Liu, R., Lalanne, L., Berry, V. "Graphene oxide as an amphiphilic material for making solar fuels."*Chemical Physics Letters* 639(2015): 82–91.
110. Wang, Ruru, Xuerui Liu, et al. "Novel electrochemical non-enzymatic glucose sensor based on 3D Au@ Pt core–shell nanoparticles decorated graphene oxide/multi-walled carbon nanotubes composite."*Microchemical Journal*174 (2022): 107061.

111. Zhang, Yin, Tapas R. Nayak, Hao Hong, et al. "Graphene: A versatile nano-platform for biomedical applications."*Nanoscale*4, no. 13 (2012): 3833–3842.
112. Shajari, Shaghayegh, Kirankumar Kuruvinashetti, Amin Komeili, et al. "The emergence of AI-based wearable sensors for digital health technology: A review."*Sensors*23, no. 23 (2023): 9498.
113. Banerjee, Amit, Chinmay Chakraborty, Anand Kumar, et al. "Emerging trends in IoT and big data analytics for biomedical and health care technologies."In *Handbook of Data Science Approaches for Biomedical Engineering*, 121–152,2020.
114. Kasoju, Naresh, et al. "Digital health: Trends, opportunities and challenges in medical devices, pharma and bio-technology." *CSI Transactions on ICT* 11, no. 1 (2023): 11–30.
115. Awad, Atheer, et al. "Connected healthcare: Improving patient care using digital health technologies." *Advanced Drug Delivery Reviews* 178 (2021): 113958.

CHAPTER 9

Advances in Glucose Electrochemical Sensors Using Nanomaterials

Dhiya P, Rejithamol R, Moumita Gangopadhyay, Santhy A, and Vishnu Vardhan Palem

9.1 INTRODUCTION

The study of nanoscience involves investigating and manipulating substances at the atomic, molecular, and macromolecular levels, which exhibit unique attributes that considerably differ from those observed at larger dimensions. Materials with a single unit size of between 1 and 100 nm are called nanomaterials, which can be nanoscale fibres, tubes, rods, or particles. A nanoparticle is defined as one that is smaller than 100 nanometres in at least one dimension. With the advancement in nanotechnology, nanomaterials are being applied in various fields, such as information transformation, electronics, textiles, cosmetics, health care, and environmental protection. Numerous electronic, computing, environmental, and biomedical industries have employed nanomaterials, namely, quantum dots, nanorods, and nanowires, due to their intriguing optical and electrical characteristics. A materials science–based approach to nanotechnology is used in nanomaterial research [1].

The characteristics of nanoscale materials differ greatly from those at the macroscale. The two main causes of nanomaterials' notable differences

DOI: 10.1201/9781003512639-9

from other materials' characteristics are their larger relative surface area and the quantum confinement effect. These elements can improve or change attributes like electrical characteristics, strength, and reactivity. The surface area of nanomaterials is shown to be higher than the same volume or mass of macro materials of the same type [2]. Because the electronic wave function in nanoparticles is constrained by the physical size of the particles, the electronic energy levels are discrete rather than continuous as they are in the bulk. Because of this phenomenon, which is known as quantum confinement, nanoparticles are also known as quantum dots (QDs) [3].

The following are typical examples of nanostructures: Self-assembled quantum dots, quantum heterostructures, gradient multilayer nanofilms, nanocages, icosahedral twins, magnetic nano chains, nanoparticles, nanopillar, nanoribbon, nanorods, nanosheet, nanoshell, nanowire, nanostructured film, nano fabrics, etc. Applications for nanostructured materials can be found in the food science, environmental, biomedical, pharmaceutical, and nanosensor fields. These substances decrease the biofouling of co-components and increase the solubility, bioavailability, and controlled release of the materials during fabrication and storage [4–7]. Furthermore, nanomaterials have attracted more attention due to their adjustable physicochemical properties, including melting point, wettability, electrical and thermal conductivity, catalytic activity, light absorption, and scattering properties [5, 6].

In this chapter, we focus on highlighting the advantages of different nanomaterials in the electrochemical sensing application for the special case of glucose as the analyte. The chapter will provide an overview of various electrochemical sensors for glucose and its relevance in our day-to-day life. A summary of types of nanomaterials used for glucose sensing is given in Figure 9.1, and the information on their existing and potential uses in glucose sensing is given in the following sections [6].

9.2 TYPES OF NANOMATERIALS

Nanomaterials can be broadly classified into four categories based on their material compositions: Composite-based nanomaterials, carbon-based nanomaterials, organic-based nanomaterials, and inorganic-based nanomaterials.

9.2.1 Carbon-Based Nanomaterials

Carbon-based nanomaterials are found in various morphologies, such as hollow tubes, spheres, or ellipsoids. Fullerenes (C_{60}), carbon nanofibres,

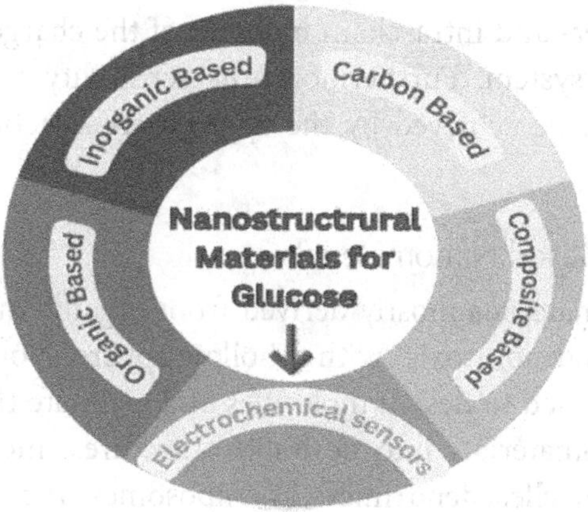

FIGURE 9.1 Different types of nanomaterials for glucose sensing.

carbon nanotubes (CNTs), carbon black, carbon onions, and graphene (Gr) are included in the carbon-based nanomaterials category. The important production methods for the fabrication of carbon-based materials, except carbon black, are chemical vapor deposition (CVD), laser ablation, and arc discharge [8]. Different types of carbon-based nanomaterials have been used for electrochemical sensing, according to a literature survey, and the contribution of carbon-based materials to glucose sensing is emphasized in this chapter [9].

9.2.2 Composite-Based Nanomaterials

Composite nanostructures include single-phase and multiphase nanoparticles, and nanomaterials that can be integrated with other nanoparticles of larger and bulk materials, such as hybrid nanofibre structures, for example metal–organic frameworks. These composites involve a combination of any metal, ceramic, or polymer bulk components with metal, organic, or carbon-based nanoparticles [10]. These nanomaterials are synthesized with varying morphologies based on the qualities needed for the intended use. The composite materials have also proved to be excellent materials in different electroanalytical applications [11, 12].

9.2.3 Inorganic-Based Nanomaterials

Metal and metal oxide–based nanomaterials are the most common examples of inorganic nanomaterials. Different inorganic-based nanomaterials like Au, Ag, etc., metal oxides such as CuO, TiO_2, and ZnO, etc., and ceramic-based nanomaterials have been widely applied in electrochemical and catalytic applications [13–16]. The inorganic nanomaterials can improve the inter- and intra-chain mobility of the charge carriers in the electrochemical system. Furthermore, the sensitivity and selectivity of the analytes can be achieved by the appropriate selection of inorganic nanomaterials.

9.2.4 Organic-Based Nanomaterials

These are nanomaterials mostly derived from organic materials and are biodegradable and non-toxic, with a hollow sphere. Non-covalent weak interactions are used to design molecules and facilitate the self-assembly of organic nanomaterials into desirable structures, including polymer nanoparticles, micelles, dendrimers, and liposomes. The spherical surface area of the nanosphere form of the organic nanomaterials can have the capacity to adsorb the analyte molecule very strongly to the electrode surface for the electrochemical reaction [13, 17].

9.3 GLUCOSE SENSORS

The human body's blood glucose (BG) concentration is a crucial diagnostic indicator for diabetes, and because diabetes death rates are high, more effective methods of BG concentration detection are required [18]. Because poorly managed BG causes over a million deaths annually, precise measurement of the concentration of BG is crucial for medical diagnosis and personal assessment. Maintaining BG concentration within the normal range and avoiding problems requires ongoing, regular monitoring [19, 20], which can significantly raise the percentage of diabetic patients who survive. Even though numerous biosensors for diabetes mellitus (DM) have been reported to date using various concepts, each with potential applications, none of them can concurrently accomplish high sensitivity, micro/nano, non-invasiveness, continuity, and low limits of detection. Hence, BG monitoring technology continues to address various challenges, offering potential opportunities for further research [21].

Several glucose sensors, such as optical and electrochemical, have been available in recent years for the regular monitoring of blood glucose levels. When compared with alternative glucose sensors, electrochemical and optical-based BG biosensors hold a dominant market share. This chapter reviews the state of the field and recent advancements, examines critical issues that must be resolved for future research, and places special emphasis on glucose biosensors and electrochemical sensors. Electrochemical sensors are extensively used in the food and agriculture industries as well as in environmental and biomedical applications [22–29]. They offer a straightforward and affordable alternative for the detection of various analytes. Their widespread use is primarily attributed to the low detection limits, resulting from the differences between the Faradaic and non-Faradaic currents, and the diverse range of electrochemical parameters, including current, voltage, net power output, and electrochemical impedance. This chapter aims to discuss the most recent developments and uses of electrochemical glucose sensors based on nanomaterials. We hope that the data provided here will stimulate more research into the development of electrochemical glucose sensors utilizing the properties of advanced types of nanomaterials [30].

In recent years, considerable attention has been directed towards nanomaterials for glucose sensing due to their expansive surface area. These nanostructures are characterized by unbalanced surface charges, thereby demonstrating heightened efficacy in the adsorption of glucose molecules. Junchao Qian et al. effectively synthesized a ZnO/CeO_2 nanocomposite by a hydrothermal process [31]. The resultant nanocomposite is evenly layered onto ultrathin petal carbon and subsequently immobilized onto the surface of a glassy carbon electrode (GCE). The electrochemical analysis of the ZnO/CeO_2-based sensor for glucose monitoring was performed using cyclic voltammetry and differential pulse voltammetry, operating within a potential of 0–1 V in a pH 7.2 phosphate buffer solution. The material exhibited excellent performance in quantifying glucose, featuring a linear range between 0.5 and 300 μM, coupled with a low detection limit (LOD) of 0.224 μM. The differential pulse voltammetry confirmed an exceptional correlation between current intensity and glucose concentration ($R^2 = 0.999$).

A sensor made of gold nanoparticles coupled with graphene nanofibres deposited on a glassy carbon electrode was reported by A.S. Shamsabadi et al. for the electrochemical sensing of glucose in a non-enzymatic way [32]. The sensor operated through the selective oxidation and reduction of glucose with the help of graphene nanofibres. The sensor provided a linear range response of 0.5–9.0 mM, a sensitivity of 1.1437 $\mu A\ mM^{-1}cm^{-2}$, a LOD of 55 μM, and a response time of the electrode of 1.7 s. Zahra Damirchi and co-workers have developed a remarkable electrochemical sensor that employs nanoparticles of Ni and Sm incorporated with double-layer MXene hydroxide for the detection of glucose from sweat using a paper-based device [33]. This magnificent design obviates the requirement for direct electrical connections between the power source and the electrode, thereby facilitating glucose detection. MXenes are one of the emerging class of two-dimensional (2D) nanomaterials having superior properties, including high conductivity, and hence, are widely applied in electrochemical applications [34]. The MXene-nanocomposite-based device demonstrates several notable features, including its ability to function without sophisticated equipment or specialized personnel, facilitating non-invasive glucose detection within physiological conditions. Through systematic optimization of various features, the measuring device enables precise detection within a linear range spanning from 10 to 200 μM, and the LOD for glucose is 3.6 μM. This flexible paper-based sensor would be a valuable device in the point-of- care diagnosis of glucose in future.

Another glucose sensor was reported by Md Abu Zahed et al. using laser-induced graphene with poly(3, 4-ethylene dioxythiophene)-poly(styrene sulfonate) nanocomposite. For the successful monitoring of glucose, this modification was further enhanced with Pt@Pd nanoparticles. The resultant biosensor demonstrated good sensitivity, working range, and remarkable selectivity towards glucose. It successfully monitored the concentration of glucose variations in sweat after meal intake and physical activity. Notably, the fabricated biosensor exhibited a remarkable current response to glucose, with a broad linear range of 10 μM to 9.2 mM [35].

Recently, Li Yu et al. developed a cost-effective, stable, environmentally friendly, and efficient non-enzymatic polyoxometalates (POM)-based hybrid electrode material for electrochemical glucose detection [36]. The compound $[Cu_4(H_2O)_2(PW_9O_{34})2]_{10}$ serves as the catalytic centre, demonstrating notable electrocatalytic performance for conventional non-enzymatic glucose sensors. Incorporating $[Cu_4(H_2O)_2(PW_9O_{34})2]_{10}$ into a carbon paste electrode (CPE) as an insoluble Cs^+ salt (Cs-CuPOM/CPE) enables effective electrocatalysis and facilitates heterogeneous electrochemical glucose sensing. Consequently, $Cs_{10}[Cu_4(H_2O)_2(PW_9O_{34})2]$ represents a promising electrode material for glucose sensors, offering enhanced stability, a broader linear range, and a lower LOD. Its rapid electron transfer kinetics contribute to its remarkable characteristics, including a wide linear range, sensitivity, and LOD of 5–1500 μM, 16.3 A $M^{-1}cm^{-2}$, and 0.99×10^{-6} M, respectively.

A non-enzymatic biosensor for the detection of glucose, which is highly selective and sensitive, was achieved by coating a cost-effective nanocomposite containing nitrogen-doped functionalized graphene (NFG), core-shell nanoparticles (Fe3O4@MIL-101-NH_2), and conductive polyaniline (PANI) on graphite sheet (GS) electrodes [37]. This sensor was developed by the research group of Mohammad Ali Ghaffarirad, and their findings

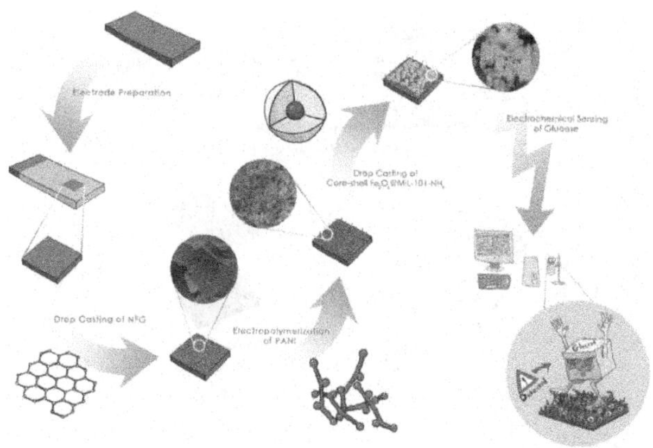

FIGURE 9.2 Schematic representation of NFG/PANI/Fe_3O_4@MIL-101-NH_2 nanocomposite for glucose detection. (Adapted from *Inorganic Chemistry Communications*, 159, M.A. Ghaffarirad, A. Sabahi, Z. Golshani, F. Manteghi, and A. Ghaffarinejad, Non-enzymatic glucose electrochemical sensor based on nitrogen-doped graphene modified with polyaniline and Fe3O4@MIL-101-NH2 nano framework, 111812, Copyright (2024), with permission from Elsevier.)

demonstrated that the iron oxide nanoparticles at the core, encased within MIL-101-NH_2 shells, enhance electron transfer and conductivity, mitigating the inherent low conductivity of the MOF (Figure 9.2). Furthermore, incorporating Fe_3O_4@MIL-101-NH_2 into the nanocomposite structure improves the sensitivity and selectivity of the determination of glucose, attributed to the presence of the amine groups and their peculiar structure. The NFG/PANI/Fe_3O_4@MIL-101-NH_2 nanocomposite provides a large surface area and catalytic properties as well as improved electron transfer, facilitating the effective oxidation of glucose, providing two linear ranges (0.5–10 μM and 100 μM to 25 mM) with a LOD of 0.3 μM. Moreover, the sensor exhibited remarkable selectivity even in the presence of various interferences.

Mahmoud Maki Ahmad and co-workers have developed a nanocomposite containing a metal-organic framework (MOF), 2,3,6,7,14,15-hexahydroxy triptycene, and Ni–P nanosheets on a nickel foam electrode [38]. The Ni–MOF (Ni–HHTT) was first synthesized by a hydrothermal method on a nickel foam (NF), and a further phosphidation process in an inert atmosphere produced Ni–P nanosheets derived from the MOF. The resultant sensor provided two concentration ranges, from 0.5 to 1700 μM and 1700 to 4100 μM, coupled with a detection limit of 0.15 μM for the monitoring of glucose. This sensor offers notable benefits, including

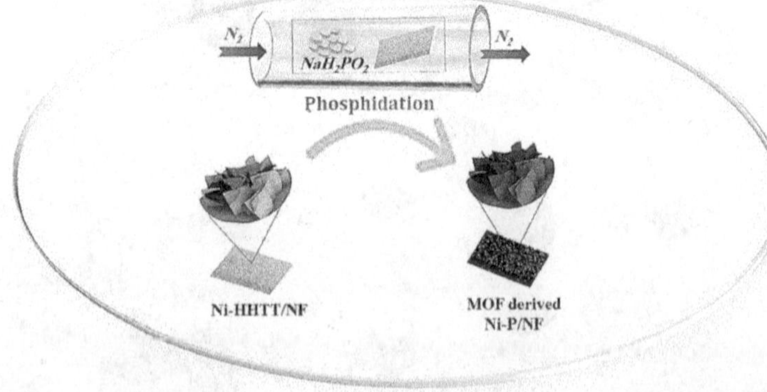

FIGURE 9.3 Graphical illustration of the synthesis of Ni–P nanosheet derived from MOF for the electrochemical sensing of glucose. (Adapted from *Microchemical Journal*, 197, M. Maki Ahmad, M. Roushani, and S. Farokhi, Ni-P nanosheets derived from a metal–organic framework containing triptycene ligand: A high-performance electrochemical sensor for glucose determination, 109737, Copyright (2024), with permission from Elsevier.)

high sensitivity, precision, rapid response time, durability, and cost-effectiveness. A graphical representation of the synthesis of Ni–P nanosheets derived from MOF for the electrochemical sensing of glucose is given in Figure 9.3.

Guangyu Wang et al. investigated the utilization of cobalt-based nanomaterials, renowned for their diverse oxidation states and robustness in alkaline environments, in glucose detection applications. Specifically, they employed a cobalt-based zeolitic imidazolate framework (ZIF-67), wherein Co^{2+} ions were coordinated with carboxyl groups of oxidized graphene oxide (GO) to generate cobalt-based nanoparticles. This integration led to the formation of a first-time cobalt-based nanomaterial to detect glucose. Hollow-Co_3O_4/GO nanocomposites were synthesized by the process of calcination at 260 °C for 4 hours. This nanomaterial was utilized as the electrochemical sensor for the determination of glucose by incorporating an impregnation method [40]. The peculiar structure of hollow-Co_3O_4/GO facilitates reoxidation reactions at the interface, which results in a significant increase in the surface area and subsequent enhancement of glucose oxidation current. Additionally, under optimized conditions, the sensitivity of the modified electrode was 1625.33 μA $mmol^{-1}cm^{-2}$, with a LOD of 0.43 μmol L^{-1} and a response time of approximately 2 s.

Qingteng Lai et al. present a method for fabricating nano Cu–Au bimetallic array microelectrodes, followed by the integration of a microfluidic channel for the electrochemical determination of glucose [20]. The microelectrodes fabricated through this technique exhibit a significantly enlarged electrode active area, which is highly advantageous for enhancing the signal in kinetically controlled glucose oxidation reactions. The result suggests that a reduced electrochemical growth driving force, when combined with a capping agent, effectively increases the Au (200) crystal plane ratio in gold particles. By utilizing flow field simulation, the optimal design of the reaction chamber in the microfluidic electrochemical chip was identified, which reduces dead space in the detection fluid flow and accelerates the sensing signal response. The obtained sensor demonstrates a linear detection range for glucose spanning from 10 to 4×10^5 μM, with a LOD of 284 nM.

B. Avinash et al. produced nanocrystalline zinc oxide (ZnO) through a bio-combustion process utilizing *Aloe vera* latex extract as fuel. The ZnO-modified carbon paste electrode was excellent at monitoring D-glucose and paracetamol in an alkaline solution by voltammetric measurements [41]. The fast responsiveness properties of the electrode material are the highlight of the work. In another work, Mir Waqas Alam et al. employed Mexican Mint gel as the fuel source to synthesize bismuth oxide nanoparticles (BONPs). They explored the catalytic properties of these nanoparticles through photocatalysis and electrochemical sensing and found excellent performance in detecting glucose and uric acid in an alkaline medium, using cyclic voltammetry measurements on the CPE/BONPs [42]. A plethora of electrochemical sensors have been reported for the electrochemical assay of glucose using different combinations of nanomaterials, and this chapter spotlights the characteristics of only a few of them. Hence, a brief description and the electrochemical performance of some of the recent nanomaterial-based electrochemical sensors for glucose monitoring is given in Table 9.1.

9.4 CONCLUSION AND FUTURE PERSPECTIVES

Scientific progress in applying nanotechnology to sensor fabrication has been accelerating, and it is an ongoing area of research. The integration of multiple functional modules into wearable devices is expected to be a trend in future research. The modules may consist of highly sensitive detection systems for fast response, self-powered systems for sensor powers, non-toxic and flexible substrates for different wearable requirements,

TABLE 9.1 Some Recent Examples of Glucose Sensors and Their Electrochemical Performance

Electrode	Modification	Linear range (μM)	LOD (μM)	Real sample	Supporting electrolyte	Reference
GCE	CB-PANI/AgNP	1000–10,000	520	–	0.1 M NaOH	[43]
Nickel foam	CuO@MCM-41	83–1500	0.016	–	0.1 M NaOH	[44]
Nickel foam	Nif/Cs/GO@Cu	5–4000	0.083	Blood serum samples	0.1 M NaOH	[45]
Screen-printed graphite	CuO/SnS_2NP	0–20,000	9.7	Lactose-free milk, ginger water, and coffee	PBS (pH 7)	[46]
Gold wire	AuNP–MIPs	0.001125–2.56	0.00125	Human serum sample	PBS (pH 7.4)	[47]
GCE	CuO/Cu2O@Cu	0-1000	0.56	Artificial sweat	0.1 M NaOH	[48]
GCE	Ni–Ca_2CuO_3	0.55–23,300	0.11	Banana juice	0.1 M NaOH	[49]
GCE	GOx@PAVE-CNTs	1.0–5000	0.36	Human urine	KNO_3	[50]

implantable drug delivery systems for better BG management, and microfluidic system control systems for trace detection. Therefore, designing advanced biosensors for the monitoring of glucose in a controlled setting and to acquire features like non-invasiveness, miniaturization, reliability, and stability is a matter of concern. More advanced nanoparticles with exceptional properties and prolonged stability in a range of conditions need to be created and fabricated shortly in order to fully harness the potential applications of nanomaterials in the realm of electrochemical sensors. The deeper advancement of electrochemical sensing applications will result from constructing electrochemical sensors that are more selective and sensitive than the analytical systems of today.

REFERENCES

1. Z. Vaseghi, A. Nematollahzadeh, Nanomaterials, in: *Green Synthesis of Nanomaterials for Bioenergy Applications*, Wiley, 2020: pp. 23–82. https://doi.org/10.1002/9781119576785.ch2.
2. F. Trotta, A. Mele, Nanomaterials: classification and properties, in: *Nanosponges*, Wiley, 2019: pp. 1–26. https://doi.org/10.1002/9783527341009.ch1.
3. P. Kambhampati, Nanoparticles, nanocrystals, and quantum dots: what are the implications of size in colloidal nanoscale materials?, *J Phys Chem Lett* 12 (2021) 4769–4779. https://doi.org/10.1021/acs.jpclett.1c00754.
4. B. Mekuye, B. Abera, Nanomaterials: an overview of synthesis, classification, characterization, and applications, *Nano Select* 4 (2023) 486–501. https://doi.org/10.1002/nano.202300038.
5. V. Singh, P. Yadav, V. Mishra, Recent advances on classification, properties, synthesis, and characterization of nanomaterials, in: *Green Synthesis of Nanomaterials for Bioenergy Applications*, Wiley, 2020: pp. 83–97. https://doi.org/10.1002/9781119576785.ch3.
6. E.C. Welch, J.M. Powell, T.B. Clevinger, A.E. Fairman, A. Shukla, Advances in biosensors and diagnostic technologies using nanostructures and nanomaterials, *Adv Funct Mater* 31 (2021). https://doi.org/10.1002/adfm.202104126.
7. K. Pathakoti, M. Manubolu, H.-M. Hwang, Nanostructures: current uses and future applications in food science, *J Food Drug Anal* 25 (2017) 245–253. https://doi.org/10.1016/j.jfda.2017.02.004.
8. O.S. Ayanda, A.O. Mmuoegbulam, O. Okezie, N.I. Durumin Iya, S.E. Mohammed, P.H. James, A.B. Muhammad, A.A. Unimke, S.A. Alim, S.M. Yahaya, A. Ojo, T.O. Adaramoye, S.K. Ekundayo, A. Abdullahi, H. Badamasi, Recent progress in carbon-based nanomaterials: critical review, *J Nanopart Res* 26 (2024) 106. https://doi.org/10.1007/s11051-024-06006-2.
9. J.N. Tiwari, V. Vij, K.C. Kemp, K.S. Kim, Engineered carbon-nanomaterial-based electrochemical sensors for biomolecules, *ACS Nano* 10 (2016) 46–80. https://doi.org/10.1021/acsnano.5b05690.

10. S. Xavier, M.K. Karuppan Perumal, R.R. Renuka, Two-dimensional nanomaterial polymer composites in biomedical applications, *Bionanoscience* (2024). https://doi.org/10.1007/s12668-024-01405-0.
11. S. Kempahanumakkagari, K. Vellingiri, A. Deep, E.E. Kwon, N. Bolan, K.H. Kim, Metal–organic framework composites as electrocatalysts for electrochemical sensing applications, *Coord Chem Rev* 357 (2018) 105–129. https://doi.org/10.1016/j.ccr.2017.11.028.
12. Y. Zuo, J. Xu, F. Jiang, X. Duan, L. Lu, H. Xing, T. Yang, Y. Zhang, G. Ye, Y. Yu, Voltammetric sensing of Pb(II) using a glassy carbon electrode modified with composites consisting of Co3O4 nanoparticles, reduced graphene oxide, and chitosan, *J Electroanal Chem* 801 (2017) 146–152. https://doi.org/10.1016/j.jelechem.2017.07.046.
13. B.H. Alshammari, M.M.A. Lashin, M.A. Mahmood, F.S. Al-Mubaddel, N. Ilyas, N. Rahman, M. Sohail, A. Khan, S.S. Abdullaev, R. Khan, Organic and inorganic nanomaterials: fabrication, properties, and applications, *RSC Adv* 13 (2023) 13735–13785. https://doi.org/10.1039/D3RA01421E.
14. R. Shashanka, B.E. Kumara Swamy, Simultaneous electro-generation and electro-deposition of copper oxide nanoparticles on glassy carbon electrode and its sensor application, *SN Appl Sci* 2 (2020). https://doi.org/10.1007/s42452-020-2785-1.
15. K. Ahmad, A. Mohammad, R. Rajak, S.M. Mobin, Construction of TiO2 nanosheets modified glassy carbon electrode (GCE/TiO2) for the detection of hydrazine, 074005 (n.d.).
16. Electrochemical sensing of ascorbic acid on ZnO-decorated reduced graphene oxide electrode, *J Biosens Bioelectron* 06 (2015). https://doi.org/10.4172/2155-6210.1000164.
17. J. Jeevanandam, A. Barhoum, Y.S. Chan, A. Dufresne, M.K. Danquah, Review on nanoparticles and nanostructured materials: history, sources, toxicity and regulations, *Beilstein J Nanotechnol* 9 (2018) 1050–1074. https://doi.org/10.3762/bjnano.9.98.
18. M. Maki Ahmad, M. Roushani, S. Farokhi, Ni-P nanosheets derived from a metal–organic framework containing triptycene ligand: A high-performance electrochemical sensor for glucose determination, *Microchem J* 197 (2024) 109737. https://doi.org/10.1016/j.microc.2023.109737.
19. G. Veerapandi, C. Sekar, Investigation of glucose concentration in banana as a function of ripening time using Ni-Ca2CuO3 based electrochemical sensor, *J Food Compos Anal* 126 (2024) 105888. https://doi.org/10.1016/j.jfca.2023.105888.
20. Q. Lai, Q. Niu, C. Zhang, N.M. Reis, M. Long, F. Wang, Z. Liu, Integrated Cu–Au stereo microelectrode arrays and microfluidic channels for the electrochemical detection of glucose, *Food Chem* 432 (2024) 137229. https://doi.org/10.1016/j.foodchem.2023.137229.
21. Z. Peng, X. Xie, Q. Tan, H. Kang, J. Cui, X. Zhang, W. Li, G. Feng, Blood glucose sensors and recent advances: a review, *J Innov Opt Health Sci* 15 (2022). https://doi.org/10.1142/S1793545822300038.

22. R. Rejithamol, S. Beena, Phenyl hydrazine and 2,4-dinitrophenyl hydrazine-based polymeric materials for the electrochemical quantification of thrombotonin, *MRS Adv* 6 (2021) 750–757. https://doi.org/10.1557/s43580-021-00116-y.
23. R. Rejithamol, S. Beena, Electrochemical quantification of pyridoxine (VB6) in human blood from other water-soluble vitamins, *Chemical Papers* 74 (2020) 2011–2020. https://doi.org/10.1007/s11696-019-01049-5.
24. R. Rejithamol, P.R. Keerthi, S. Beena, A novel disposable pencil graphite electrode for the voltammetric determination of cysteamine, *Mater Today Proc* (2019). https://doi.org/10.1016/j.matpr.2019.07.504.
25. R. Rejithamol, R.G. Krishnan, S. Beena, Disposable pencil graphite electrode decorated with a thin film of electro-polymerized 2,3,4,6,7,8,9,10-octahydropyrimido [1,2-a] azepine for simultaneous voltammetric analysis of dopamine, serotonin and tryptophan, *Mater Chem Phys* 258 (2021). https://doi.org/10.1016/j.matchemphys.2020.123857.
26. R. Rejithamol, R.G. Krishnan, S. Beena, Electrochemical quantification of L-tryptophan via molecular imprinted pyromellitic acid polymer-based indium tin oxide electrode, *J Electrochem Soc* 167 (2020) 117507. https://doi.org/10.1149/1945-7111/aba33e.
27. R. Rejithamol, S. Adwaitha, G. Darsana, M.V. Sreenidhi, R. Rajeswary, M. Abhirami, R. Reshma, Fabrication of pyromellitic acid polymer modified electrochemical sensor for the determination of tryptophan, *Asian J Chem* 34 (2022) 1526–1530. https://doi.org/10.14233/ajchem.2022.23705.
28. C. Devu, S.J. Kaveri, P. Dhiya, M.S. Aiswarya, K.C. Seetha Lakshmi, R. Rejithamol, In situ polymerization of meta nitro benzoic acid on graphite electrode as a mediator for the electrochemical oxidation of resorcinol, *J Appl Electrochem* 54 (2024) 13–24. https://doi.org/10.1007/s10800-023-01944-6.
29. R. Rajamani, D.C., S.P.J., D.V., A.P.S., M.K., V. Balaraman, Biopolymer supported electroanalytical methods for the determination of biomolecules and food additives – a comprehensive perspective, *Analytical Methods* 15 (2023) 2886–2904. https://doi.org/10.1039/D3AY00459G.
30. J. Baranwal, B. Barse, G. Gatto, G. Broncova, A. Kumar, Electrochemical sensors and their applications: a review, *Chemosensors* 10 (2022) 363. https://doi.org/10.3390/chemosensors10090363.
31. J. Qian, Y. Wang, J. Pan, Z. Chen, C. Wang, J. Chen, Z. Wu, Yangyue, Nonenzymatic glucose sensor based on ZnO–CeO2 whiskers, *Mater Chem Phys* 239 (2020) 122051. https://doi.org/10.1016/j.matchemphys.2019.122051.
32. A.S. Shamsabadi, H. Tavanai, M. Ranjbar, A. Farnood, M. Bazarganipour, Electrochemical non-enzymatic sensing of glucose by gold nanoparticles incorporated graphene nanofibers, *Mater Today Commun* 24 (2020). https://doi.org/10.1016/j.mtcomm.2020.100963.
33. Z. Damirchi, A. Firoozbakhtian, M. Hosseini, M.R. Ganjali, Ti3C2/Ni/Sm-based electrochemical glucose sensor for sweat analysis using bipolar electrochemistry, *Microchimica Acta* 191 (2024). https://doi.org/10.1007/s00604-024-06209-3.

34. A.G. Lekshmi, R. Rejithamol, A. Santhy, A. Raman, A. Aparna, A. Saritha, Surface terminations and surface functionalization strategies of MXenes, in: *MXene Reinforced Polymer Composites: Fabrication, Characterization and Applications*, 2024. https://doi.org/10.1002/9781119901280.ch3.
35. M.A. Zahed, S.C. Barman, P.S. Das, M. Sharifuzzaman, H.S. Yoon, S.H. Yoon, J.Y. Park, Highly flexible and conductive poly(3,4-ethylene dioxythiophene)-poly(styrene sulfonate) anchored 3-dimensional porous graphene network-based electrochemical biosensor for glucose and pH detection in human perspiration, *Biosens Bioelectron* 160 (2020) 112220. https://doi.org/10.1016/j.bios.2020.112220.
36. L. Yu, X. Ma, X. Cao, J. Zhao, Nanostructured polyoxometalate-based heterogeneous electrode materials for electrochemical sensing of glucose, *Inorg Chem* 63 (2024). https://doi.org/10.1021/acs.inorgchem.3c04596.
37. M.A. Ghaffarirad, A. Sabahi, Z. Golshani, F. Manteghi, A. Ghaffarinejad, Non-enzymatic glucose electrochemical sensor based on nitrogen-doped graphene modified with polyaniline and Fe3O4@MIL-101-NH2 nano framework, *Inorg Chem Commun* 159 (2024) 111812. https://doi.org/10.1016/j.inoche.2023.111812.
38. M. Maki Ahmad, M. Roushani, S. Farokhi, Ni-P nanosheets derived from a metal–organic framework containing triptycene ligand: A high-performance electrochemical sensor for glucose determination, *Microchem J* 197 (2024). https://doi.org/10.1016/j.microc.2023.109737.
39. M. Maki Ahmad, M. Roushani, S. Farokhi, Ni-P nanosheets derived from a metal–organic framework containing triptycene ligand: A high-performance electrochemical sensor for glucose determination, *Microchem J* 197 (2024) 109737. https://doi.org/10.1016/j.microc.2023.109737.
40. G. Wang, L. Li, X. Han, W. Gao, C. Wang, Nickel foam-decorated hollow-Co3O4/GO nanocomposites: A highly sensitive non-enzymatic electrochemical sensor for glucose detection, n.d. https://ssrn.com/abstract=4717133.
41. B. Avinash, C.R. Ravikumar, N. Basavaraju, B. Abebe, T.N. Kumar, S.N. Manjula, H.C.A. Murthy, Facile green synthesis of zinc oxide nanoparticles: Its photocatalytic and electrochemical sensor for the determination of paracetamol and D-glucose, *Environ Funct Mater* 2 (2023) 133–141. https://doi.org/10.1016/j.efmat.2024.01.002.
42. M.W. Alam, N. Allag, M. Utami, M. Waheed-Ur-Rehman, M. Al Saleh Al-Othoum, S. Sadaf, Facile green synthesis of α-bismuth oxide nanoparticles: Its photocatalytic and electrochemical sensing of glucose and uric acid in an acidic medium, *J Compos Sci* 8 (2024) 47. https://doi.org/10.3390/jcs8020047.
43. Q. Dong, H. Ryu, Y. Lei, Metal oxide based non-enzymatic electrochemical sensors for glucose detection, *Electrochim Acta* 370 (2021). https://doi.org/10.1016/j.electacta.2021.137744.
44. Z.D. Arif, M. Hassan, M. Abdullah, W. Miran, M.A. Nasir, S. Batool, M.A. Baig, U. Liaqat, An electrochemical sensor based on copper oxide nanoparticles loaded on a mesoporous MCM-41 for non-enzymatic detection of glucose, *Ceram Int* 50 (2024) 12614–12620. https://doi.org/10.1016/j.ceramint.2024.01.171.

45. F. Shi, S. Hu, J. Li, F. Wang, N. Chen, Glucose and UA sensing based on Cu nanoparticle decorated Nif/GO flexible electrode, *Microchimica Acta* 191 (2024). https://doi.org/10.1007/s00604-023-06066-6.
46. S. Kondee, W. Pon-On, W. Siriwatcharapiboon, A. Tuantranont, C. Wongchoosuk, CuO/SnS2 nanoparticles on PEDOT:PSS for nonenzymatic electrochemical glucose sensors, *ACS Appl Nano Mater* 7 (2024) 6722–6735. https://doi.org/10.1021/acsanm.4c01093.
47. E. Sehit, J. Drzazgowska, D. Buchenau, C. Yesildag, M. Lensen, Z. Altintas, Ultrasensitive nonenzymatic electrochemical glucose sensor based on gold nanoparticles and molecularly imprinted polymers, *Biosens Bioelectron* 165 (2020). https://doi.org/10.1016/j.bios.2020.112432.
48. X. Ban, J. Li, W. Sun, A. Sun, H. Li, Z. Yang, P. Pan, J. He, R. Zhang, Y. Zhao, A highly sensitive non-enzymatic glucose electrode based on truncated octahedral CuO-modified Cu2O@Cu composite, *Microchem J* 205 (2024). https://doi.org/10.1016/j.microc.2024.111221.
49. G. Veerapandi, C. Sekar, Investigation of glucose concentration in banana as a function of ripening time using Ni-Ca2CuO3 based electrochemical sensor, *J Food Compos Anal* 126 (2024). https://doi.org/10.1016/j.jfca.2023.105888.
50. S. Xu, Y. Zhang, Y. Zhu, J. Wu, K. Li, G. Lin, X. Li, R. Liu, X. Liu, C.P. Wong, Facile one-step fabrication of glucose oxidase loaded polymeric nanoparticles decorating MWCNTs for constructing glucose biosensing platform: Structure matters, *Biosens Bioelectron* 135 (2019) 153–159. https://doi.org/10.1016/j.bios.2019.04.017.

CHAPTER 10

How Nanomaterial-Based Sensors Contribute to a More Sustainable Environment

Varun Donnakatte Neelalochana, Sahana Kamanna Metry, Jubate Mohanty, Sandeep Shadakshari, Hareesha Nagarajappa, Sanjay Ballur Prasanna, Santhosh Arehalli Shivamurthy, Manjunatha Jamballi Gangadharappa, Eleonora Tomasino, Paolo Scardi, and Narges Ataollahi

10.1 INTRODUCTION

10.1.1 Environmental Sustainability and Sensor Technology

In the face of escalating environmental challenges, harnessing innovative technologies is crucial for preserving the health of our planet. Environmental sustainability and sensor technology are increasingly intertwined, as the latter offers groundbreaking solutions for monitoring, managing, and mitigating ecological threats [1]. This synergy can

DOI: 10.1201/9781003512639-10

revolutionize our approach to environmental stewardship, enabling data-driven decision-making and propelling us towards a more sustainable future. To ensure the planet's health and well-being, environmental sustainability is crucial. This involves implementing resource-efficient practices that don't disrupt natural cycles. Sensor technology is key to effective environmental monitoring, providing essential data to address ecological concerns [2]. Nanomaterial-based sensors, with their enhanced properties, represent a significant advancement in this field, offering new opportunities to address environmental challenges [3, 4]. This stands out as one of the leading technologies, widely regarded by policymakers, scientists, and industry leaders as a crucial innovation that will drive economic prosperity and sustainable development.

Some key aspects of environmental sustainability with nanomaterial-based sensors include:

a) ***Green synthesis:*** this involves using environmentally friendly methods and natural reagents for nanomaterial production, which reduces the need for hazardous chemicals and minimizes waste [5, 6].

b) ***Energy efficiency:*** The development of NMs and processes that prioritize energy conservation and minimize harmful emissions is vital for maintaining the planet's ecological balance [7]. A promising approach involves nanostructure-decorated interfaces, which offer a simplified and energy-efficient alternative to traditional deposition methods. These interfaces can be easily constructed without complex self-assembly, requiring fewer starting materials and streamlining the manufacturing process. This makes them an attractive option for sensor and energy applications, where sustainability and efficiency are paramount [8].

c) ***Sensitivity and selectivity:*** Sensitivity is the ability to quantitatively determine a small amount of a target from a sample [9, 10]. Selectivity refers to a sensor's ability to distinguish the intended analyte from other interfering species. Nanomaterial-based sensors can be designed to detect environmental contaminants at very low concentrations with high specificity, allowing early detection and prevention of environmental damage [11].

d) ***Portability and field-deployability:*** Nanomaterial-based sensors offer a groundbreaking solution for environmental monitoring, providing compact, lightweight, and real-time detection capabilities that eliminate the need for cumbersome laboratory equipment. These innovative sensors enable continuous, in-field monitoring of environmental conditions, such as air and water quality, temperature, and humidity, with rapid response times, high sensitivity, and low power consumption [12].

e) ***Sustainability of NMs:*** The creation of eco-friendly NMs, sourced from nature, offers a sustainable solution to reduce the environmental footprint of sensor production and deployment. By leveraging biodegradable and non-toxic materials, the ecological effects of sensor manufacturing and usage can be significantly mitigated, paving the way for a more environmentally responsible approach to sensor technology [13, 14].

10.1.2. Importance of Environmental Sustainability

Environmental sustainability is vital for safeguarding the well-being of both present and future generations by preserving the integrity of the planet's ecosystems [15]. This concept centres on adopting responsible practices that interact harmoniously with the environment, preventing the depletion or degradation of natural resources, and ensuring the long-term health and quality of our planet's ecosystems for generations to come.

a) ***Preservation of natural resources:*** To preserve natural resources like air, land, and water for future generations, sustainable practices are helpful. Overuse of these resources can result in their scarcity and loss of biodiversity, which will harm human existence and the ecological balance [16].

b) ***Mitigation of climate change:*** Sustainable practices aid in the fight against climate change by lowering greenhouse gas emissions [17]. Deforestation, industrial emissions, and the use of fossil fuels are some of the main causes of climate change, which puts biodiversity, sea levels, and weather patterns at serious risk [18].

c) ***Economic benefits:*** There are substantial financial advantages to environmental sustainability. Sustainable approaches frequently cut waste and increase efficiency, which lowers costs. Additionally, funding sustainable businesses and technologies can boost employment and the economy [19].

d) ***Health and well-being:*** Environments that are sustainable benefit human health and well-being. Air pollution, water pollution, and pollution in general are associated with several health problems, such as cancer, heart disease, and respiratory disorders. We can save healthcare expenses and enhance public health outcomes by preserving the environment [20].

e) ***Biodiversity conservation:*** Ecosystem resilience and stability depend on biodiversity. It offers a wide range of ecological services that are critical to agriculture and human existence, like pollination, nutrient cycling, and pest control. Sustainable practices ensure that these benefits will continue to be provided by helping to safeguard animals and their habitats [21].

f) ***Intergenerational responsibility:*** Using sustainable methods is a way for us to achieve our obligation to the next generation. It promotes a sense of stewardship and moral obligation towards the environment by guaranteeing that our children and grandchildren inherit a planet that can fulfil their needs and aspirations [22].

10.1.3 Role of Sensor Technology in Environmental Monitoring

Developing an environmental monitoring system is essential for maintaining the sustainability of the ecosystem [23]. The implementation of sensor technologies has greatly enhanced the efficiency and precision of environmental monitoring [24, 25]. Although there are still many obstacles to be solved, sensor technology has significantly improved environmental monitoring capabilities [26, 27]. Sensor technology plays a crucial role in environmental monitoring by collecting data on various environmental parameters, such as temperature, humidity, air quality, water quality, and more. These sensors are designed to detect and measure physical, chemical, and biological parameters, providing real-time data that help in understanding and managing the environment effectively [28].

Air quality monitoring: Metal oxide semiconductors are widely used in chemical sensors that rely on electrical transduction principles. These sensors are especially useful for detecting a wide range of gaseous analytes, such as harmful gases and environmental pollutants [29]. Particularly, NMs and nanoparticles (NPs) have demonstrated significant potential in reducing air pollution [30]. Growing amounts of carbon dioxide (CO_2) contribute to climate change and global warming, harming the environment as well as human health. As a result, CO_2 detection is crucial for both environmental and public health [31]. A study conducted by Jamila Boudaden et al. demonstrated that amine-functionalized silica NPs can accurately measure CO_2 levels within the range of 400 to 2000 ppm. These particles function well at temperatures below 60 °C and remain reliable even in humid environments with up to 80% relative humidity [32]. CuO nanowires synthesized by Jae-Hun Kim et al. by a thermal oxidation method have been used as gas sensors for the detection of carbon monoxide (CO); these sensors have shown the ability to detect CO gas concentrations ranging from 1 to 10 ppm when operating at an optimal temperature of 300 °C [33]. Moreover, these NMs are inherently recyclable, which further enhances their environmental benefits [30].

Water quality monitoring: Water pollution can occur accidentally or intentionally, and if the right steps aren't taken at once, the effects may be severe [34]. The quality of drinking water is impacted by the steadily rising rate of pollution, which also has an impact on aquatic systems and the biological cycle [35]. In addition to heavy metal contamination, which is becoming, if not already, a major concern for human health worldwide [36], there are also growing concerns about the environmental and health impacts of persistent hazardous pollutants, including dyes, pesticides, fertilizers, heavy metals, and their degradation products or derivatives [37]. The majority of traditional sensor systems either do not have user-friendly on-site monitoring features or can only monitor one parameter at a time [38]. An outstanding way to overcome this obstacle is by using a completely integrated electrochemical sensor array, due to their benefits, which include affordability, ease of use, and quick analysis times. Electro-analytical approaches have garnered significant interest lately [39]. Yashas S R et al. developed a biosensor, Gr/PPyNT/$SrCuO_2$/lac, to detect 2,4-dichlorophenol (2,4-DCP) in water. This enzyme-based sensor combines perovskite micro-seeds and Polypyrrole nanotube (PPyNT)-tethered laccase. It achieved a 99% recovery rate with minimal interference and a detection limit of 0.1878 μM [40]. The construction of an Sr@FeNi–S/GnRbs electrode, where GnRbs refer to graphene nanoribbons, produced an effective Flutamide sensor with a broad linear range of 20–450 nM and a low detection limit (20 nM) due to the beneficial interaction of the materials used [41].

10.1.4 Overview of Nanomaterial-Based Sensors

10.1.4.1 Nanomaterials

The physical and chemical characteristics of NMs are heavily influenced by factors such as their size, shape, and other properties. Establishing a single, universally accepted definition of NMs has proven to be a complex task, and the scientific community continues to engage in discussions to establish a rigorous and comprehensive definition for these materials [42, 43].

A nanomaterial is a broad term that encompasses nano-objects and nanostructured materials. It refers to any material that possesses at least one external dimension or an internal structure on the nanoscale, typically within the size range of 100 nanometres or less. NMs can take various forms, composed of discrete functional parts operating at the nanoscale level. Compared with the same material without nanoscale features, NMs often exhibit novel and unique characteristics due to their diminutive size and high surface-area-to-volume ratio [44].

To fully comprehend and value the wide range of variations among NMs, it is necessary to establish some form of classification system. NMs can be categorized according to their dimensional properties, ranging from zero-dimensional (0D) and one-dimensional (1D) to three-dimensional (3D) structures (Figure 10.1) [43, 45].

10.1.4.2. Properties of Nanomaterials

In the field of nanomaterial-based detection techniques, most of the mechanisms rely on the unique properties exhibited by NMs. These properties include their high surface-area-to-volume ratio, fluorescent behaviour, magnetic characteristics, and highly sensitive dielectric properties. By capitalizing on these distinctive nanoscale features, researchers have developed innovative detection methods that offer enhanced sensitivity and performance compared with conventional approaches [47, 48]. Unique properties of NPs are as follows:

a) NPs possess a unique characteristic: As their size decreases within the range of 1 to 100 nanometres, their physical behaviour transitions from being governed by classical physics to exhibiting quantum mechanical effects [43].

b) At the incredibly small scale of atomic clusters, an intriguing phenomenon known as negative thermal capacity can occur [49]. This behaviour arises from the distinction between temperature, which indicates kinetic energy, and thermal capacity, which relates to total energy. When a portion of kinetic energy transforms into potential energy, a decrease in temperature is observed. This phenomenon can

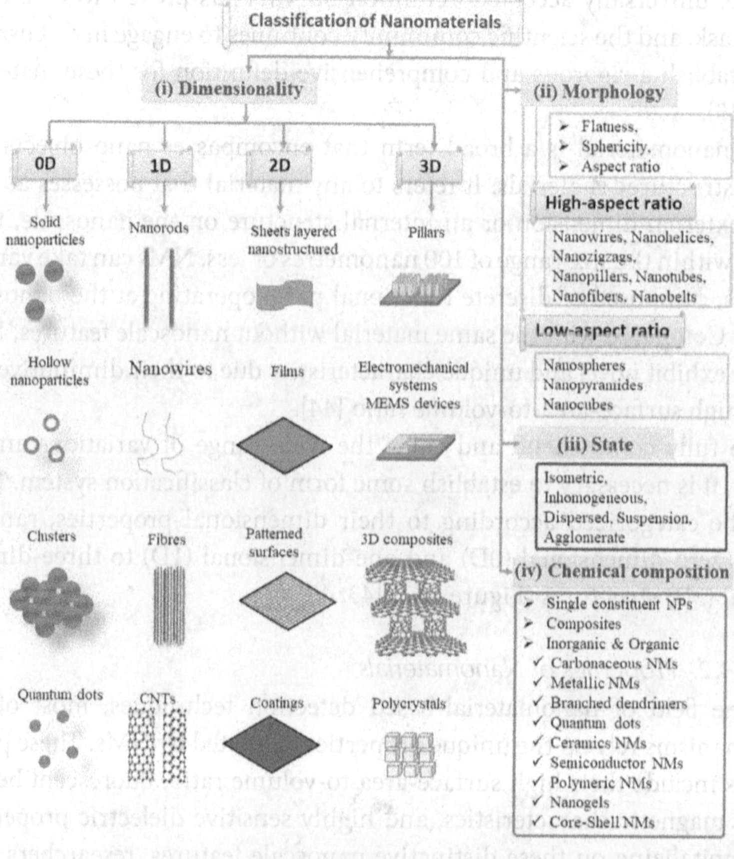

FIGURE 10.1 **Classification of NMs based on dimensions and their examples.** (From Saleh, T.A., *Environ. Technol. Innov.*, 20, 101067, 2020.)

be understood by examining the structural differences between the S8 ring and S8 linear chain formations of sulfur. Upon heating the S8 ring, the vibrational and rotational motions increase in amplitude, leading to a rise in temperature. However, when heat is localized to break a bond and convert the hot S8 ring into the colder S8 linear chain structure, the temperature drops, despite the energy input, due to the conversion of kinetic energy into potential energy during the structural change [50].

c) When materials are scaled down to the nanometre range, their electronic properties can differ significantly from those of their bulk counterparts. As an illustration, boron in its bulk form is not classified as a metal, yet when arranged in a two-dimensional network structure known as borophene, it exhibits remarkable metallic properties characteristic of an exceptional 2D metal [42, 51].

d) NPs exhibit enhanced thermal properties compared with their fluid counterparts due to their large surface area, which facilitates direct heat transfer on the material's surface. The thermal characteristics of materials can be gradually improved by increasing the concentration of added metal oxides, such as SiO_2, to the polycarbonate matrix. The presence of these metal oxides promotes stronger interactions between the NPs and polymers while also imposing constraints on polymer chain formation. Incorporating nanofillers with inherently high thermal conductivity can influence the overall thermal behaviour of NMs. Ultimately, the thermal properties of NPs are governed by several factors, including their substantial surface area, mass concentration, the ratio of energetic atoms within the NPs, and the volume fraction of dispersed NPs [52, 53].

e) NPs exhibit mechanical behaviours that are distinct from those of their bulk material counterparts due to surface effects and quantum phenomena that arise at the nanoscale. As an illustration, conventional FeAl powder comprised of microparticles larger than 4 micrometres exhibits brittle characteristics. In contrast, ultrafine FeAl alloy powder consisting of NPs displays an advantageous combination of strength, ductility, and enhanced plasticity [54]. These unique mechanical properties observed in NPs are believed to originate from the diverse interaction forces that occur between the NPs themselves or between the NPs and their surrounding surfaces [55–57].

f) Electrons are described by the quantum confinement effects in terms of potential wells, valence bands, conduction bands, and electron energy band gaps [58]. The "dimensionality" of a system in the nanometre range is associated with quantum size effects, as the quantum confinement effect is only noticeable when the particle's size is too small to be connected to the electron's wavelength [59].

g) The optical properties of metal NPs, such as silver, gold, and aluminium, can be understood through surface plasmon phenomena, where the free electrons within the NPs resonate at a frequency that corresponds to the wavelength of the incident light [60].

10.1.4.3 Synthesis Route for Nanomaterials

There are two well-known methods for synthesizing NMs. Both top-down and bottom-up methods are used. The raw materials are mechanically transformed into nanoscale particles in a top-down approach. Bottom-up procedures use co-precipitation or self-assembly techniques to assemble tiny particles into NMs [61]. The primary distinction between top-down and bottom-up methods for nanoparticle preparation lies in the starting material [62].

Top-down methods: These begin with bulk materials, which are then reduced to NPs through various physical, chemical, and mechanical processes. In contrast, bottom-up methods start with individual atoms or molecules, which are assembled to form NPs [62].

- **Mechanical milling/grinding procedures:** A top-down method that has been extensively used for the production of different NMs, nanograins, nanoalloys, nanocomposites, and nano-quasicrystalline materials is high-energy ball milling [63]. During ball milling, kinetic energy from a grinding medium is transferred to a material that is undergoing reduction, resulting in the production of NMs through mechanical attrition [64].
- The lithographic process is a printing technique that makes use of a perfectly smooth plate or stone [65]. An innovative technique for creating nanoscale materials or patterns is called nanoimprint lithography. Lithography projects an image with the desired pattern onto a surface using visible or ultraviolet (UV) light, X-rays, electrons, or ions [66].
- Various laser-based methods, including laser ablation, laser vaporization, pulsed laser deposition (PLD), laser–chemical vapor deposition (LCVD), and so on, are employed to create materials at the nanoscale that possess precise, regulated dimensions [67]. Reagent costs are modest, yield is reasonably high, and the laser ablation technique does not require molecular precursors (Figure 10.2) [68].

Bottom-up approach: This is a widely employed technique for synthesizing vast amounts of colloids with a variety of sizes and shapes. Through this process, organic molecules, atoms, and ions combine to form larger structures. Moreover, this process produces homogeneous NPs with a

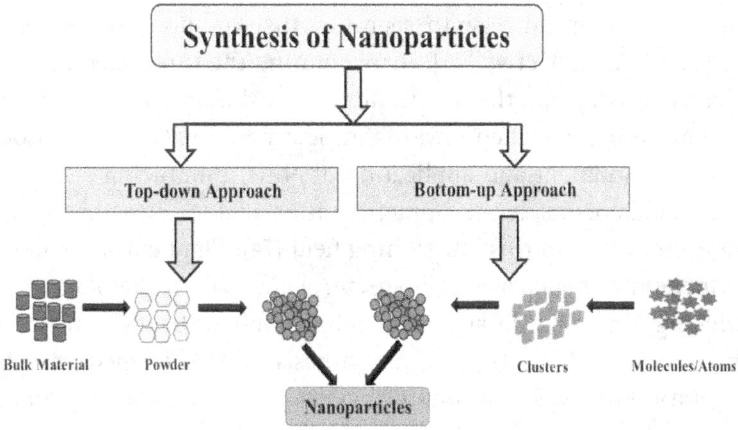

FIGURE 10.2 Bottom-up and top-down approaches in the synthesis of carbon-based NMs. (From Habiba, K. et al., *Manufacturing Nanostructures*, UK: One Central Press (OCP) 2014.)

narrow size distribution, a constant shape, and fewer surface imperfections [70].

- Using the inexpensive and straightforward **coprecipitation** approach, one can control size and size distribution by adjusting the relative rates of nucleation and growth during the synthesis process [71].
- The **sol–gel process** is a wet chemical approach for producing various nanostructures, especially metal oxide NPs [72]. It consists of the solution-borne condensation, hydrolysis, and thermal breakdown of metal alkoxides or precursors.
- Making NMs (dots, clusters, colloids, rods, wires, thin films) is incredibly easy with electrodeposition and electroless deposition. One easy technique to create NPs is to anodize aluminium to create a porous oxide structure [64].

10.1.4.4 Green Synthesis of Nanomaterials

Currently, chemical and physical methods are the primary techniques employed for large-scale NM production. However, these approaches pose several drawbacks, including detrimental environmental effects, potential health hazards, high energy consumption, and substantial costs associated with their implementation. These limitations have prompted the exploration of alternative synthesis routes that are more sustainable, eco-friendly, and cost-effective [73]. Incorporating the fundamental concepts of green chemistry into the rapidly advancing domain of nanoscience has become an imperative step towards mitigating potential risks associated with the development and application of NMs. Embracing environmentally conscious principles and practices is crucial for ensuring a sustainable trajectory within this burgeoning field [74]. Plant extracts, abundant in phytochemicals such as flavonoids, terpenoids, and alkaloids, can serve as reducing and capping agents, enabling their utilization in the green synthesis of NPs [73]. The green synthesis approach encompasses the incorporation of the 12 guiding principles of green chemistry and engineering, with the aim of producing NMs through environmentally benign and sustainable processes that prioritize safety considerations throughout their lifecycle (Figure 10.3) [75, 76].

A diverse array of plant extracts, derived from various parts such as leaves, flowers, fruits, and roots, have been employed in the synthesis of silver NPs (AgNPs). One notable example involves the utilization of pomelo peel extract, as reported by Nguyen. In their study, the peel extract underwent pretreatment with a citric acid solution for 2 hours at 85 °C, followed by mixing with a silver nitrate ($AgNO_3$) solution. Subsequent exposure of this mixture to sunlight facilitated the formation of AgNPs ranging from 20 to 30 nanometres in size, as characterized by transmission emission microscopy (TEM) analysis [76, 78]. Another study by Santhosh et al. presents the synthesis of AgNPs using *Euphorbia geniculate* leaf extract

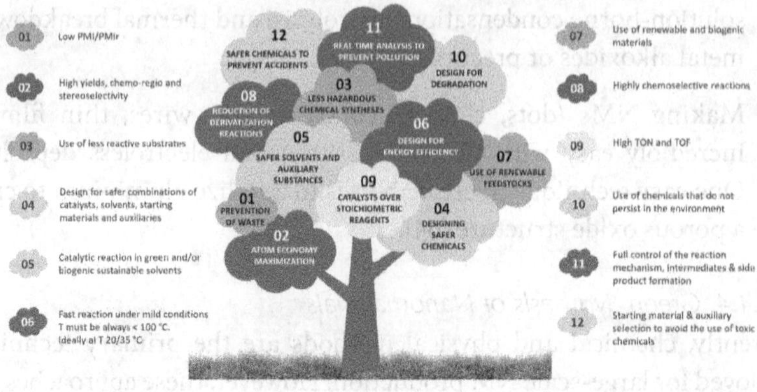

FIGURE 10.3 The 12 principles of green chemistry proposed by Anastas and Warner. (From Anastas, P.T. and Warner, J.C., *Green Chemistry: Theory and Practice*, Oxford University Press, 2000.)

(EugeAgNPs) and examines their effectiveness in catalytically degrading methyl orange dye and detecting Hg^{2+} ions in the solutions [79].

Agricultural waste products are being utilized as precursor materials for the synthesis of silicon NPs (SiNPs). A key advantage of employing agricultural waste lies in its abundant availability after each harvesting season, making nanoparticle synthesis methods that leverage these resources highly cost-effective compared with other approaches. Rice hull, a waste byproduct of the agriculture industry, is rich in silica content. Numerous studies have documented the successful use of rice hulls as a source for producing high-quality SiNPs. Notably, Jansomboon et al. pioneered the synthesis of silica microparticles and NPs by exploiting rice hull as the starting material [80, 81].

10.1.4.5 Principle of Nanomaterial-Based Sensors

A device that analyses substances by transforming recorded data into visible or quantifiable signals is called a chemical sensor [82, 83]. An electrochemical sensor is a self-contained, integrated device that converts information from electrochemical reactions into signals ready for analysis. It is designed to deliver specific quantitative or semi-quantitative analytical data using a chemical recognition element (receptor) that is directly in contact with the electrode [84]. The electrochemical sensing process employs a three-electrode system comprising a sensing electrode (working electrode, WE) that can be functionalized with diverse materials to enhance sensitivity and lower detection limits, a reference electrode (RE), and a counter electrode (CE), where the operating principle involves the binding of analytes (heavy metals, pesticides, or pathogenic bacteria) onto receptors, modulating the physicochemical signal, which is then transduced into an electrical signal with variations in current, voltage, resistance, conductance, impedance, intensity, phase of electromagnetic radiation, temperature, or mass, amplified and displayed for analyte quantification, with nanostructures like nanowires, nanorods, NPs, quantum dots, or nanotubes acting as intermediaries between the analytes and the transducer combined with the receptor [47, 85, 86].

Some commonly used NMs in various sensor applications are:

Graphene: Graphene is utilized in sensors to detect the presence of harmful pesticides and contaminants in food. Its high sensitivity and large surface area make it ideal for identifying trace amounts of these substances, ensuring food safety and quality [4].

Gold NPs (AuNPs): AuNPs are used for the detection of pharmaceutical compounds. Their ability to be functionalized with various molecules enhances their selectivity, allowing precise identification of drugs and other medicinal compounds [87].

CNTs: CNTs are employed in sensors to detect toxic gases and heavy metals. Their excellent electrical conductivity and high aspect ratio contribute to their high sensitivity and rapid response time in detecting hazardous substances [88].

AgNPs: AgNPs are effective in sensing biological molecules such as DNA and proteins. Their strong plasmonic properties facilitate enhanced detection sensitivity, which is crucial in medical diagnostics and biological research [89].

Zinc Oxide NPs (ZnONPs): ZnONPs are used to monitor environmental pollutants in air and water. Their high reactivity and surface area enable the detection of various pollutants, aiding in environmental protection and pollution control [90].

Conductive polymers: Polymers like polyaniline, polypyrrole, and poly(3,4-ethylenedioxythiophene) (PEDOT) are intrinsically conductive and can be used to modify electrode surfaces. They are often used in sensors for the detection of ions, gases, and biological molecules due to their ability to change conductivity upon interaction with the analyte [15, 91].

Nanocomposites: Combinations of different NMs, such as metal NPs embedded in CNTs or graphene, can be used to create sensors with enhanced properties. These nanocomposites can provide improved sensitivity, selectivity, and stability compared with individual NMs [92, 93].

Quantum dots: Quantum dots are applied in biological imaging to visualize cells and molecular structures. Their unique optical properties, such as size-tuneable emission and high brightness, make them ideal for detailed imaging in biomedical applications [94].

10.1.5 Need for Nanomaterial-Based Sensors and Their Advantages

The integrity and security of our food supply are jeopardized by a multitude of contaminants [95]. These include microbial pollutants such as pathogenic bacteria, fungi, and viruses, as well as chemical contaminants like pesticides, veterinary drugs, hormones, and growth promoters. Physical contaminants, ranging from plastic fragments to metal pieces, hair, stones, or plant stalks, also pose a threat. Additionally, the use of growth-promoting feed additives like β-agonists and antibiotics, the presence of food allergens, and the occurrence of natural toxins contribute to the compromised state of food quality and safety [96, 97]. Contaminants of emerging concern refer to pollutants that are not currently included in established monitoring and evaluation programmes. These contaminants are the subject of ongoing research to assess the potential health risks they pose and their ecotoxicity, and to gather data on their prevalence and activities within ecosystems. Based on the findings from such investigations, these emerging contaminants may be identified as candidates for future regulatory measures and oversight [98]. To assess food quality and safety, various analytical techniques and methods are currently employed. These include advanced chromatographic techniques like high-performance liquid chromatography (HPLC) coupled with mass spectrometry (MS), as well as nuclear magnetic resonance (NMR) spectroscopy [99]. Molecular biology techniques such as polymerase chain reaction (PCR) combined with gas chromatography (GC) and MS are also utilized. Additionally, immunoassay-based methods, including the enzyme-linked immunosorbent assay (ELISA), are widely adopted for conducting comprehensive food quality and safety evaluations [100]. While conventional analytical techniques for detecting pollutants are confined to laboratory settings, they face limitations for on-site field testing due to their reliance on rigorous sample pre-processing, costly equipment, and the need for trained personnel, posing challenges for deployment in resource-constrained environments lacking adequate infrastructure [101].

Employing materials at the nano-dimensional scale offers numerous enhancements in analytical characteristics, such as increased sensitivity, faster response times, improved selectivity, and greater robustness, which highlight the significant benefits of using NMs over micromaterials in creating advanced and highly efficient analytical tools [102]. Especially, carbon NMs provide distinct benefits across various fields, including a high surface-to-volume ratio, excellent electrical conductivity, chemical stability, biocompatibility, and strong mechanical strength and also possess rapid electron transfer rates, high aspect ratios, and resistance to fouling [103, 104]. The use of NMs provides powerful analytical techniques with advantages such as simple instrumentation, moderate cost, portability, and lower detection limits compared with conventional methods[105].

Electrochemical sensors have emerged as the market leaders and are among the most widely adopted sensor types due to their advantageous features, including exceptionally low limits of detection (reaching picomolar levels), rapid response times, and cost-effective sensing apparatus [106, 107]. The design of electrochemical sensors spans a wide range, from leading-edge devices to fully integrated wearable platforms [108]. The key attribute that renders chemical sensors valuable is their capability to provide accurate, real-time information about the chemical composition of their surroundings [109].

10.1.5.1. Limitations

10.1.5.1.1 Potential Environmental and Health Risks Associated with Nanomaterials

- **Toxicity:**

  Health risks: NMs, due to their small size, can penetrate biological membranes and accumulate in various organs, posing potential health risks. Some NMs, such as certain metal NPs, have been shown to cause cytotoxicity and genotoxicity.

  Environmental impact: The release of NMs into the environment, whether during production, usage, or disposal, can pose risks to ecosystems. NPs can accumulate in soil, water, and air, potentially affecting flora and fauna.

- **Regulatory Challenges:**

  Lack of regulations: The rapid development of NMs has outpaced the establishment of comprehensive regulations. This creates uncertainty in terms of safety standards, environmental impact assessments, and risk management protocols.

  Testing and approval: Rigorous testing and approval processes are required to ensure the safety of NMs. Developing standardized testing protocols for toxicity and environmental impact remains a challenge.

10.1.6 Environmental Challenges

The rapid pace of industrialization, intensified agricultural operations, and unabated urban expansion have contributed to the release of numerous toxic and hazardous substances into the environment. These harmful contaminants can infiltrate natural water sources, compromising the safety of drinking water supplies and consequently, posing grave risks to human health and well-being [110]. According to estimates, the detrimental effects of pollution across air, water, and soil contribute to a staggering 9 million premature deaths annually on a global scale. This alarming figure translates to one in every six deaths worldwide, exacting an enormous economic toll that amounts to trillions of dollars in associated costs [111]. The water environment has been impacted by a wide range of environmental pollutants, with their concentrations spanning from nanograms per millilitre to milligrams per millilitre. Remarkably, the diversity of these pollutants encompasses more than 50 different varieties, underscoring the severity and complexity of the contamination challenge [112].

10.1.6.1 Essential Role of Nanomaterial-Based Sensors in Accurate Environmental Monitoring

- ***Miniaturization and portability:*** NMs enable sensors to be smaller, which makes them more easily deployable in the field and is crucial for real-time environmental contamination monitoring [11].
- ***Quick quantification and detection of environmental contaminants:*** NMs can offer quick signal response times, which are essential for prompt mitigation and intervention plans [113].
- Food safety and public health are protected by the great sensitivity and precision with which nanomaterial sensors can identify pollutants in food products [114].
- ***Environmental quality assessment:*** By monitoring and reducing environmental pollution and its effects on ecosystems and human health, nanomaterial sensors can evaluate the quality of the air, water, and soil [115, 116].
- ***Precision agriculture:*** Nanosensors can detect and measure plant diseases, pesticides, nutrients, water content, and soil humidity [117].

10.1.6.2 Several Pivotal Areas Emerge as Critical Focal Points for Future Research and Development Endeavours within the Nanosensor Domain

Materials innovation: Continuous exploration and synthesis of novel NMs exhibiting enhanced properties and functionalities will be of paramount importance. Research efforts will concentrate on

TABLE 10.1 Comparison of Nanomaterial-Based Sensors with Traditional Sensor Technologies

Parameter	Traditional Sensors	Nanomaterial-Based Sensors
Sensitivity	Lower, due to limited surface area	Higher, due to large surface area
Selectivity	Often limited, requires complex modifications	High, through functionalization
Response Time	Slower	Faster, due to rapid interactions
Portability	Generally larger and less portable	Highly portable
Reproducibility	Variable	High, due to uniform properties

discovering materials that offer exceptional stability, biocompatibility, and environmental sustainability [118, 119].

Convergence with emerging technologies: The fusion of nanosensor technologies with cutting-edge advancements such as artificial intelligence, the Internet of Things (IoT), and blockchain could catalyse a revolutionary transformation in data analysis, security, and sensor networking capabilities. Future work will likely investigate how these integrations can pave the way for more intelligent, efficient, and interconnected sensing systems [120, 121].

Enhancing sensing capabilities: Efforts to augment the sensitivity and selectivity of nanosensors for specific applications, particularly in the early detection of diseases or environmental pollutants, will be crucial to unlocking their full potential [122].

10.1.7 Challenges and Limitations

10.1.7.1. Challenges Associated with Nanomaterial-Based Sensors

a) ***Scalability:***

Production scale: One of the primary challenges is scaling up the production of NMs while maintaining their unique properties. Many synthesis methods are labour-intensive, time-consuming, and not easily adaptable for large-scale production.

Uniformity and consistency: Ensuring uniform size, shape, and functionalization of NMs across large batches is difficult, which can impact the performance consistency of sensors.

b) ***Cost:***

High production costs: The synthesis and functionalization processes for NMs can be expensive, especially for high-purity or specialized NMs. This increases the overall cost of nanomaterial-based sensors.

Materials cost: Some NMs, such as gold NPs, involve high material costs, further contributing to the expense.

c) ***Reliability:***

Long-term stability: Nanomaterial-based sensors may suffer from degradation over time, affecting their sensitivity and selectivity. Environmental factors such as temperature, humidity, and exposure to chemicals can impact their long-term stability.

Reproducibility: Achieving reproducible results across different sensors and batches remains a challenge due to variations in nanomaterial properties.

d) ***Integration with existing technologies:***

Compatibility issues: Integrating nanomaterial-based sensors with existing electronic systems and devices can be complex. Ensuring compatibility and effective signal transduction is essential for reliable performance.

Standardization: The lack of standardized protocols for nanomaterial synthesis, functionalization, and sensor fabrication can lead to variability in sensor performance and hinder widespread adoption [123].

10.1.7.2 Emerging Trends and Future Directions in Nanomaterial-Based Sensor Technology

a) **Smart and flexible sensors:**

Wearable devices: The development of flexible, wearable sensors using NMs is gaining momentum. These sensors can monitor various physiological parameters in real time, offering applications in healthcare and fitness.

Smart packaging: Nanomaterial-based sensors are being integrated into smart packaging solutions to monitor the freshness and quality of food products, ensuring safety and reducing waste.

b) **Environmental monitoring:**

Air quality sensors: Advances in NMs are enabling the creation of highly sensitive and selective sensors for monitoring air pollutants, contributing to improved air quality management in urban areas.

Water quality sensors: Emerging nanomaterial-based sensors can detect contaminants in water at extremely low concentrations, providing valuable tools for ensuring clean and safe water supplies.

c) **IoT integration:**

IoT-enabled sensors: The integration of nanomaterial-based sensors with IoT platforms is revolutionizing the way data is collected, analysed, and utilized. This enables remote monitoring and real-time decision-making across various applications.

Big data and AI: The combination of nanomaterial-based sensors with big data analytics and artificial intelligence is facilitating predictive maintenance, anomaly detection, and advanced data interpretation.

d) **Biomedical applications:**

Point-of-care diagnostics: Nanomaterial-based sensors are being developed for point-of-care (POC) diagnostic devices, enabling rapid and accurate detection of diseases and biomarkers at low cost.

Drug delivery systems: Researchers are exploring the use of nanomaterial-based sensors in smart drug delivery systems that can release therapeutics in response to specific physiological triggers.

e) **Sustainability and green technologies:**

Eco-friendly nanomaterials: The focus is shifting towards the development of eco-friendly and biodegradable NMs to minimize environmental impact and enhance sustainability.

Renewable energy: Nanomaterial-based sensors are being employed in renewable energy technologies, such as solar cells and fuel cells, to optimize performance and efficiency.

Nanomaterial-based sensors are at the forefront of technological innovation, offering solutions to pressing challenges in various fields. Ongoing research efforts are addressing current limitations related to scalability, cost, reliability, and integration. The future of nanomaterial-based sensor technology is promising, with emerging trends pointing towards smart and flexible sensors, environmental monitoring, IoT integration, biomedical applications, and sustainable solutions. As these advancements continue to unfold, nanomaterial-based sensors will play an increasingly vital role in creating a more sustainable and technologically advanced future [122].

10.1.8 Future Perspectives

Nanotechnology unveils an exciting frontier, paving the way for transformative advancements in the realms of biomechanics and biomedicine. The true potential of NMs will be shaped by their strategic design, the development of cutting-edge instrumentation, and the expansion of their applications in the medical field, unlocking novel opportunities for innovation and progress [124].

Portable and user-friendly diagnostic or therapeutic instruments intended for use beyond conventional healthcare facilities like hospitals or clinics are termed point-of-care devices [125]. Their compact design and ability to deliver swift results make them invaluable assets for a wide range of medical applications [126]. POC devices encompass various categories, such as diagnostic tools for detecting specific diseases or conditions, monitoring devices that track vital signs or other physiological parameters, and therapeutic devices geared towards delivering targeted treatments [127]. The integration of carbon NMs' exceptional properties with the remarkable selectivity offered by molecularly imprinted polymers (MIPs) has facilitated the development of innovative sensor platforms. These MIP sensors, constructed by leveraging carbon NMs, have demonstrated enhanced sensitivity, enabling their application in the detection of agricultural and veterinary drug residues, illicit additives, processing contaminants, and other major pollutants present in food products [128]. Electrochemiluminescence (ECL) biosensors represent a unique fusion of electrochemistry and spectroscopy techniques, combining their respective strengths to offer high sensitivity, simplicity, and rapid analysis capabilities [129]. At the core of ECL biosensor development lies the synthesis of electrochemiluminescent materials that exhibit high efficiency, remarkable stability, and excellent biocompatibility. Over the past decade, significant advancements have been made in leveraging ECL biosensors for food analysis applications, enabling ECL to emerge as a powerful tool for the ultrasensitive detection of a wide range of analytes. These integrated biosensors hold great promise, merging electrochemistry with spectroscopy to provide a robust platform for diverse analytical needs [130].

10.2 CONCLUSION

Nanomaterial-based sensors are poised to play a pivotal role in the global effort towards environmental sustainability. Their unparalleled sensitivity and selectivity enable precise monitoring of environmental pollutants, ensuring cleaner air, water, and soil. In the realm of healthcare, these sensors facilitate early disease detection and personalized medicine, contributing to improved health outcomes and reduced healthcare costs. Furthermore, their applications in renewable energy technologies and smart packaging solutions are driving advancements in energy efficiency and waste reduction. By providing real-time, accurate data, nanomaterial-based sensors empower policymakers, industries, and individuals to make informed decisions that promote environmental stewardship and sustainable practices. Their integration with IoT platforms and big data analytics further enhances their potential to address complex environmental challenges through predictive maintenance, anomaly detection, and advanced data interpretation.

REFERENCES

1. Hu J, Chen J, Li P, Yan J, Wang H. Systematic review of socially sustainable and community regeneration: research traits, focal points, and future trajectories. *Buildings*. 2024;14. https://doi.org/10.3390/buildings14040881
2. Heo JH, Sung M, Trung TQ, Lee Y, Jung DH, Kim H, et al. Sensor design strategy for environmental and biological monitoring. *EcoMat*. 2023;5:1–26. https://doi.org/10.1002/eom2.12332
3. Darwish MA, Abd-Elaziem W, Elsheikh A, Zayed AA. Advancements in nanomaterials for nanosensors: a comprehensive review. *Nanoscale Adv*. 2024:4015–46. https://doi.org/10.1039/d4na00214h
4. Sanjay BP, Sandeep S, Santhosh AS, Karthik CS, Varun DN, Swamy NK, et al. Unprecedented 2D GNR-CoB nanocomposite for detection and degradation of malachite green – A computational prediction of degradation pathway and toxicity. *Chemosphere*. 2022;287:132153. https://doi.org/10.1016/j.chemosphere.2021.132153
5. Wiederrecht GP, Bachelot R, Xiong H, Termentzidis K, Nominé A, Huang J, et al. Nanomaterials and sustainability. *ACS Energy Lett*. 2023;8:3443–9. https://doi.org/10.1021/acsenergylett.3c01303
6. Verma A, Gautam S, Bansal K, Prabhakar N, Rosenholm J. Green nanotechnology: advancement in phytoformulation research. *Medicines*. 2019;6:39. https://doi.org/10.3390/medicines6010039

7. Corsi I, Winther-Nielsen M, Sethi R, Punta C, Della Torre C, Libralato G, et al. Ecofriendly nanotechnologies and nanomaterials for environmental applications: key issue and consensus recommendations for sustainable and ecosafe nanoremediation. *Ecotoxicol Environ Saf.* 2018;154:237–44. https://doi.org/10.1016/j.ecoenv.2018.02.037
8. Rajeev A, Yin L, Kalambate PK, Khabbaz MB, Trinh B, Kamkar M, et al. Nano-enabled smart and functional materials toward human well-being and sustainable developments. *Nanotechnology.* 2024;35. https://doi.org/10.1088/1361-6528/ad4dac
9. Zhang Y, Guo Y, Xianyu Y, Chen W, Zhao Y, Jiang X. Nanomaterials for ultrasensitive protein detection. *Adv Mater.* 2013;25:3802–19. https://doi.org/10.1002/adma.201301334
10. Varun DN, Hareesha JGMN, Mallu SSP, Prinith CSKNS. Simple and sensitive electrochemical analysis of riboflavin at functionalized carbon nanofiber modified carbon nanotube sensor. *Monatshefte Für Chemie - Chem Mon.* 2021;152:1183–91. https://doi.org/10.1007/s00706-021-02839-y
11. Willner MR, Vikesland PJ. Nanomaterial enabled sensors for environmental contaminants Prof Ueli Aebi, Prof Peter Gehr. *J Nanobiotechnology.* 2018;16:1–16. https://doi.org/10.1186/s12951-018-0419-1
12. Mohamed EF, Awad G. Development of nano-sensor and biosensor as an air pollution detection technique for the foreseeable future. *Curr Opin Anal Chem.* 2022;99. https://doi.org/10.1016/bs.coac.2021.11.003
13. Ahmed H. Nanosensors for environmental monitoring and protection: applications of nanosensors in environmental monitoring. 2023;1:1–5.
14. Ravikumar SB, Mallu TA, Subbareddy S, Shivamurthy SA, Neelalochana VD, Shantakumar KC, et al. Detection of 4-nitrophenol and nitrofurantoin. *J Mater Chem B.* 2024. https://doi.org/10.1039/d3tb03054g
15. Neelalochana VD, Tomasino E, Di Maggio R, Cotini O, Scardi P, Mammi S, et al. Anion exchange membranes based on chemical modification of recycled PET bottles. *ACS Appl Polym Mater.* 2023. https://doi.org/10.1021/acsapm.3c01391
16. Costanza R, D'Arge R, de Groot R, Farber S, Grasso M, Hannon B, et al. The value of the world's ecosystem services and natural capital. *Nature.* 1997;387:253–60.
17. Devi KP, Chaturvedi H. An overview of nanotechnology in water treatment applications and combating climate change. In: *Water conservation in the era of global climate change*; 2021. p. 191–212.
18. Driga AM, Drigas AS. Climate change 101: how everyday activities contribute to the ever-growing issue. *Int J Recent Contrib Eng Sci IT.* 2019;7:22–31.
19. Porter ME, Van Der Linde C. Toward a new conception of the environment-competitiveness relationship. *Corp Environ Responsib.* 2017;9:61–82. https://doi.org/10.1257/jep.9.4.97
20. Landrigan PJ, Fuller R, Acosta NJR, Adeyi O, Arnold R, Baldé AB, et al. The Lancet Commission on pollution and health. *Lancet.* 2018;391:462–512.
21. Cardinale BJ, Duffy JE, Gonzalez A, Hooper DU, Perrings C, Venail P, et al. Biodiversity loss and its impact on humanity. *Nature.* 2012;486:59–67. https://doi.org/10.1038/nature11148

22. Norton BG. *Sustainability: a philosophy of adaptive ecosystem management*. University of Chicago Press; 2005.
23. Sardà R, O'Higgins T, Cormier R, Diedrich A, Tintoré J. A proposed ecosystem-based management system for marine waters: linking the theory of environmental policy to the practice of environmental management. *Ecol Soc*. 2014;19:1–14.
24. Jawad HM, Nordin R, Gharghan SK, Jawad AM, Ismail M. Energy-efficient wireless sensor networks for precision agriculture: a review. *Sensors*. 2017;17:1781.
25. Itagi SA, Manjunatha JG, Madikeri M, Charithra PM, Sandeep S, Karthik CS, et al. Fast and facile voltammetric detection of acetaminophen at poly(DL-...) modified electrode. *Open Chem Eng J*. 2021:31–40. https://doi.org/10.2174/1874123102115010031
26. Ramadhani D. The role of sensor technology in the development of environmental monitoring systems: a systematic review of the literature. *Sciences and Applied Journal*.2024;1:1–8.
27. Hareesha N, Soumya DM, Manjunatha JG, Rohit RN, Manikanta P, Varun DN, et al. Honeycomb polypore biomass-derived activated porous carbon nanosheets/graphite/nafion composite: green and sensitive electrocatalyst for nanomolar detection of Hg^{2+} ions and water-splitting reactions. *J Environ Chem Eng*. 2024;12:113584. https://doi.org/10.1016/j.jece.2024.113584
28. Ho CK, Robinson A, Miller DR, Davis MJ. Overview of sensors and needs for environmental monitoring. *Sensors*. 2005;5:4–37.
29. Steinhauer S. Gas sensors based on copper oxide nanomaterials: a review. *Chemosensors*. 2021;9:1–20. https://doi.org/10.3390/chemosensors9030051
30. Mamun MA Al, Yuce MR. Recent progress in nanomaterial enabled chemical sensors for wearable environmental monitoring applications. *Adv Funct Mater*. 2020;30:1–43. https://doi.org/10.1002/adfm.202005703
31. Bharati K, Tiwari PR, Singh RP, Singh A, Yadav BC, Kumar S. Cobalt-doped praseodymium ortho ferrite as a promising nanomaterial for carbon dioxide gas sensing. *J Mater Chem C*. 2023;11:15581–90.
32. Boudaden J, Klumpp A, Endres H-E, Eisele I. Capacitive CO_2 sensor. *Proceedings*. 2017;1:472.
33. Kim J-H, Katoch A, Choi S-W, Kim SS. Growth and sensing properties of networked p-CuO nanowires. *Sens Actuators B Chem*. 2015;212:190–5.
34. Yaroshenko I, Kirsanov D, Marjanovic M, Lieberzeit PA, Korostynska O, Mason A, et al. Real-time water quality monitoring with chemical sensors. *Sensors*. 2020;20:3432.
35. Kanoun O, Lazarević-Pašti T, Pašti I, Nasraoui S, Talbi M, Brahem A, et al. A review of nanocomposite-modified electrochemical sensors for water quality monitoring. *Sensors*. 2021;21:4131.
36. Schweitzer L, Noblet J. Chapter 3.6 - Water contamination and pollution. In: Török B, Dransfield TBT-GC, editors. *Green chemistry: an inclusive approach*. Elsevier; 2018. p. 261–90. https://doi.org/10.1016/B978-0-12-809270-5.00011-X

37. Rathi BS, Kumar PS, Vo D-VN. Critical review on hazardous pollutants in water environment: occurrence, monitoring, fate, removal technologies and risk assessment. *Sci Total Environ.* 2021;797:149134. https://doi.org/10.1016/j.scitotenv.2021.149134
38. Alam AU, Clyne D, Jin H, Hu NX, Deen MJ. Fully integrated, simple, and low-cost electrochemical sensor array for in situ water quality monitoring. *ACS Sensors.* 2020;5:412–22. https://doi.org/10.1021/acssensors.9b02095
39. Karimi-Maleh H, Beitollahi H, Senthil Kumar P, Tajik S, Mohammadzadeh Jahani P, Karimi F, et al. Recent advances in carbon nanomaterials-based electrochemical sensors for food azo dyes detection. *Food Chem Toxicol.* 2022;164:112961. https://doi.org/10.1016/j.fct.2022.112961
40. Yashas SR, Sandeep S, Shivakumar BP, Swamy NK. A matrix of perovskite micro-seeds and polypyrrole nanotubes tethered laccase/graphite biosensor for sensitive quantification of 2,4-dichlorophenol in wastewater. *Anal Methods.* 2019;11:4511–9. https://doi.org/10.1039/C9AY01468C
41. Santhosh AS, Sahana KM, Sandeep S, Kumar PNP, Alsaiari NS, Katubi KM, et al. Synthesis and application of a 0D/2D nanocomposite for the nanomolar level detection of an antiandrogen drug. *New J Chem.* 2022. https://doi.org/10.1039/D2NJ01967A
42. Baig N, Kammakakam I, Falath W, Kammakakam I. Nanomaterials: a review of synthesis methods, properties, recent progress, and challenges. *Mater Adv.* 2021;2:1821–71. https://doi.org/10.1039/d0ma00807a
43. Sudha PN, Sangeetha K, Vijayalakshmi K, Barhoum A. Nanomaterials history, classification, unique properties, production and market. In: *Emerging applications of nanoparticles and architecture nanostructures.* Elsevier Inc.; 2018. https://doi.org/10.1016/B978-0-323-51254-1.00012-9
44. Scientific Committee on Consumer Products. Preliminary opinion on nanomaterials in cosmetic products. *In Vivo (Brooklyn).* 2007;95:1–56.
45. Buzea C, Pacheco II, Robbie K. Nanomaterials and nanoparticles: sources and toxicity. *Biointerphases.* 2007;2:MR17–71. https://doi.org/10.1116/1.2815690
46. Saleh TA. Nanomaterials: classification, properties, and environmental toxicities. *Environ Technol Innov.* 2020;20:101067. https://doi.org/10.1016/j.eti.2020.101067
47. Prajapati S, Padhan B, Amulyasai B, Sarkar A. Nanotechnology-based sensors. In: *Nanomaterials for biosensors.* Elsevier Inc.; 2020. https://doi.org/10.1016/B978-0-12-816897-4.00011-4
48. Meng R, Zhu Q, Long T, He X, Luo Z, Gu R, et al. The innovative and accurate detection of heavy metals in foods: a critical review on electrochemical sensors. *Food Control.* 2023;150:109743. https://doi.org/10.1016/j.foodcont.2023.109743
49. Schmidt M, Haberland H. Phase transitions in clusters. *C R Phys.* 2002;3:327–40.
50. Khanna VK. Nanomaterials and their properties. In: *Nanotechnology.* Springer; 2016. p. 25–41. https://doi.org/10.1007/978-81-322-3625-2_3

51. Mannix AJ, Zhou X-F, Kiraly B, Wood JD, Alducin D, Myers BD, et al. Synthesis of borophenes: anisotropic, two-dimensional boron polymorphs. *Science*. 2015;350. https://doi.org/10.1126/science.aad1080
52. Jeon H, Lee K. Effect of gold nanoparticle morphology on thermal properties of polyimide nanocomposite films. *Colloids Surf A Physicochem Eng Asp*. 2019;579:123651. https://doi.org/10.1016/j.colsurfa.2019.123651
53. Shashanka R, Chaira D, Kumara Swamy BE. Electrocatalytic response of duplex and yittria dispersed duplex stainless steel modified carbon paste electrode in detecting folic acid using cyclic voltammetry. *Int J Electrochem Sci*. 2015;10:5586–98. https://doi.org/10.1016/s1452-3981(23)17279-8
54. Rajendrachari S, Shrinivas Mahale R, Adimule V, Gulen M, Mahesh V, Basavegowda N. Fabrication of mechanically alloyed high entropy alloys (20-Fe-20Cr-20Ni-20Mn-20Ti) modified carbon paste electrode sensor for the cyclic voltammetric determination of methyl orange. *Inorg Chem Commun*. 2024;164:112428. https://doi.org/10.1016/j.inoche.2024.112428
55. Joudeh N, Linke D. Nanoparticle classification, physicochemical properties, characterization, and applications: a comprehensive review for biologists. *J Nanobiotechnology*. 2022;20:1–29. https://doi.org/10.1186/s12951-022-01477-8
56. Wu Q, Miao WS, Zhang Y Du, Gao HJ, Hui D. Mechanical properties of nanomaterials: a review. *Nanotechnol Rev*. 2020;9:259–73. https://doi.org/10.1515/ntrev-2020-0021
57. Pithawalla YB, El-Shall MS, Deevi SC, Ström V, Rao KV. Synthesis of magnetic intermetallic FeAl nanoparticles from a non-magnetic bulk alloy. *J Phys Chem B*. 2001;105:2089–90. https://doi.org/10.1021/jp002354i
58. Neikov OD. Chapter 4 - Nanopowders. In: Neikov OD, Naboychenko SS, Murashova IV, Gopienko VG, Frishberg IV, Lotsko DV, editors. *Handbook of non-ferrous metal powders*. Elsevier; 2009. p. 80–101. https://doi.org/10.1016/B978-1-85617-422-0.00004-5
59. Asha AB, Narain R. Chapter 15 - Nanomaterials properties. In: Narain R, editor. *Polymeric sensors and nanotechnology*. Elsevier; 2020. p. 343–59. https://doi.org/10.1016/B978-0-12-816806-6.00015-7
60. Makkar P, Ghosh NN. A review on the use of DFT for the prediction of the properties of nanomaterials. *RSC Adv*. 2021;11:27897–924.
61. Paramasivam G, Palem VV, Sundaram T, Sundaram V, Kishore SC, Bellucci S. Nanomaterials: synthesis and applications in theranostics. *Nanomaterials*. 2021;11:3228.
62. Jamkhande PG, Ghule NW, Bamer AH, Kalaskar MG. Metal nanoparticles synthesis: an overview on methods of preparation, advantages and disadvantages, and applications. *J Drug Deliv Sci Technol*. 2019;53:101174.
63. Prasad Yadav T, Manohar Yadav R, Pratap Singh D. Mechanical milling: a top-down approach for the synthesis of nanomaterials and nanocomposites. *Nanosci Nanotechnol*. 2012;2:22–48. https://doi.org/10.5923/j.nn.20120203.01
64. Arole VM, Munde SV. Fabrication of nanomaterials by top-down and bottom-up approaches – an overview. *JAAST Mater Sci (Special Issue)*. 2014;1:2–89.

65. Bashir S, Liu J. Overviews of synthesis of nanomaterials. In: *Advanced nanomaterials and their applications in renewable energy*. Elsevier; 2015. p. 51–115.
66. Fu X, Cai J, Zhang X, Li W-D, Ge H, Hu Y. Top-down fabrication of shape-controlled, monodisperse nanoparticles for biomedical applications. *Adv Drug Deliv Rev*. 2018;132:169–87.
67. Larosi MB, García JV, Rodríguez AR. Laser synthesis of nanomaterials. *Nanomaterials*. 2022;12:10–3. https://doi.org/10.3390/nano12172903
68. Jiang Z, Li L, Huang H, He W, Ming W. Progress in laser ablation and biological synthesis processes: "Top-down" and "bottom-up" approaches for the green synthesis of Au/Ag nanoparticles. *Int J Mol Sci*. 2022;23. https://doi.org/10.3390/ijms232314658
69. Habiba K, Makarov V, Weiner B, Morell G. Fabrication of nanomaterials by pulsed laser synthesis. 2014. p. 263–91. https://doi.org/10.13140/RG.2.2.16446.28483
70. Ahmed M. Nanomaterial synthesis. *Nanomaterials for sustainable energy and environmental remediation*. Elsevier Inc.; 2020. https://doi.org/10.1016/B978-0-12-816806-6.00016-9
71. Houshiar M, Zebhi F, Razi ZJ, Alidoust A, Askari Z. Synthesis of cobalt ferrite ($CoFe_2O_4$) nanoparticles using combustion, coprecipitation, and precipitation methods: a comparison study of size, structural, and magnetic properties. *J Magn Magn Mater*. 2014;371:43–8.
72. Bokov D, Turki Jalil A, Chupradit S, Suksatan W, Javed Ansari M, Shewael IH, et al. Nanomaterial by sol-gel method: synthesis and application. *Adv Mater Sci Eng*. 2021;2021:5102014.
73. Gupta D, Boora A, Thakur A, Gupta TK. Green and sustainable synthesis of nanomaterials: recent advancements and limitations. *Environ Res*. 2023;231:116316. https://doi.org/10.1016/j.envres.2023.116316
74. Varma RS. Greener approach to nanomaterials and their sustainable applications. *Curr Opin Chem Eng*. 2012;1:123–8. https://doi.org/10.1016/j.coche.2011.12.002
75. Halfadji A, Naous M, Rajendrachari S, Ceylan Y, Ceylan KB, Shekar PVR. Effective investigation of electro-catalytic, photocatalytic, and antimicrobial properties of porous CuO nanoparticles green synthesized using leaves of *Cupressocyparis leylandii*. *J Mol Struct*. 2024;1301:137318. https://doi.org/10.1016/j.molstruc.2023.137318
76. Noah NM, Ndangili PM. Green synthesis of nanomaterials from sustainable materials for biosensors and drug delivery. *Sensors Int*. 2022;3:100166. https://doi.org/10.1016/j.sintl.2022.100166
77. Anastas PT, Warner JC. *Green chemistry: theory and practice*. Oxford University Press; 2000. https://doi.org/10.1093/oso/9780198506980.001.0001
78. Nguyen VT. Sunlight-driven synthesis of silver nanoparticles using pomelo peel extract and antibacterial testing. *J Chem*. 2020;2020:6407081. https://doi.org/10.1155/2020/6407081

79. Santhosh AS, Sandeep S, Swamy NK. Green synthesis of nano silver from *Euphorbia geniculata* leaf extract: investigations on catalytic degradation of methyl orange dye and optical sensing of Hg^{2+}. *Surf Interfaces*. 2019;14:50–4.
80. Karande SD, Jadhav SA, Garud HB, Kalantre VA, Burungale SH, Patil PS. Green and sustainable synthesis of silica nanoparticles. *Nanotechnol Environ Eng.* 2021;6:1–14. https://doi.org/10.1007/s41204-021-00124-1
81. Jansomboon W, Boonmaloet K, Sukaros S, Prapainainar P. Rice hull micro and nanosilica: synthesis and characterization. *Key Eng Mater.* 2017;718:77–80. https://doi.org/10.4028/www.scientific.net/KEM.718.77
82. Fraden J. *Handbook of modern sensors*. 3rd ed. Springer; 1999.
83. Fraden J. Physical principles of sensing. In: *Handbook of modern sensors: physics, designs, and applications*. 2nd ed.; Springer, New York, NY; 2004. p. 37–121.
84. Hulanicki A, Glab S, Ingman F. Chemical sensors: definitions and classification. *Pure Appl Chem*. 1991;63:1247–50.
85. Manjunatha JG, Deraman M, Basri NH, Nor NSM, Talib IA, Ataollahi N. Sodium dodecyl sulfate modified carbon nanotubes paste electrode as a novel sensor for the simultaneous determination of dopamine, ascorbic acid, and uric acid. *Comptes Rendus Chim*. 2014;17:465–76. https://doi.org/10.1016/j.crci.2013.09.016
86. Pushpanjali PA, Manjunatha JG, Hareesha N, Girish T, Al-Kahtani AA, Tighezza AM, et al. Electrocatalytic determination of hydroxychloroquine using sodium dodecyl sulphate modified carbon nanotube paste electrode. *Top Catal.* 2022. https://doi.org/10.1007/s11244-022-01568-8
87. Mohammed IA, Al-Gawhari FJ. Gold nanoparticle: synthesis, functionalization, enhancement, drug delivery and therapy: a review. *Syst Rev Pharm.* 2020;11:888–910. https://doi.org/10.31838/srp.2020.6.127
88. Amrutha BM, Manjunatha JG, Bhatt AS, Hareesha N, Al-Kahtani AA, Tighezza AM, et al. Electrocatalytic analysis of diclofenac in the presence of dopamine at surface amplified voltammetric sensor based on poly glycine modified carbon nano tube paste electrode. *Top Catal.* 2022. https://doi.org/10.1007/s11244-022-01567-9
89. Shashanka R, Kumara Swamy BE. Biosynthesis of silver nanoparticles using leaves of *Acacia melanoxylon* and their application as dopamine and hydrogen peroxide sensors. *Phys Chem Res.* 2020;8:1–18. https://doi.org/10.22036/pcr.2019.205211.1688
90. Rajendrachari S, Taslimi P, Karaoglanli AC, Uzun O, Alp E, Jayaprakash GK. Photocatalytic degradation of Rhodamine B (RhB) dye in wastewater and enzymatic inhibition study using cauliflower-shaped ZnO nanoparticles synthesized by a novel one-pot green synthesis method. *Arab J Chem.* 2021;14:103180. https://doi.org/10.1016/j.arabjc.2021.103180
91. Tomasino E, Mukherjee B, Neelalochana VD, Scardi P, Ataollahi N. Computational modeling of hydrated polyamine-based anion exchange membranes via molecular dynamics simulation. *J Phys Chem C.* 2024;128:623–34. https://doi.org/10.1021/acs.jpcc.3c07118

92. Charithra MM, Manjunatha JG, Al-Kahtani AA, Tighezza AM, Ataollahi N. Electroanalysis of epinephrine using polymerized carbon nanotube composite sensor. *Top Catal.* 2022. https://doi.org/10.1007/s11244-021-01560-8
93. Shashanka R. Investigation of optical and thermal properties of CuO and ZnO nanoparticles prepared by *Crocus sativus* (saffron) flower extract. *J Iran Chem Soc.* 2021;18:415–27. https://doi.org/10.1007/s13738-020-02037-3
94. Wagner AM, Knipe JM, Orive G, Peppas NA. Quantum dots in biomedical applications. *Acta Biomater.* 2019;94:44–63. https://doi.org/10.1016/j.actbio.2019.05.022
95. Modi B, Timilsina H, Bhandari S, Achhami A, Pakka S, Shrestha P, et al. Current trends of food analysis, safety, and packaging. *Int J Food Sci.* 2021;2021:9924667
96. Adampourezare M, Hasanzadeh M, Hoseinpourefeizi MA, Seidi F. Iron/iron oxide-based magneto-electrochemical sensors/biosensors for ensuring food safety: recent progress and challenges in environmental protection. *RSC Adv.* 2023;13:12760–80. https://doi.org/10.1039/d2ra07415j
97. Jiang L, Hassan MM, Ali S, Li H, Sheng R, Chen Q. Evolving trends in SERS-based techniques for food quality and safety: a review. *Trends Food Sci Technol.* 2021;112:225–40. https://doi.org/10.1016/j.tifs.2021.04.006
98. Adeniji TM, Stine KJ. Nanostructure modified electrodes for electrochemical detection of contaminants of emerging concern. *Coatings.* 2023;13. https://doi.org/10.3390/coatings13020381
99. Agriopoulou S, Stamatelopoulou E, Varzakas T. Advances in analysis and detection of major mycotoxins in foods. *Foods.* 2020;9:518.
100. Ayerdurai V, Cieplak M, Kutner W. Molecularly imprinted polymer-based electrochemical sensors for food contaminants determination. *TrAC Trends Anal Chem.* 2023;158:116830. https://doi.org/10.1016/j.trac.2022.116830
101. Dhamu VN, Poudyal D, Samson M, Paul A, Muthukumar S, Prasad S. Review—Environmental biosensors for agro-safety based on electrochemical sensing mechanism with an emphasis on pesticide screening. *ECS Sensors Plus.* 2023;2:024601. https://doi.org/10.1149/2754-2726/acde5d
102. Arduini F, Cinti S, Scognamiglio V, Moscone D. Nanomaterial-based sensors. In: *Handbook of nanomaterials in analytical chemistry.* Elsevier; 2020. p. 329–59.
103. Yang C, Denno ME, Pyakurel P, Venton BJ. Recent trends in carbon nanomaterial-based electrochemical sensors for biomolecules: a review. *Anal Chim Acta.* 2015;887:17–37.
104. Rajendrachari S, Basavegowda N, Vinaykumar R, Narsimhachary D, Somu P, Lee MJ. Electrocatalytic determination of methyl orange dye using mechanically alloyed novel metallic glass modified carbon paste electrode by cyclic voltammetry. *Inorg Chem Commun.* 2023;155:111010. https://doi.org/10.1016/j.inoche.2023.111010
105. Chen A, Chatterjee S. Nanomaterials based electrochemical sensors for biomedical applications. *Chem Soc Rev.* 2013;42:5425–38. https://doi.org/10.1039/C3CS35518G

106. Adimule V, Sharma K, Lakshminarayana P, Rajendrachari S, Keri R, Alharthi FA. Fabrication of carbon nanospheres functionalized Sm_2O_3/Co_3O_4 novel hybrid nanocomposites for selective and sensitive amperometric detection of hydrogen peroxide. *J Appl Electrochem.* 2024;54:323–40. https://doi.org/10.1007/s10800-023-01972-2
107. Fu L, Karimi-Maleh H. Leveraging electrochemical sensors to improve efficiency of cancer detection. *World J Clin Oncol.* 2024;15:360.
108. Zakashansky JA. *High surface area, wrinkled gold electrodes for high-sensitivity electrochemical sensing in physiological environments* [dissertation]. University of California, Irvine; 2022.
109. Thakur A, Kumar A. Recent trends in nanostructured carbon-based electrochemical sensors for the detection and remediation of persistent toxic substances in real-time analysis. *Mater Res Express.* 2023;10. https://doi.org/10.1088/2053-1591/acbd1a
110. Ivanišević I. The role of silver nanoparticles in electrochemical sensors for aquatic environmental analysis. *Sensors.* 2023;23. https://doi.org/10.3390/s23073692
111. Barhoum A, Hamimed S, Slimi H, Othmani A, Abdel-Haleem FM, Bechelany M. Modern designs of electrochemical sensor platforms for environmental analyses: principles, nanofabrication opportunities, and challenges. *Trends Environ Anal Chem.* 2023;38:e00199. https://doi.org/10.1016/j.teac.2023.e00199
112. He Q, Wang B, Liang J, Liu J, Liang B, Li G, et al. Research on the construction of portable electrochemical sensors for environmental compounds quality monitoring. *Mater Today Adv.* 2023;17:100340. https://doi.org/10.1016/j.mtadv.2022.100340
113. Lei J, Ju H. Signal amplification using functional nanomaterials for biosensing. *Chem Soc Rev.* 2012;41:2122–34.
114. Wang Y, Duncan TV. Nanoscale sensors for assuring the safety of food products. *Curr Opin Biotechnol.* 2017;44:74–86.
115. Xu K, Fu C, Gao Z, Wei F, Ying Y, Xu C, et al. Nanomaterial-based gas sensors: a review. *Instrum Sci Technol.* 2018;46:115–45.
116. Wang L, Ma W, Xu L, Chen W, Zhu Y, Xu C, et al. Nanoparticle-based environmental sensors. *Mater Sci Eng R Reports.* 2010;70:265–74.
117. Beegum S, Das S. 17 - Nanosensors in agriculture. In: Ghosh S, Thongmee S, Kumar A, editors. *Advances in nanosensors for food analysis.* Woodhead Publishing; 2022. p. 465–78. https://doi.org/10.1016/B978-0-323-91908-1.00012-2
118. Ahmed SF, Mofijur M, Rafa N, Chowdhury AT, Chowdhury S, Nahrin M, et al. Green approaches in synthesising nanomaterials for environmental nanobioremediation: technological advancements, applications, benefits and challenges. *Environ Res.* 2022;204:111967.
119. Ahmad F, Salem-Bekhit MM, Khan F, Alshehri S, Khan A, Ghoneim MM, et al. Unique properties of surface-functionalized nanoparticles for bioapplication: Functionalization mechanisms and importance in application. *Nanomaterials.* 2022;12:1333.

120. Nikhat A, Yusuf P. The Internet of Nano Things (IoNT): existing state and future prospects. *GSC Adv Res Rev.* 2020;5:131–50.
121. Pratama A, Rahayu S. Biosensors enabled by nanotechnology: advancing precision agriculture through IoT monitoring. *J Intell Connect Emerg Technol.* 2024;9:25–39.
122. Darwish MA, Abd-Elaziem W, Elsheikh A, Zayed AA. Advancements in nanomaterials for nanosensors: a comprehensive review. *Nanoscale Adv.* 2024. https://doi.org/10.1039/d4na00214h
123. Bakker E, Telting-Diaz M. Electrochemical sensors. *Anal Chem.* 2002;74:2781–800.
124. Gandhi M, Amreen K. Emerging trends in nanomaterial-based biomedical aspects. *Electrochem.* 2023;4:365–88. https://doi.org/10.3390/electrochem4030024
125. Rink S, Baeumner AJ. Progression of paper-based point-of-care testing toward being an indispensable diagnostic tool in future healthcare. *Anal Chem.* 2023;95:1785–93.
126. Wang K, Meng X, Yan X, Fan K. Nanozyme-based point-of-care testing: revolutionizing environmental pollutant detection with high efficiency and low cost. *Nano Today.* 2024;54:102145.
127. Moulahoum H, Ghorbanizamani F, Beduk T, Beduk D, Ozufuklar O, Guler Celik E, et al. Emerging trends in nanomaterial design for the development of point-of-care platforms and practical applications. *J Pharm Biomed Anal.* 2023;235:115623. https://doi.org/10.1016/j.jpba.2023.115623
128. Chi H, Liu G. Carbon nanomaterial-based molecularly imprinted polymer sensors for detection of hazardous substances in food: recent progress and future trends. *Food Chem.* 2023;420:136100. https://doi.org/10.1016/j.foodchem.2023.136100
129. Mwanza C, Ding S-N. Newly developed electrochemiluminescence based on bipolar electrochemistry for multiplex biosensing applications: a consolidated review. *Biosensors.* 2023;13:666.
130. Zhou J, Lv X, Jia J, Din ZU, Cai S, He J, et al. Nanomaterials-based electrochemiluminescence biosensors for food analysis: Recent developments and future directions. *Biosensors.* 2022;12. https://doi.org/10.3390/bios12111046

CHAPTER 11

Limitations, Challenges, and Future Aspects of Electrochemical Sensing Devices

Daniel Yaezu Tiba, Willian Venancio Lemos Belino, João Henrique Astolfi Fereira, Mariana Munik da Silva, Laura Chaluleu Lieb, and Thiago Cruz Canevari

11.1 INTRODUCTION

Electrochemical sensors are devices that use electrical properties to measure any changes in the chemical system or force these and convert them into measurable data in a computer. They can calculate the concentration of the given analyte. The device is generally built in the following order: A receptor that recognizes the target analyte (conventional electrode or modified electrode), the transducer that converts the receptor's signal into an electrical signal, and an amplifier that augments the received signal to ease the interpretation for the system's operator. Thus, an electrochemical sensor can convert the concentration of different species in solution into an electrical signal formed by increasing or decreasing the current

DOI: 10.1201/9781003512639-11

generated at the electrode–solution interface. Bare conventional electrodes, such as gold, platinum, ITO (indium–tin oxide), and glassy carbon (GC) electrodes, were the first to be used to determine different species in solution. However, their sensitivity and selectivity were minimal, mainly due to the poisoning of the surface by the analysed species. Murray and collaborators solved this problem with the emergence of chemically modified electrodes, in which these conventional electrodes served only as a substrate to be modified with different species. Compared with other analytical techniques, electrochemical sensors have both advantages and disadvantages. The benefits are the ability to be miniaturized (point of care), in situ analyses, ease of surface renewal, ease of surface modification with other species (chemically modified electrodes), and low cost when compared with other analytical techniques, such as chromatography and mass spectrometry. Disadvantages include a lower detection limit compared with some analytical methods, surface poisoning by some species that bind irreversibly, the ability to analyse only in solution, poor electrical contact that generates noise, and solvent effect in the analyses, among others. However, some of the disadvantages mentioned can be resolved by developing electrodes modified with different nanomaterials and/or nanostructures. In this way, modified electrochemical sensors can achieve greater sensitivity, excellent selectivity, and the ability to determine different species at low concentrations (approximately picomolar) individually or simultaneously, in addition to excellent stability.

The first to idealize the concept of nanomaterials and nanotechnology was Richard Feynman, who is considered one of the fathers of nanotechnology, in his lecture titled "There is Plenty of Room at the Bottom" [1].

The possibility of finding new properties of materials on nanometric scales (10^{-9} m) was mentioned. As well as Feynman, Richard Zsigmondy and Norio Taniguchi also contributed to developing this new field, as they used the concepts of "nanometre" and "nanotechnology" for the first time. After a few years, fullerene and carbon nanotubes (CNTs) were discovered for the first time by Kroto and Iijima [2, 3], respectively, boosting the creation of facilities for research, development, and application of new nanotechnologies [4].

As mentioned before, nanomaterials are materials with at least one of their dimensions on the nanometric scale, specifically between 1 and 100 nm. As well as this classification, a nanomaterial must also present enhanced or new properties compared with bulk materials on a micrometric scale or above. Furthermore, nanomaterials can be classified according to the number of dimensions on the nanoscale, being classified as zero-dimensional (0D) nanomaterials (such as quantum dots, nanoparticles, and fullerene) when all three dimensions are in the nanoscale; one-dimensional (1D) (such as CNTs or multi-walled carbon nanotubes [MWCNTs], nanowires, nanoneedles) when two dimensions are in nanometric scale; and two-dimensional (2D) (such as graphene, graphene oxide [GO], reduced graphene oxide [rGO], nanoplates, nanosheets) when one dimension is in nanometric scale [5–7].

Synthesizing nanomaterials is of interest due to their novel or enhanced properties. Many nanomaterials exhibit high reactivity due to incomplete bonds on their surfaces, which are associated with a large superficial area. Additionally, their conductive properties can change; a material that is an insulator, semi-conductor, or conductor in bulk form may exhibit different properties at the nanoscale. Other unique properties may also emerge, which are often intrinsic to the type of material [8], such as carbon, metallic, silicon, or polymeric-based nanomaterials.

Moreover, a nanomaterial property can be further enhanced by combining multiple nanomaterials or even bulk materials, producing novel nanohybrids or nanocomposites, as nanohybrids can be defined by a combination of more than one nanomaterial in which they are connected by strong chemical bonds, such as ionic or covalent bonds. As for the nanocomposites, they are composed of at least one nanomaterial, the others being bulk materials or other materials, connected by weak chemical bonds such as Van der Waals forces [9, 10].

Nanomaterials, nanohybrids, or composites can be used in multiple areas, such as reaction catalysts, fuel cells, membrane filtration, drug delivery systems, electronic devices, polymers or cement reinforcement, and sensing. However, in this chapter, the discussion will be focused on sensing, more specifically on electrochemical sensing [11–15].

There are a few classes of electrochemical sensors: Amperometric, impedimetric, potentiometric, and voltammetric. The amperometric sensor type works in a fixed potential, and the current is recorded; generally, for research purposes, the current is recorded as a function of time, called chronoamperometry; mass transfer phenomena occur on the electrode surface, and the analyte is oxidized or reduced at a given potential. Impedimetric sensors are usually applied to immunosensors, using alternate currents to measure charge transfer resistance on the electrode surface. They can precisely distinguish when the target analyte is bonded to the antibodies. The potentiometric sensors do not displace current within the electrochemical cell; they measure the potential at a given current, normally 0 A, to measure the zero-charge potential; for example, pH probes are potentiometric sensors, and the indicator electrode is a proton ion-selective electrode, the value of which changes the zero-charge potential. Voltametric sensors use the potential to reduce or oxidize a particular analyte. The voltammogram is a graph that shows this phenomenon as current versus potential, indicated by valleys or peaks for reduction or oxidation, respectively [16].

This chapter will discuss nanomaterials that support working electrodes and solve many limitations. They have increased electrocatalytic properties and area, increasing the sensor sensitivity towards the analyte and focusing on voltammetric, amperometric, and impedimetric techniques.

11.2 CARBON-BASED NANOMATERIALS

Carbon is an exciting material to study because it naturally has multiple structures. This versatility is due to its allotropy, which allows the material to arrange its atoms differently in the unit cell.

Its most conventional forms are diamond and coal (bulk graphite). Diamond has an sp^3-hybridized tetrahedral structure. Each carbon atom bonds with four other carbon atoms without free electrons in this structure. Graphite has an sp^2-hybridized two-dimensional hexagonal structure. In this structure, each carbon atom is bonded to three other carbon atoms through sigma bonds and a weaker π bond formed by the free electron in the p orbital.

The π bond formed in graphite reduces its mechanical resistance compared with that of diamond. However, the delocalized electron of this π bond makes graphite an excellent electrical conductor along the plane (two-dimensional).

These carbon materials obtained in their nanometric form have structures similar to their massive analogs (sp^2 and sp^3 hybridization) yet with better and different properties than the conventional material (bulk) provided by the nanometric size of at least one of their dimensions. These properties include high mechanical strength, high conductivity, and photoluminescence.

These carbon nanomaterials can be applied in sensors and biosensors that use electrochemical techniques to analyse samples without needing prior treatment [17]. Sensors based on carbon nanomaterials such as rGO and carbon dot (Cdot) may improve the sensitivity of the analysis, dynamic range, and response time compared with conventional carbon electrodes due to their properties

Supramolecular nanotechnology is a field of science that combines nanotechnology and nanomaterials with supramolecular chemistry, which is responsible for the interaction between materials (intermolecular forces). It becomes possible to integrate other nanomaterials with carbon nanomaterials to obtain a hybrid nanomaterial with new electrical, optical, and catalytic properties [18] and specific properties, making sensors more sensitive to a particular analyte.

Laccase/polypyrrole-3-carboxylic acid/antimony pentoxide (Sb_2O_5)-hybridized rGO deposited on a glassy carbon electrode is a promising material for constructing an electrochemical biosensor detecting bisphenol A [17].

The hybrid nanomaterial favours the anchoring of a biological species (laccase) responsible for detection on the surface of the vitreous electrode through the presence of polypyrrole and its cross-links [17]. At the same time, rGO functionalized with Sb_2O_5 favours the electrochemical potential of the biosensor.

Bisphenol A is a chemical compound classified as an endocrine disruptor capable of altering the production of natural hormones even at low concentrations. The use of electrochemical techniques increases the limit of detection and the sensitivity of the method for this substance, in addition to not requiring prior treatment of the sample to be analysed [17].

This electrochemical biosensor using the differential pulse voltammetry (DPV) technique to detect bisphenol A presented excellent responses, such as a theoretical detection limit of 9.9 nmol L^{-1} in the linear range of 0.1–1.0 μmol and excellent substance selectivity even when using ordinary water (tap) containing different substances such as 17β-oestradiol, oestriol, progesterone, catechol, hydroquinone, ascorbic acid (AA), and dopamine [17].

The hybrid biosensor based on polypyrrole-3-carboxylic/rGO/Sb_2O_5 and laccase enzyme is an excellent solution and can be used in basic analytical conditions for presenting determinations of bisphenol A in water, with a recovery rate ranging from 99.6% to 101% [17].

Cdot is another carbon nanomaterial hybridized with palladium nanoparticles (PdNPs) and silica. It can be used with an electrochemical biosensor, which has the advantage of simultaneous analysis of dopamine and serotonin.

These hormones are known as "happiness hormones," acting as neurotransmitters. Changes in the body can cause various cognitive and motor difficulties, requiring practical and reliable analysis of their levels. Combining several nanomaterials (hybridization) to form a unique nanostructure improves the performance of the electrochemical sensor, such as reliability and speed of analysis, and provides the possibility of multiple functions, such as simultaneous analysis [18].

The use of PdNps favours the transfer of electrons at the interface between the electrode and the analyte due to the increase in electrochemical potential. The Cdot 0D nanomaterial also helps in electron transfer and anchoring of other chemical species (functionalization) due to its easy modification. Silica is widely used in constructing carbon-based hybrid electrochemical sensors because its porosity facilitates the encapsulation of different species and protects them from the solvent effect and chemical resistance [18].

The carbon-based hybrid electrode managed to overcome the challenge of simultaneous analysis of dopamine and serotonin; due to its high analytical sensitivity, there were no significant interferences from ascorbic acid and uric acid (UA), which tend to inactivate the electrodes through adsorption when analysing real samples [18].

The PdNPs/Cdot/silica nanostructure effectively provided the electrode with excellent electrocatalytic activity for simultaneously determining dopamine and serotonin with a detection limit of 36 and 33 nmol L^{-1}, respectively [18].

Another simultaneous analysis made possible by a carbon-based nanostructure is the determination of nicotinamide adenine dinucleotide (NADH) and dopamine without suffering the interference of AA and UA in determining the analytes [19].

NADH is a coenzyme found in all cells, and its primary function is to produce energy for the body. Its levels may change when the body has Parkinson's disease or HIV or is in contact with narcotics such as cocaine, and this, together with the change in dopamine levels, can interfere with cognition, mood, libido, and motor functions [19].

To increase Cdot's excellent electrochemical properties and avoid interference from AA and UA, doping (functionalization) was carried out with fluorine (F) and sulfur (S) and the use of copper oxide nanoparticles (CuONPs), forming a hybrid nanostructure in the construction of the electrode. In this way, making a non-enzymatic electrode with high analytical sensitivity is possible [19].

Biosensors based on carbon nanomaterials are promising due to their high analytical sensitivity, the possibility of simultaneous analysis without interference with the results, and easy functionalization.

Through supramolecular nanotechnology, this nanomaterial can become a hybrid. It provides modular structures that require only the researcher's curiosity to achieve a new objective.

11.3 SILICA-BASED NANOMATERIALS

Silica is one of the major components of the earth [20]. The silica matrix in the earth's crust, generally called silica, is formed of silicon oxide (SiO_2). The standard method of fabricating the matrix is sol–gel. This method follows two main steps, hydrolysis and condensation, which allow the control of the shape and size of the silica produced.

One of the most influential parameters for silica production is the charge of species based on the pH value of the synthesis. There are two routes for this: Acid and alkali. When the synthesis solution is below the isoelectric point (around pH 2), the charges are most favourable, and with the decrease in pH, the density of charges increases. There are negative charges between pH 2 and 7, and the tendency is to interact by electrostatic and hydrogen bonding. Above pH 7, the charges are highly negative, and the electrostatic interactions dominate. Some known sol–gel production techniques, such as the Stöber method, use alkali synthesis and are widely used for nanohybrid production. Silica matrix can be applied in several fields, such as membranes for filtration, chromatographic studies, and electrochemical sensors [21].

Silica-based sensors are attractive materials for the electrochemical field because of their porous network, fast mass transport, and organic–functional binding. The sol–gel production method enables their composition, size, structure, and morphology to be modified for specific applications [22]. The literature describes various applications of electrochemical sensors, such as for neurotransmitters, coenzymes, antibiotics, and pesticides.

Bavatharani (2020) [23] has developed a nanohybrid of graphene decorated with polydiphenylamine (PDPA) and silicon oxide for dopamine detection. A fluorine-doped tin oxide (FTO) electrode was modified with graphene nanosheets decorated with the PDPA, aiming for an electrochemically active layer for amperometric detection. The detection limit was 100 nmol L^{-1}, and the result was stable in the presence of interferents (glucose and acetaminophen).

Canevari (2011) [24] synthesized a nanohybrid with silica (SiO_2), tin oxide (SnO_2), and antimony oxide (Sb_2O_5) to detect the coenzyme NADH by an indirect method using Meldola's blue. The combination of the silica oxide nanomaterial and Meldola's blue is because of the electroactivity of the dye, which modifies the surface of the nanomaterial to enable detection at a low potential. The NADH oxidation in the presence of the electroactive dye produces a signal for electrochemical analysis. A chronoamperometric method was used to determine the coenzyme, and the limit of detection obtained by the nanomaterial was 150 nmol L^{-1}; the presence of interferents (dopamine and ascorbic acid) does not change the function of the modified electrode.k

Antibiotics are among the highest public health issues, and monitoring is necessary. Ceftriaxone sodium (CFX) is a third-generation cephalosporin, a semisynthetic antibiotic, and acts in many bacterial infections. The mechanism of action is in the bacteria's cell wall, preventing the wall's division and formation. Approximately 45% to 60% of an intravenous dose is excreted in the urine within 48 hours of application. Detection is required; techniques like electrochemical detection, primarily because of the simple and rapid determination and low cost of production, show benefits compared with others [25].

Silva (2024) [26] has developed a nanohybrid with silica, zirconium oxide, and nitrogen-doped carbon quantum dots in a screen-printed electrode for CFX detection. The innovation in his technique is producing a screen-printed electrode with a vinyl sheet and carbon ink printer. For CFX detection, square wave voltammetry (SWV) was used, and the limit of detection obtained was 0.2 nmol $L-^{-1}$. In the presence of glucose, uric acid, ascorbic acid, and caffeine interferents, the electrode presented good separation of oxidation signals, and in the presence of cationic interferents (Na^+, K^+, Ca^+, Mg^{2+}, Cu^{2+}, Pb^{2+}, Mn^{2+}, Zn^{2+}, Co^{2+}), the results continued to exhibit the selectivity of the electrode. The recovery test demonstrated 95% to 100% in synthetic urine and 95% to 105% in urine samples; in tap water, the recovery was 98% to 105%.

Fenitrothion is an organophosphate pesticide extensively used in crops to combat pest infestations. Because of the P=O bonding, its main action in the human body is on the nervous system. Some techniques for fenitrothion detection exist, but most require sample preparation or pretreatment; electrochemical analysis does not. The direct application of the analyte gives this technique an advantage over others.

Canevari (2016) [27] produced a modified glassy carbon electrode with silica, multi-walled carbon nanotubes, and ruthenium phthalocyanine (RuPc) for fenitrothion detection. Ruthenium phthalocyanine is a metal phthalocyanine (MPc) with redox and catalytic properties. The detection was by DPV, with a limit of detection of 1.62 μmol L^{-1}; the maximum residue limit (MRL) determined by Codex Alimentarius in fruits is 3.61 μmol L^{-1}. An interferent study was carried out with malathion and chlorpyrifos, other organo-phosphate pesticides, and the result demonstrated that their presence is not harmful for the electrode. An orange juice analysis was also performed, with spiked sample for better results, and displayed the same results as the controlled analyses.

11.4 METAL-BASED NANOMATERIALS

As well as their morphology, an essential aspect of nanoparticles is their composition. Nanoparticles can be synthetically formed from a wide range of elements, with metal ions imparting additional beneficial properties during the formation process. These properties are increasingly being harnessed in new technologies and processes. A variety of chemical and physical techniques can be employed for the synthesis of metal-based nanoparticles. However, these methods often face significant challenges, including the use of harmful solvents, the generation of hazardous by-products, and high energy consumption.

Metal-based nanoparticles have a broad spectrum of applications, including food storage, photonics, information storage, electronic and optical detection systems, therapeutics, diagnostics, photovoltaics, and electrochemical sensors. Consequently, there is a critical need to develop environmentally friendly methods for synthesizing these nanoparticles. Some of the most commonly used metal-based nanoparticles include aluminium (Al), aluminium oxide (Al_2O_3), gold (Au), silver (Ag), copper (Cu), iron (Fe), iron oxide (Fe_2O_3, Fe_3O_4), manganese (Mn), manganese oxide (MnO), silicon dioxide (SiO_2), titanium dioxide (TiO_2), zinc (Zn), zinc oxide (ZnO), and other metal oxides such as ceria (CeO_2), nickel oxide (NiO), and zirconia (ZrO_2) [28–30].

Glyphosate is a herbicide widely used in agriculture to control weeds. However, due to its extensive use, concerns have been raised about its potential health and environmental impacts, making it crucial to develop sensitive and accurate methods for its detection. Electrochemical sensors offer a promising approach to glyphosate detection due to their high sensitivity, selectivity, and potential for on-site monitoring.

Tiba et al (2023) [31] have developed an electrochemical sensor for glyphosate based on copper oxide (CuO) nanostructures with a nanoneedle shape obtained by direct reaction with nitrogen-doped carbon quantum dots (N-CQDs). The CuO nanoneedles provide a high surface area and excellent catalytic properties, while the N-CQDs enhance the sensor's electronic conductivity and stability.

The study employs electrochemical techniques such as cyclic voltammetry (CV) and DPV. The resulting electrochemical sensor exhibits high sensitivity with a detection limit of 0.1 μM. Additionally, the sensor demonstrates a rapid response time and excellent reproducibility, making it well suited for practical applications in environmental monitoring.

The innovative approach of using CuO nanoneedles in combination with N-CQDs for the electrochemical detection of glyphosate highlights the potential of this sensor as a powerful tool for environmental analysis. It also contributes to developing effective and reliable monitoring methods for glyphosate, thereby addressing the growing concerns regarding its presence in the environment.

Nitric oxide (NO) is a vital signalling molecule in various physiological processes, including vasodilation, neurotransmission, and immune response. However, abnormal NO levels are associated with several pathological conditions, making its accurate detection crucial for medical diagnostics and research. Electrochemical sensors are particularly suitable for NO detection due to their high sensitivity, rapid response, and potential for miniaturization.

The study by Maluta et al. (2014) [32] focused on the development of a suitable electrochemical sensor for the detection of nitric oxide. The sensor uses MWCNTs decorated with spherical gold nanoparticles (AuNPs). The combination of MWCNTs and AuNPs provides a large surface area, excellent electrical conductivity, and enhanced catalytic activity, which are essential for the efficient detection of NO.

The electrochemical techniques used in this study include CV and amperometry. The developed electrochemical sensor demonstrates high sensitivity and a low detection limit for nitric oxide. Specifically, it achieves a detection limit of 5 nM, allowing the detection of deficient concentrations of NO. The sensor also shows a fast response time and good reproducibility, making it suitable for practical medical diagnostics and environmental monitoring applications.

The use of MWCNTs adorned with spherical Au nanoparticles in detecting NO electrochemically represents an innovative approach that underscores the sensor's potential as a robust tool for nitric oxide detection. This research significantly advances the development of dependable methods for monitoring NO, meeting the increasing demand for precise detection across diverse applications.

Fenitrothion is an organophosphate pesticide widely used in agriculture to control pests. Its presence in food products, such as orange juice, poses potential health risks due to its toxicity. Therefore, sensitive and accurate methods for detecting fenitrothion in food samples are essential for food safety and regulatory compliance. Electrochemical sensors offer a promising approach to pesticide detection due to their high sensitivity, rapid response, and potential for on-site analysis.

Ferreira et al (2024) [33] focused on the development of an ultrasensitive electrochemical sensor for the determination of fenitrothion in orange juice. The sensor utilizes a gold-printed electrode modified with a nanoarchitecture consisting of silver nanoparticles (AgNPs), C-dots, and MWCNTs. This nanoarchitecture enhances the surface area, conductivity, and catalytic activity of the electrode, improving the sensitivity and selectivity of the sensor towards fenitrothion.

Electrochemical impedance spectroscopy (EIS) is the primary technique employed in this study to detect fenitrothion. The developed electrochemical sensor demonstrates ultra-high sensitivity with a low detection limit for fenitrothion in orange juice. Specifically, it achieves a detection limit of 0.48 nmol L^{-1}, enabling the detection of trace amounts of fenitrothion with exceptional accuracy. The sensor also shows excellent reproducibility and stability, making it suitable for practical food quality control and environmental monitoring applications.

Fenitrothion is a potent pesticide whose accurate detection in food products is crucial for ensuring consumer safety and regulatory compliance. The use of AgNP/C-dots/MWCNT nanoarchitecture in the electrochemical sensor highlights its potential as an effective tool for ultra-sensitive detection of fenitrothion. This study contributes significantly to advancing methods for pesticide detection in food matrices, addressing the critical need for reliable analytical techniques in food safety.

Biotin, also known as vitamin B7, is an essential nutrient involved in numerous metabolic processes, including the synthesis of fatty acids and the metabolism of amino acids and glucose. It is vital for cell growth and skin, hair, and nail health. Accurate biotin detection is essential in various fields, such as clinical diagnostics, nutrition, and pharmaceutical research. Electrochemical methods are highly effective for detecting biotin due to their high sensitivity, specificity, and rapid response.

Zhan et al. (2023) present the development of an ultrasensitive electrochemical sensor to detect biotin. Zhan et al (2023) [34] presented the development of an ultrasensitive electrochemical sensor for the detection of biotin. The sensor utilizes heme conjugated with AuNPs to enhance its sensitivity and performance. Combining heme and AuNPs provides a large surface area, excellent electrical conductivity, and catalytic activity, which is crucial for effectively detecting biotin.

The primary electrochemical techniques used in this study are CV and DPV. The developed electrochemical sensor demonstrates ultra-high sensitivity and a low detection limit of 0.0016 μmol L^{-1}, giving high accuracy in detecting deficient concentrations of biotin. The sensor also shows excellent reproducibility and stability, making it suitable for practical clinical diagnostics and nutritional analysis applications.

Biotin is an important vitamin whose accurate detection is essential for various health-related applications. The innovative use of heme conjugated with gold nanoparticles in the electrochemical sensor highlights its potential as a powerful tool for ultrasensitive biotin detection.

Electrochemical sensors that utilize metal-based nanoparticles demonstrate remarkable sensitivity and selectivity in detecting various analytes, such as pesticides, vitamins, and signalling molecules. For example, electrochemical sensors for detecting glyphosate, nitric oxide, fenitrothion, and biotin have shown excellent performance due to the unique properties of the metal-based nanoparticles in their construction.

11.5 CHALLENGES AND FUTURE ASPECTS

With the emergence of the Internet of Things (IoT) and additive manufacturing, the development of electrochemical devices is becoming increasingly advanced, especially regarding sensitivity, application areas, and durability. Through additive manufacturing, it is possible to build/print more sensitive electrodes that can contain different nanostructures already inserted. IoT enables real-time monitoring of these electrochemical sensors in various media, expanding the spectrum of applications. In addition to IoT and additive manufacturing, printed and paper electrodes are revolutionizing the development of electrochemical sensors in terms of usability, ease of modification, sensitivity, and miniaturization. These advances, combined with nanotechnology, can further increase the use of electrochemical sensors in different areas, further advancing conventional analytical techniques.

11.6 CONCLUSIONS

Many limitations in the development of electrochemical devices can be solved by a combination of electrode modification, mainly with nanostructures, and the use of different electrochemical techniques, such as cyclic voltammetry, differential pulse voltammetry, and electrochemical impedance spectroscopy to achieve low detection limits and high sensitivity. The development of sensors using innovative nanoarchitectures, such as those incorporating gold nanoparticles, carbon dots, and multi-walled carbon nanotubes allied to miniaturization, is leading to potentially useful electrochemical devices to be employed in advancing analytical methods for environmental monitoring, food safety, and clinical diagnostics.

REFERENCES

1. R.P. Feynman, *Plenty of Room at the Bottom*, n.d.
2. K.H. W, H.J. R, H.W. Kroto, *This Week's Citation Classic ® C 60-The Third Man*, n.d.
3. S. Iijima, Helical microtubules of graphitic carbon, *Nature* 354 (1991) 56–58. https://doi.org/10.1038/354056a0.
4. J.E. Hulla, S.C. Sahu, A.W. Hayes, Nanotechnology: History and future, *Hum Exp Toxicol* 34 (2015) 1318–1321. https://doi.org/10.1177/0960327115603588.
5. R.C. Sanfelice, A. Pavinatto, D.S. Corrêa, *Capítulo 1 Introdução À Nanotecnologia 1.1 Breve Histórico Sobre A Nanotecnologia E Nanoci-Ência*, n.d.
6. J.N. Tiwari, R.N. Tiwari, K.S. Kim, Zero-dimensional, one-dimensional, two-dimensional and three-dimensional nanostructured materials for advanced electrochemical energy devices, *Prog Mater Sci* 57 (2012) 724–803. https://doi.org/10.1016/j.pmatsci.2011.08.003.
7. S.K. Tiwari, V. Kumar, A. Huczko, R. Oraon, A. De Adhikari, G.C. Nayak, Magical allotropes of carbon: Prospects and applications, *Crit Rev Solid State Mater Sci* 41 (2016) 257–317. https://doi.org/10.1080/10408436.2015.1127206.
8. E. Roduner, Size matters: Why nanomaterials are different, *Chem Soc Rev* 35 (2006) 583–592. https://doi.org/10.1039/b502142c.
9. F. De Clippel, M. Dusselier, S. Van De Vyver, L. Peng, P.A. Jacobs, B.F. Sels, Tailoring nanohybrids and nanocomposites for catalytic applications, *Green Chem* 15 (2013) 1398–1430. https://doi.org/10.1039/c3gc37141g.
10. J.A. Ferreira, L.L. Name, L.C. Lieb, D.Y. Tiba, M.M. da Silva, A.C. Oliveira, T.C. Canevari, Carbon dots hybrid nanostructure-based electrochemical sensors: applications in determining different species in a real sample, *Curr Nanosci* 20 (2024) 31–46. https://doi.org/10.2174/1573413719666221221095901.

11. W.C. Chong, E. Mahmoudi, Y.T. Chung, C.H. Koo, A.W. Mohammad, K.F. Kamarudin, Improving performance in algal organic matter filtration using polyvinylidene fluoride–graphene oxide nanohybrid membranes, *Algal Res* 27 (2017) 32–42. https://doi.org/10.1016/j.algal.2017.08.023.
12. S. Mu, J. Yue, Y. Wang, C. Feng, Electrical, piezoresistive and electromagnetic properties of graphene reinforced cement composites: A review, *Nanomaterials* 11 (2021). https://doi.org/10.3390/nano11123220.
13. N. Zhang, W. She, F. Du, K. Xu, Experimental study on mechanical and functional properties of reduced graphene oxide/cement composites, *Materials* 13 (2020). https://doi.org/10.3390/ma13133015.
14. X. Li, W. Xu, Y. Xin, J. Yuan, Y. Ji, S. Chu, J. Liu, Q. Luo, Supramolecular polymer nanocomposites for biomedical applications, *Polymers (Basel)* 13 (2021) 1–17. https://doi.org/10.3390/polym13040513.
15. A. Hulanicki, S. Glab, F. Ingman, Chemical sensors: Definitions and classification, *Pure Appl Chem* 63 (1991) 1247–1250. https://doi.org/10.1351/pac199163091247.
16. J. Baranwal, B. Barse, G. Gatto, G. Broncova, A. Kumar, Electrochemical sensors and their applications: A review, *Chemosensors* 10 (2022) 363. https://doi.org/10.3390/chemosensors10090363.
17. F.H. Cincotto, J. Oliveira Fernandes, A.C.O. Travassos, D.Y. Tiba, A.M.A.B. Medeiros, R.E. Santelli, A. Sánchez, R. Villalonga, J.M. Pingarón, T.C. Canevari, Sensitive electrochemical biosensor for bisphenol A based on laccase immobilized on polypyrrole-3-carboxylic/Sb2O5/reduced graphene oxide hybrid nanomaterial, *Electroanalysis* 35 (2023). https://doi.org/10.1002/elan.202300086.
18. J.H.A. Ferreira, R.M. Peres, M. Nakamura, H.E. Toma, T.C. Canevari, PdNPs/carbon dots/silica hybrid nanostructures: The development of an electrochemical sensor for simultaneous determination of dopamine and serotonin in real samples, *J Nanopart Res* 25 (2023). https://doi.org/10.1007/s11051-022-05659-1.
19. O.P.L. de Souza, D.Y. Tiba, J.H.A. Ferreira, L.C. Lieb, T.C. Canevari, Non-enzymatic biosensor based on F,S-doped carbon dots/copper nanoarchitecture applied in the simultaneous electrochemical determination of NADH, dopamine, and uric acid in plasma, *Analyst* 149 (2024) 2728–2737. https://doi.org/10.1039/d3an02239k.
20. R.K. Iler, *The Chemistry of Silica*, John Wiley & Sons, Inc., 1979.
21. S.H. Wu, H.P. Lin, Synthesis of mesoporous silica nanoparticles, *Chem Soc Rev* 42 (2013) 3862–3875. https://doi.org/10.1039/c3cs35405a.
22. A. Walcarius, Silica-based electrochemical sensors and biosensors: Recent trends, *Curr Opin Electrochem* 10 (2018) 88–97. https://doi.org/10.1016/j.coelec.2018.03.017.
23. C. Bavatharani, E. Muthusankar, Z.A. Alothman, S.M. Wabaidur, V.K. Ponnusamy, D. Ragupathy, Ultra-high sensitive, selective, non-enzymatic dopamine sensor based on electrochemically active graphene decorated Polydiphenylamine-SiO2 nanohybrid composite, *Ceram Int* 46 (2020) 23276–23281. https://doi.org/10.1016/j.ceramint.2020.06.054.

24. T.C. Canevari, R.C.G. Vinhas, R. Landers, Y. Gushikem, SiO2/SnO2/Sb2O5 microporous ceramic material for immobilization of Meldola's blue: Application as an electrochemical sensor for NADH, *Biosens Bioelectron* 26 (2011) 2402–2406. https://doi.org/10.1016/j.bios.2010.10.020.
25. H. M. Owens, A. K. Dash, Ceftriaxone sodium: Comprehensive profile, *Profiles Drug Subst Excip Relat Methodol* 30 (2003) 21–57. https://doi.org/10.1016/S0099-5428(03)30002-4.
26. F.W.L. Silva, L.L. Name, D.Y. Tiba, B.F. Braz, R.E. Santelli, T.C. Canevari, F.H. Cincotto, High sensitivity, low-cost, and disposability: A novel screen-printed electrode developed for direct electrochemical detection of the antibiotic ceftriaxone, *Talanta* 266 (2024). https://doi.org/10.1016/j.talanta.2023.125075.
27. T.C. Canevari, T.M. Prado, F.H. Cincotto, S.A.S. Machado, Immobilization of ruthenium phthalocyanine on silica-coated multi-wall partially oriented carbon nanotubes: Electrochemical detection of fenitrothion pesticide, *Mater Res Bull* 76 (2016) 41–47. https://doi.org/10.1016/j.materresbull.2015.12.007.
28. M.J. Sweet, A. Chesser, I. Singleton, Review: Metal-based nanoparticles; size, function, and areas for advancement in applied microbiology, in: *Advances in Applied Microbiology*, Academic Press Inc., 2012: pp. 113–142. https://doi.org/10.1016/B978-0-12-394381-1.00005-2.
29. K.N. Thakkar, S.S. Mhatre, R.Y. Parikh, Biological synthesis of metallic nanoparticles, *Nanomedicine* 6 (2010) 257–262. https://doi.org/10.1016/j.nano.2009.07.002.
30. A.M. Schrand, M.F. Rahman, S.M. Hussain, J.J. Schlager, D.A. Smith, A.F. Syed, Metal-based nanoparticles and their toxicity assessment, *Wiley Interdiscip Rev Nanomed Nanobiotechnol* 2 (2010) 544–568. https://doi.org/10.1002/wnan.103.
31. D.Y. Tiba, L.L. Name, R. Landers, T.C. Canevari, Copper oxide nanostructures with nanoneedles shape obtained by direct reaction with nitrogen-doped carbon quantum dots: development of an electrochemical sensor to glyphosate, *J Mater Sci* 58 (2023) 12569–12583. https://doi.org/10.1007/s10853-023-08827-3.
32. J.R. Maluta, T.C. Canevari, S.A.S. Machado, Sensitive determination of nitric oxide using an electrochemical sensor based on MWCNTs decorated with spherical Au nanoparticles, *J Solid State Electrochem* 18 (2014) 2497–2504. https://doi.org/10.1007/s10008-014-2505-0.
33. J.H.A. Ferreira, A.M.A.B. Medeiros, R.M. Peres, T.C. Canevari, Ultrasensitive determination of fenitrothion pesticide in orange juice by gold-printed electrode modified with AgNP/Carbon Dot/MWCNT Nanoarchitecture employing electrochemical impedance spectroscopy, *Food Anal Methods* 17 (2024) 825–833. https://doi.org/10.1007/s12161-024-02614-9.
34. T. Zhan, X.Z. Feng, Y.Y. Cheng, G.C. Han, Z. Chen, H.B. Kraatz, Electrochemical sensor for ultrasensitive sensing of biotin based on heme conjugated with gold nanoparticles and its electrooxidation mechanism, *Food Chem* 429 (2023). https://doi.org/10.1016/j.foodchem.2023.136997.

CHAPTER 12

Advantages of Nanomaterial-Modified Electrochemical Sensors over Conventional Electrochemical Sensors

Gangadhar M Kanaginahal, Sateeshkumar Kanakannavar, Prashant Kakkamari, D. Nagaraju, and H C Ananda Murthy

12.1 INTRODUCTION TO ELECTROCHEMICAL SENSORS

Electrochemical sensors use electrochemical reactions between electrodes and analytes to generate qualitative or quantitative outputs. These sensors are divided into three categories: Potentiometric, conductometric, and amperometric/voltammetric. They have great sensitivity and selectivity, making them perfect for detecting particular molecules such as NADH. An electrochemical sensor consists of two main components: A transducer, which turns biological processes into electrical signals, and a sensing device, which is typically a catalyst. Electrochemical sensors are frequently utilized because of their portability, fast response time,

DOI: 10.1201/9781003512639-12

and adaptability. They are critical in applications such as blood glucose monitoring, where redox reactions at the working electrode generate detectable electrical signals. Advances in materials, such as conductive metal–organic frameworks (cMOFs), have improved the performance of these sensors, increasing their conductivity and stability and broadening their applications in a variety of disciplines. An electrochemical sensor is made up of three electrodes: The working electrode, which conducts the electrocatalytic reaction and is enhanced by nanomaterials; the counter electrode, which finishes the circuit and keeps the flow of electrons constant; and the reference electrode, which ensures that the working electrode potential is accurately applied [1, 2]. Electrochemical sensors are available in a variety of configurations, including benchtop systems and fully integrated wearable devices. Their principal duty is to provide real-time, accurate information on chemical composition without interfering with the sample. These sensors use a recognition layer on a transduction element to generate electrical signals in response to target analyte interactions. Various varieties are appropriate for environmental monitoring, depending on the analyte, sample matrix, and sensitivity requirements (Figure 12.1) [3, 4].

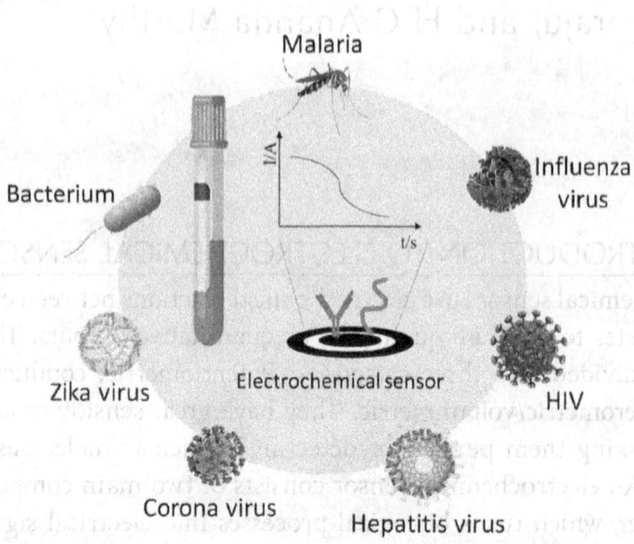

FIGURE 12.1 Schematic representation of electrochemical sensor working. (From Baranwal, J. et al., *Chemosensors*, 10, 2022.)

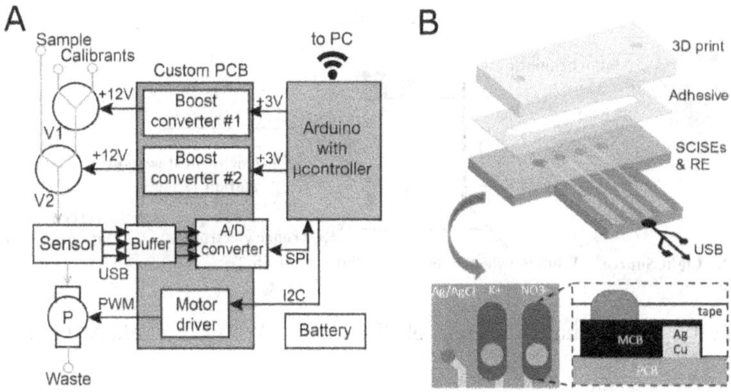

FIGURE 12.2 Potentiometric sensor used in PCB systems. (A) Block diagram of PCB Circuit. (B) Multiplexed sensor with microfluid integration. (From Zhang, Z. et al., *Chemosensors*, 11, 2023.)

12.2 TYPES OF ELECTROCHEMICAL SENSOR

A. **Potentiometric**

When there is no current flowing across two electrodes, potentiometric sensors measure the potential difference between them and are often used to calculate the concentration of a solution component using the Nernst equation. These sensors frequently use polymeric membrane ion-selective electrodes or field effect transistors to detect potential changes. This approach is widely used for physiological electrolyte testing, since it provides information on ion activity irrespective of sample volume. While symmetrical topologies give stability, they limit downsizing, resulting in asymmetrical systems where the membrane interfaces directly with a solid phase (Figure 12.2).

B. **Photoelectrochemistry**

Photoelectrochemistry, a branch of electrochemistry, investigates the interaction of light with photoactive materials, thereby promoting the conversion of light into electricity and the interconversion of electrical and chemical energy. Photoactive materials absorb photons and form electron–hole pairs. These charge carriers travel to electrodes, producing electrical signals, and in the electrolyte, they initiate redox processes for energy conversion (Figure 12.3) [6, 7].

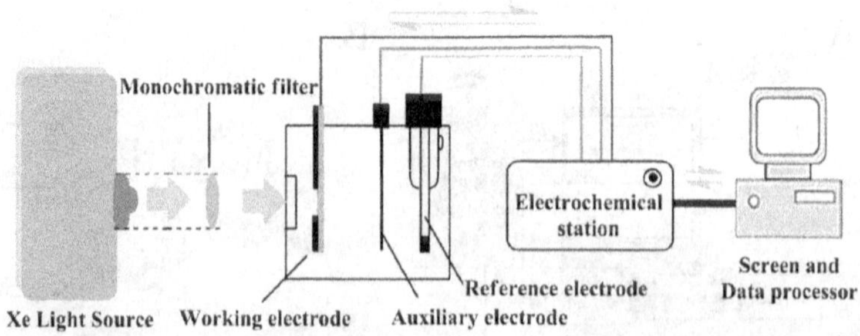

FIGURE 12.3 Schematic diagram of PEC. (From Wang, Y. et al., *Nanomaterials*, 14, 2024.)

C. **Electrogenerated chemiluminescence**

Electrogenerated chemiluminescence (ECL) biosensing relies on three main mechanisms: Ion annihilation, coreactant processes, and hot electron-induced cathodic reactions. In the ion annihilation process, radical cations and anions are formed and destroyed, releasing light from the excited species. The coreactant technique employs a single potential scan to generate ECL via interactions between luminophores and oxidizing or reducing molecules. ECL is useful for portable sensors because it gives a direct optical reading that does not require an external light source. This combination of electrochemical and spectroscopic approaches makes ECL biosensing extremely successful for real-time, low-cost, and visually detectable signal detection (Figure 12.4) [9, 10].

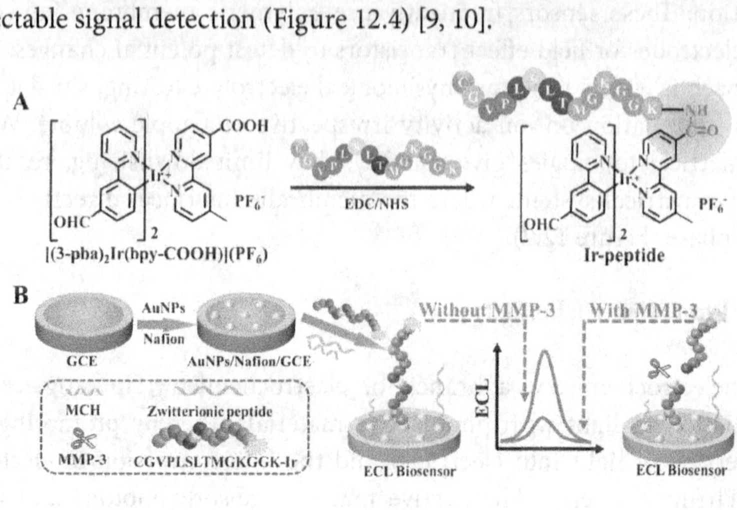

FIGURE 12.4 Representation of peptide-based ECL sensor. (A) Synthesis of Ir-peptide. (B) Fabrication of the peptide-based ECL biosensor. (From Qian, M. et al., *Biosensors*, 14, 2024.)

12.3 NEED FOR NANOMODIFIED ELECTROCHEMICAL SENSORS AND WORKING PRINCIPLE OF TYPES OF NANO-ELECTROCHEMICAL SENSOR

ECS have been utilized to offer accurate and rapid details about a particular environment's composition of chemicals. They display a good estimate that is linearly proportional to the adsorbed analyte. This is utilized to identify key contaminants through targeted on-site detection. These sensors are made up of a transducer coated in a layer involving chemical or organic recognition. The biomolecular recognition is transformed into a physical value, such as mass or electrical resistance, by the transducer. Because of the sensor element's limited lateral extension, which becomes closer to the length scale of the physical event that the transducer principle is based on, the transducer is vulnerable to edge effects at the nanoscale. The sensitivity and selectivity of ECS that have been nanomodified have improved the detectability of biological or chemical materials [12].

12.4 TYPES OF NANO-ELECTROCHEMICAL SENSOR

12.4.1 Electrochemical Impedance Spectroscopy

A sensor that applies sinusoidal signals across a broad frequency range and tracks input response. By default, the system generates a linear curve with an output signal that is exactly proportionate to the input. Reactance and resistance, or impedance, are produced when a current flows through a sensor by its resistance to flow. The material properties are defined collectively by varying the frequency and the impedance at each frequency. An electrochemical system is fed a tiny voltage or current overlaid on a DC signal, and its response is monitored over a range of frequencies (Figure 12.5) [13, 14].

12.4.1.1 Application

12.4.1.1.1 Determination of Ocular Diseases from Tears

ECS protein-based biomarkers utilize tears to detect ocular diseases. The process is painless; the tears are collected instantaneously, and the results are provided on the spot, Additionally, the approach does not require any extra detection molecules to identify the presence of bacteria or parasites that are the result of occlusal allergies [16, 17]. A setup comprising a 5 mM ferri/ferrocyanide redox couple in a 0.1 M KCl buffer mixture was exposed to a 5 mV AC signal with a small amplitude. After agitation in a tube at 13,000 rpm, 10 mg of tear was recovered, yielding a density of 1.005 g. Amyloid concentrations were determined by measuring the resistance

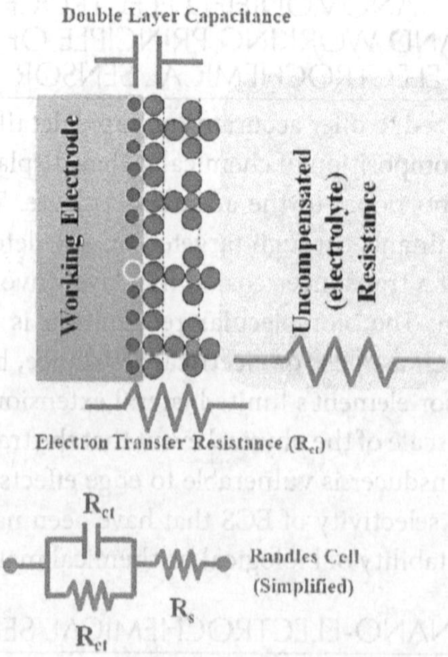

FIGURE 12.5 Schematic diagram of EIS. (From Magar, H.S. et al., *Sensors*, 21, 2021.)

caused by β-amyloid at frequencies ranging from 0.02 Hz to 100 kHz. Aβ levels ranged from 1 pg/mL to 1 ng/mL, with good sensitivity and a low identification limit. The Aβ measurement's resistance ratio ranged from 2.4 to 3.2, resulting in a concentration of 10 pg/mL. This suggests that there was a significant variation in the Aβ ratio between the youngest and oldest age groups [18, 19].

12.4.1.1.2 Determination of Corrosive Behaviour in Metals

In determining the corrosive behaviour of materials, EIS has shown better results. The working electrode consists of copper single crystals (Cu SCs), the counter electrode is a platinum grid, and the reference electrode is a calomel electrode (SCE) saturated with KCl. These were arranged in a conventional three-electrode setup for SC Cu(100)/Cu(111) plates. To conduct EIS studies, a fixed AC signal of 10 mV was modified to frequencies ranging from 10^5 to 10^{-2} Hz. It was concluded that accumulated oxides on Cu(111) SC were causing a decrease in diffusion. Absolute impedance ($|Z|f = 0.01$ Hz) can be used to assess the corrosion resistance of materials at a low frequency. The histogram clearly revealed that Cu(100) had a stronger impedance than Cu(111) at various locations, which corresponds to fluctuations in Nyquist's capacitance loop. After 10D open circuit potential, Cu(100) had a higher impedance of 13,115 Ω cm² (Figure 12.6) [20, 21].

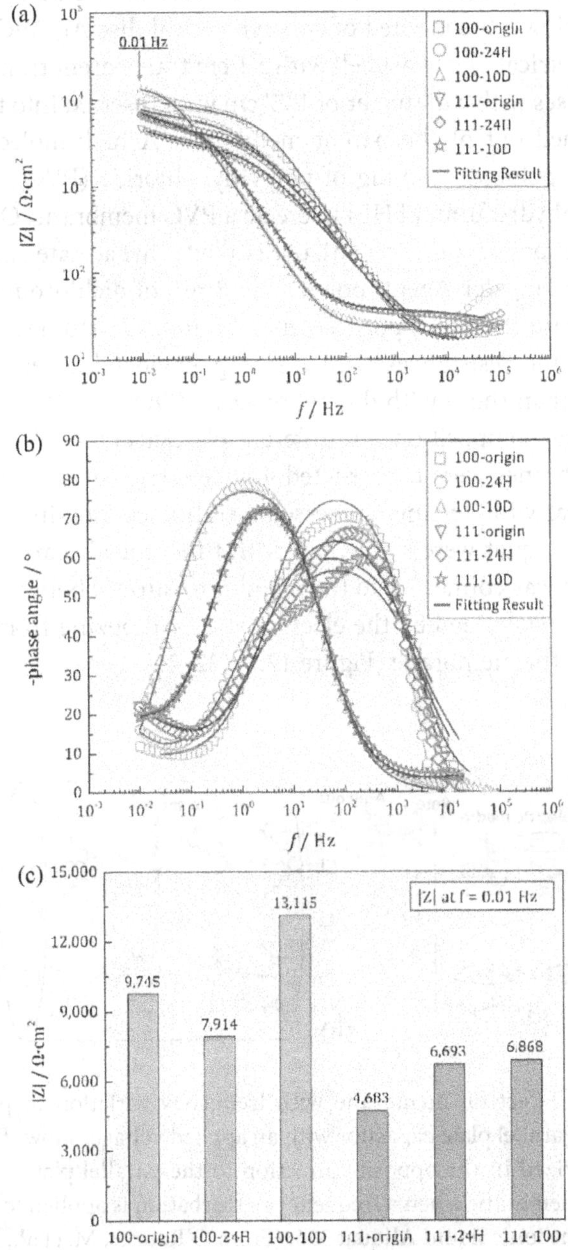

FIGURE 12.6 Histogram of EIS test impedance results.

12.4.1.1.3 Determination of Polymer Inclusion Membrane in Aliquat

Measurements were conducted using a four-electrode arrangement in saturated KCl (Analar) over a frequency range of 0.1–1 MHz, utilizing reagents such as potassium chloride and hydrochloric acid. The impedance measuring cell consisted of two cylindrical disc Ag/AgCl electrodes and a symmetrical plexiglass cell with a 1 cm transparent transmembrane diameter. Discs with a diameter of 1.77 cm were inserted into the cell after being punched out of the parent membrane. A high-molecular-weight selectophore polymer, 250 mg of polyvinyl chloride (PVC), was diluted in 8 mL tetrahydrofuran (THF) to create a PVC membrane. One hundred milligrams of polystyrene in 8 mL of distilled ethyl acetate and 250 mg of CTA (Chain Transfer Agent) polymer in 8 mL of dichloromethane were used to prepare the other polymer membranes. EIS studies showed that bulk membrane resistances with 2-nitrophenyloctyl ether (NPOE) are much lower than those with dioctyl sebecate (DOS) for PVC membranes with the same composition but different plasticizers. PVC, Aliquat, and NPOE membranes were constructed for dielectric constant comparison. During the early EIS testing, processed membranes containing significant amounts of Aliquat were stable, but with time, they became brittle. The phenomenon was connected to the membrane surface being "oily," which might also be explained by the electrostatic fluid oozing from the outermost layer of the membrane (Figure 12.7) [22, 23].

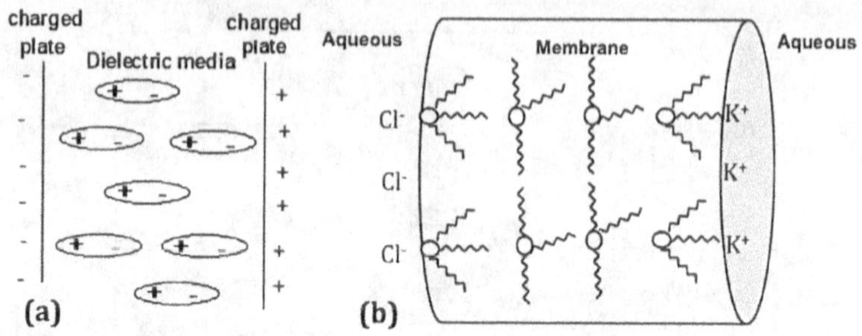

FIGURE 12.7 Effect on membrane with frequency variation in parallel plate capacitor. (a) Parallel plate capacitor with an applied voltage, shows the dielectric medium polarized in the opposite direction to the parallel plates; and (b) The effect on the membrane when a frequency perturbation is applied to the internal dynamics of the ionic liquid Aliquat 336. (From O'Rourke, M. et al., *Membranes*, 1, 132–148, 2011.)

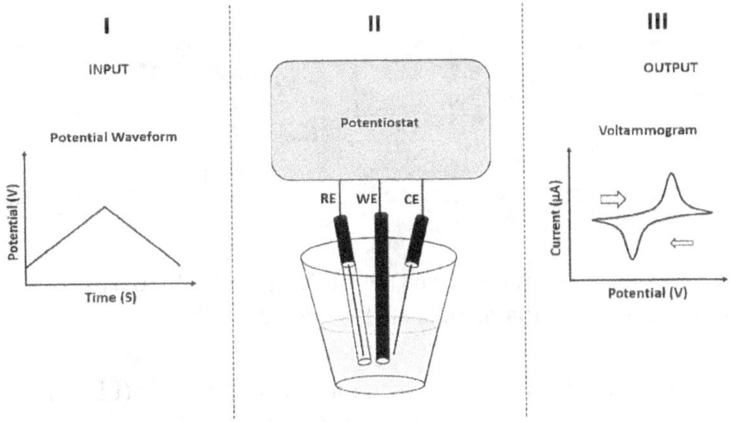

FIGURE 12.8 Schematic diagram of cyclic voltammetry setup. (From Wang, H.-W. et al., *Signals*, 2, 138–158, 2021.)

12.4.2 Cyclic Voltammetry

The reduction and oxidation processes of molecular species are studied by an electrochemical procedure in which potential is linearly swept in relation to time and the current measurement is taken. Scan rate (ms^{-1}) is the slope of voltage change with respect to time. The rate of electron transfer, the materials' electrochemical reactivity, and the measurement scan rate are used to gauge these outcomes. When a scan is performed with a voltage range of E1 to E2, the flow ceases at E1 and returns to E1, which is why the scan is referred to as cyclic. Further data on the voltage, reversibility, and cyclability of redox processes can be obtained by performing multiple cycles (Figure 12.8) [24, 25].

12.4.2.1 APPLICATION

12.4.2.1.1 Determination of Organic Contaminants in Water

Since identifying organic contaminants is so difficult, cyclic voltammetry (CV) has produced higher-quality results. Using deep learning and the CV technique, the benzoquinone (BQ) and hydroquinone (HQ) metabolites of benzene were identified from water sources. Screen-printed electrodes (SPEs) that are inexpensive and disposable, such as graphite ink or single- or multiwalled carbon nanotubes (SWCNTs or MWCNTs), were used as the working, counter, and reference electrodes. As an optional component, hydroquinone was utilized to construct the calibration curve that determined the SPE's sensitivity. The I–V curves recorded at multiple levels of concentration enable both analyte categorization and sensor "calibration."

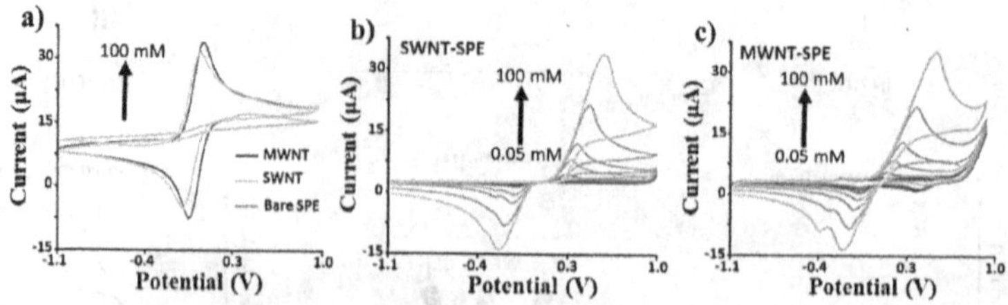

FIGURE 12.9 CV voltagrams of varying working electrode: a) bare, b) SWCNT, c) MWCNT. (From Molinara, M. et al., *Sensors*, 22, 2022.)

In addition, potassium ferri/ferrocyanide $[Fe(CN)_6]^{4-/3-}$ (PF) was examined as a standard electroactive species (Figure 12.9).

Deep learning was applied to the CV findings, revealing hydroquinone and benzoquinone to have extremely similar electroactivity and as a result, identical voltammetry footprints. This suggests that the platform is more important in determining the results than the working electrode [27, 28].

12.4.2.1.2 DETERMINATION OF LEVODOPA FOR TREATMENT OF PARKINSON'S DISEASE

Phosphate buffer solution was mixed with powdered shell and ultrasonicated, forming a electrochemical solution. Utilizing three screen printed carbon sensors, one working electrode composed of carbon electrode, mesoporous carbon, and ordered mesoporous carbon, a reference electrode (Ag/AgCl) and a counter electrode (a Pt wire) were utilized to determine the percentage of levodopa (L-DOPA) in food supplements. For L-DOPA detection, a potential range of −0.4 to +1.3 V was used with scan speeds ranging from 0.1 to 1.0 V s^{-1} (Figure 12.10).

Measurement of food samples such as Now, Haya Labs, and Bio Raw were originally disseminated in the solution for supporting electrolytes (PBS 0.1 M, pH = 7.0) (Figure 12.11).

OSPE (On Site Printed Electrodes) electrodes revealed both reduction and oxidation peaks associated with the existence of L-DOPA, and the coefficient of variation was lower than 3.5% for all experiments, confirming the high accuracy of the procedures [29, 30].

12.4.2.1.3 MEASUREMENT OF CAFFEINE IN DIETARY SUPPLEMENTS

Sodium diphosphate, phosphoric acid, and caffeine (10–3 M) were combined to create a solution of phosphate buffer (PBS). Pt wire as a counter

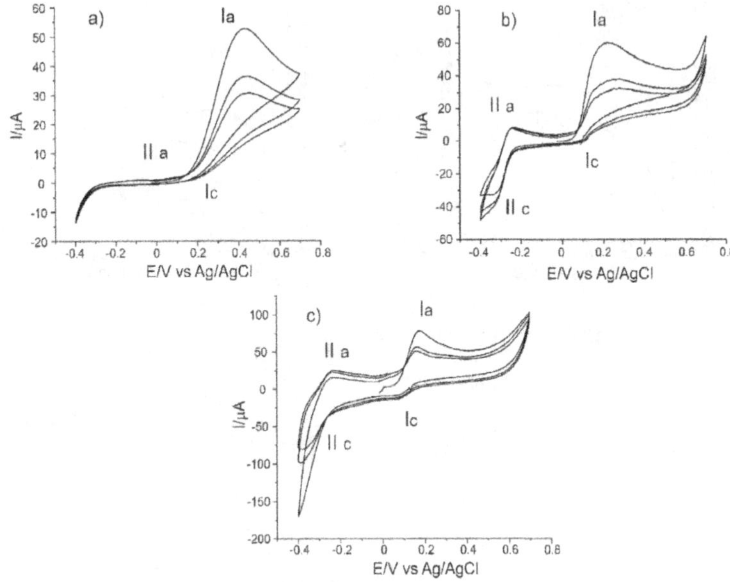

FIGURE 12.10 Cyclic voltammograms obtained from Carbon Screen-Printed Electrode (CSPE), Modified Screen-Printed Electrode (MSPE), and On-Site Printed Electrode (OSPE). Cyclic voltammograms obtained from (a) Carbon Screen-Printed Electrode (CSPE), (b) Modified Screen-Printed Electrode (MSPE), and (c) On-Site Printed Electrode (OSPE). (From Dăscălescu, D. and Apetrei, C., *Sensors*, 21, 2021.)

electrode and an Ag or AgCl substitute electrode were used to maintain a rate of scanning of 0.1 V s^{-1} with carbon nanofibre (CNF), MWCNT, and carbonaceous screen printed electrodes (SPE) as the electrolyte. After being dissolved in 50 millilitres of PBS solution, food supplements such as Active Detox, DVR-Stem Glycemo, and Green Tea underwent five cycles within the ideal potential range (−0.4 to +1.3 V) (Figure 12.12).

The strongest anodic and cathodic peaks on the CNF-modified electrode indicate the maximum susceptibility for caffeic acid identification. CNFs' tubular nanostructure gives them a higher ratio of surface area to volume than MWCNTs, which includes additional edge-plane imperfections on the outermost layer as well as an arranged pattern of graphene sheets. Features of CNFs include a greater rate of electron transfer, a larger specific surface area, enhanced interfacial adsorption, and superior electrocatalytic capabilities. Catechol has the greatest impact on the quantitative measurement of caffeic acid, whereas gallic, L-ascorbic, uric, ferulic, and vanillic acids have less of an impact [31, 32].

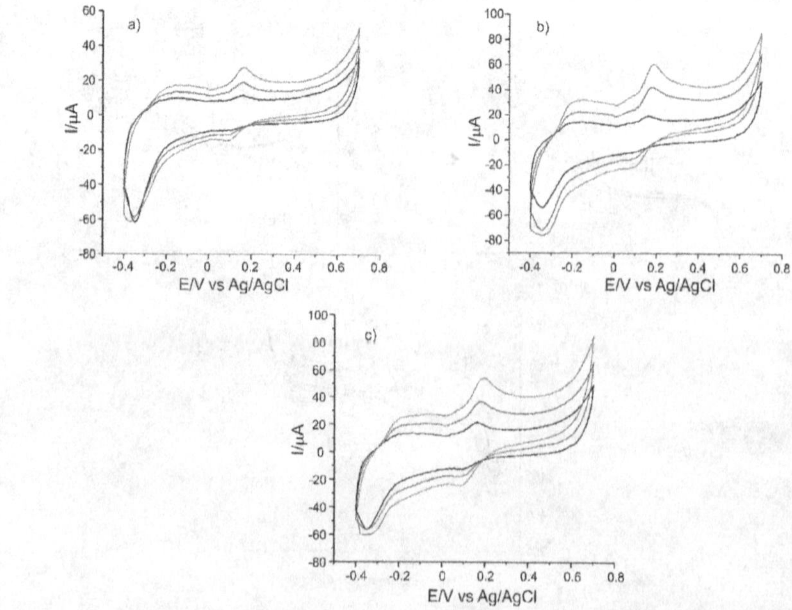

FIGURE 12.11 Cyclic voltammograms of food samples. (a) Bio Raw Foods, (b) Haya Labs, (c) Now Foods, of different concentrations. (From Dăscălescu, D. and Apetrei, C., *Sensors*, 21, 2021.)

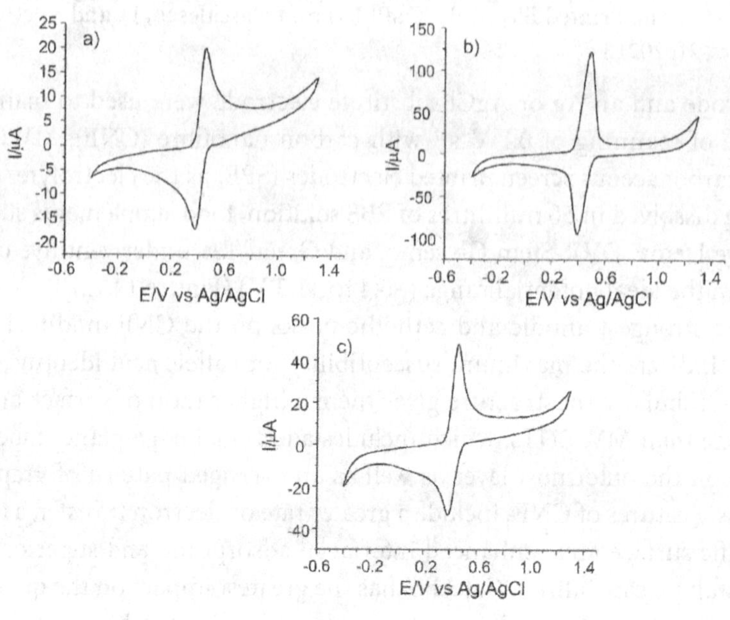

FIGURE 12.12 CV graphs of CSPE, MWCNT SPE, and CNF SPE in 10^{-3} M solution of caffeic acid. (a) CSPE, (b) MWCNT SPE, and (c) CNF SPE. (From Bounegru, A.V. and Apetrei, C., *Chemosensors*, 8, 2020.)

12.5 KEY PERFORMANCE ELECTROCHEMICAL CHARACTERISTICS OF NANOMODIFIED VERSUS CONVENTIONAL ELECTROCHEMICAL SENSORS

Nanomodified electrochemical sensors have been regarded as instantaneous composition meters that operate in real time. These are far more easily installed and operated than spectrometric or chromatographic methods. Compared with optical biosensors, nanomodified ECS biosensors provide inexpensive, label-free assessment of heterogeneous samples. The nanomodified method requires no pre-treatment and uses a smaller volume of sample than the standard method. Comparing nanomodified ECS with standard ECS, measuring biomolecules and characterizing the biosensor system did not require a secondary molecule for measurement [16, 33].

Methods like mass spectrometry and gas chromatography have been employed to identify the pollutants in water. Glass electrodes were altered to include an enzymatic layer, which allowed the detection of carbaryl and azinphosethyl when the electrode was incubated with an enzyme substrate and pesticide. Electrochemical sensors are based on the electrocatalytic response produced by the active electrode material(s) towards electrochemical alterations of the analyte, which are based on the direct electrochemistry of pesticides. One issue with carbon's chemical inertness in electrochemistry is that electrocatalytic conversion and the ensuing detection/quantification require a contact between the analyte and the electrode material (Figure 12.13) [23].

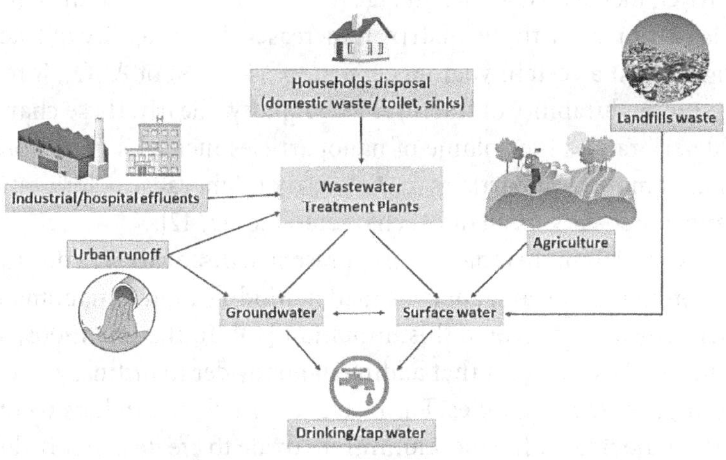

FIGURE 12.13 Pollutants in water sources. (From Kanoun, O. et al., *Sensors*, 21, 2021.)

Graphene quantum dots, on the other hand, enabled additional limit of detection (LOD) improvements with a reduction of 8.23×10^{-8} μM. An ultra-low LOD of just 1.7×10^{-8} μM for acetamiprid was demonstrated using an impedimetric apt sensor that immobilized an aptamer using a decorated MWCNT-reduced graphene oxide nanoribbon. When compared with detection limits obtained using enzyme-based biosensors, these are significantly lower [14, 34].

12.6 MECHANICAL CHARACTERIZATION

A high surface-to-volume ratio, superior electrical properties, stability in chemicals, biological compatibility, and robust durability are just a few of the many unique advantages that nanomaterials provide. Consequently, nanomaterials are often used as reinforcing elements in sensing applications [35, 36]. Because of their characteristics, nanomaterials including graphene, SWCNTs, and MWCNTs have attracted more attention, distinct shapes and a broad range of applications, including sensing devices, medication administration, and storage of energy. By forming an intragranular or intergranular structure, the reinforcement of nanomaterials with a prevalent substance will improve the grain structure to some degree. This improves the grain boundary, resulting in enhanced mechanical properties [37–39]. Different nanoparticles present varied mechanical characteristics, so comparing the durability of natural nanomaterials with that of nanoceramic materials is irrational (Figure 12.14).

The change in the durability of fractures and rigidity as the amount of nano-Al_2O_3 increases is shown in Figure 12.14 [40, 41]. When the nanomaterial is reinforced with the matrix, it increases the durability of fractures and rigidity. At a certain volume percentage (4 vol.%) of Al_2O_3, it reaches its maximum durability of fractures and rigidity. Clearly, these characteristics deteriorate as the volume of nanoparticles increases. Therefore, the addition of more nanoparticles to the matrix might cause deterioration in the nanocomposite's mechanical characteristics [2, 42].

Nanomaterials in their native form presented disadvantages due to poor dispersion in organic and aqueous media. Modification of nanomaterials was performed to overcome this limitation [43]. In the late 1900s, scientists proposed the concept that adding nanopowder to ordinary materials might improve their qualities. For instance, British researchers combined zirconium dioxide with nano-aluminium oxide to create a highly durable substance [44]. Also, it has been shown that the modification of cermet with nano-TiN enhanced the fracture toughness, bending strength, and

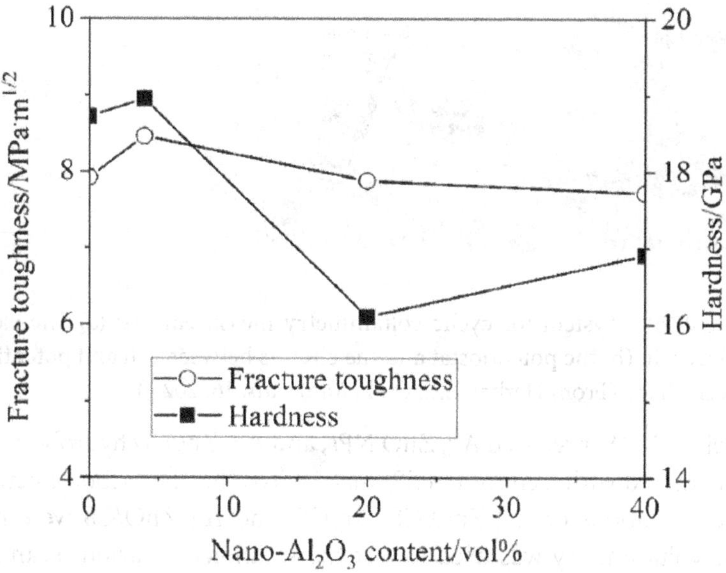

FIGURE 12.14 The effect of the content of nano-Al_2O_3 on fracture toughness and hardness. (From Wu, Q. et al., 9(1), 259–273, 2020.)

hardness by 7%, 25%, and 4%, respectively, as compared with the general cermet.

12.7 ELECTROCHEMICAL CHARACTERIZATION

The electrochemical characteristics of materials adsorbed on an electrode are assessed using cyclic voltammetry. Over time, the working electrode's potential difference (p.d.) varies linearly. Repeated voltage cycles are the result of its returning to the beginning potential after reaching a certain value. A graph of working and counter electrodes against the applied voltage gives CV. The cathodic current increases until the analyte's reduction potential is reached and then decreases as the analyte is depleted. In reversible electrochemical reactions, an anodic current appears during the reverse scan, indicating re-oxidation of the analyte. Redox reactions of electroactive species in solution, adsorption and desorption processes mediated by electric potentials, and double layer charging are typical characteristics of CV. The sum of the capacitive and faradaic currents in CV equals the overall current density [45–47].

In order to modify the surface of screen-printed carbon electrodes (SPCEs) and examine the electrochemical behaviour of the typical redox system of a 10 mM potassium ferrocyanide–0.1 M buffer solution (BS),

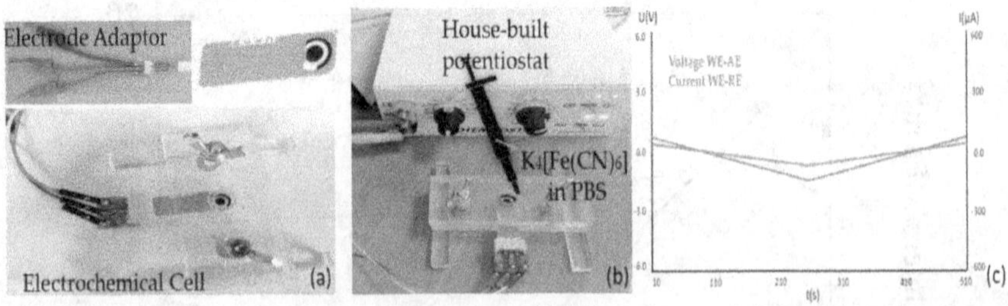

FIGURE 12.15 System for cyclic voltammetry measurements: (a) The electrochemical cell; (b) the potentiostat and the circuits between cell and potentiostat; (c) check chart. (From Herbei, E.E. et al., *Materials*, 16, 2023.)

Herbei et al. [48] prepared Ag, ZnO NPs, and Ag–ZnO by hydrolysing zinc acetate mixed with a chitosan (CS) matrix. To alter the carbon electrode's surface, solutions of CS, ZnO/CS, Ag/CS, and Ag–ZnO/CS were made. Cyclic voltammetry was used to test the solutions at various scan rates, ranging from 0.02 to 0.7 V s^{-1}. A homemade potentiostat (HBP) was used for the CV (Figure 12.15). The impact of altering the scan rate was demonstrated by the measured electrodes' cyclic voltammetry. The anodic and cathodic peak strength is influenced by variations in scan rate. With Ia = 22 μA and Ic = −25 μA, both anodic and cathodic current levels are larger at 0.1 V s^{-1} than they are at 0.06 V s^{-1} (Ia = 10 μA and Ic = −14 μA).

12.8 APPLICATIONS

12.8.1 Molecularly Imprinted Polymer (MIP)–Based Electrochemical Sensors for Ferritin Detection

Because of their excellent stability, quick electropolymerization time, and high selectivity for the target analyte, MIP-based electrochemical sensors have attracted a lot of interest. Interestingly, sensors based on MIPs have become popular for use in both clinical and environmental settings (Figure 12.16) [49].

Using a fluorescence sensor that had been altered with imprinted nanoparticles, Li et al. [50] created a test strip for ferritin detection. They used the sol-gel approach to create the imprinted nanoparticles after deciding on SiO^2 as the core material (Figure 12.16).

They prevented electron or energy transmission between green and red CdTe by making the green and red imprinted nanoparticles independently. Because these imprinted nanoparticles were light-transmissive, fluorescence emission could be seen with the naked eye. Visual identification was made easier by the progressive change in fluorescence colour

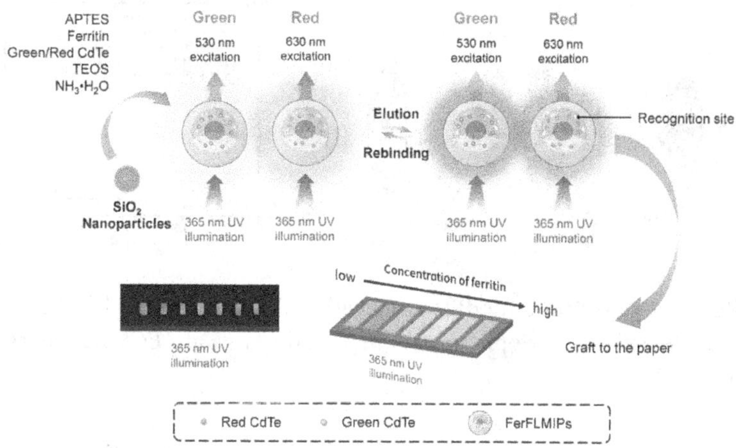

FIGURE 12.16 Preparation of imprinted nanoparticle-modified fluorescence sensor for ferritin detection. (From Saylan, Y. et al., *Processes*, 12, 2024.)

from green to red as the ferritin concentration rose after the nanoparticles were put to the test paper.

12.8.2 NANOPARTICLE-MODIFIED SENSOR FOR THE DETECTION OF DOPAMINE, URIC ACID, AND ASCORBIC ACID

Blood and urine are examples of biological fluids that contain dopamine (DA), ascorbic acid (AA), and uric acid (UA). AA is essential for metabolic functions, abnormal DA levels can cause brain abnormalities, and the kidneys release UA in urine, which can cause imbalances that lead to diseases, including Lesch–Nyhan syndrome and hyperuricaemia [51, 52]. Using electrochemically produced FeO^4 nanoparticles, Gaya et al. [53] created a basic electrochemical sensor to identify the aforementioned components. In order to make the sensor, they combined the nanoparticles with carbon Vulcan, NAFION, and isopropanol to create an ink. Using cyclic voltammetry and square wave voltammetry, it was shown that the detection linear ranges for DA, UA, and AA were 10–100 μM, 20–160 μM, and 1050–2300 μM, respectively, with detection limits of 4.5 μM, 14 μM, and 95 μM.

12.8.3 NANOPARTICLE-MODIFIED SENSOR FOR ANTICANCER DRUG OXOPLATIN

Cis-dichlorodiamminedihydroxoplatin (IV), often known as oxoplatin (OXP), is frequently used to treat cancer. OXP and its metabolites also

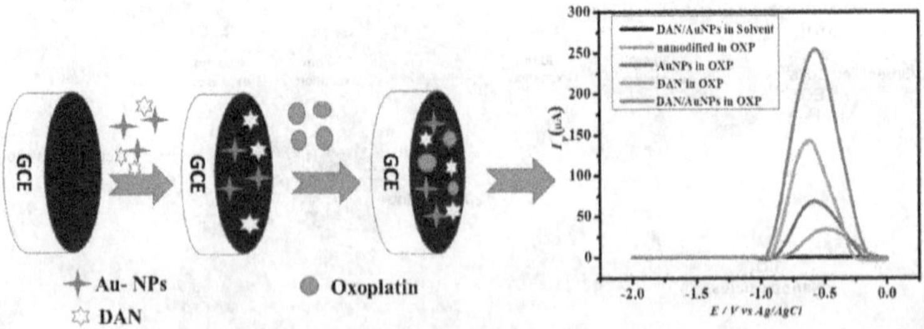

FIGURE 12.17 Electrode modification with DAN/AuNPs enables the electrochemical detection of oxoplatin. (From Zahid, A. et al., *ECS Advances*, 2(4), 040506, 2023.)

make their way to wastewater treatment facilities after excretion. Therefore, it is essential to identify OXP in wastewater in order to safeguard aquatic life. Zahid et al. [54] modified the surface of a glassy carbon electrode (GCE) with a suitable modifier to enhance its sensing characteristics. For the sensitive detection of OXP, a commonly used anticancer medication, a combination of gold nanoparticles (AuNPs) and the surfactant 1-(2, 4-initrophenyl)-dodecanoylthiourea (DAN) was utilized as a modifier. A 0.05 µm alumina slurry was used to polish the 0.07 cm^2 GCE in this work, and it was then washed with doubly distilled water. Using a micropipette, a 10 µl droplet of DAN/Au nanoparticles (1:1 ratio) was placed on the GCE and left to dry. After that, as shown in Figure 12.17, the DAN/AuNPs/GCE was put into the electrochemical cell for more research. With a limit of detection (LOD) of 1.45×10^{-11} M, the results demonstrated that under ideal circumstances, the DAN/AuNPs/GCE demonstrated exceptional sensitivity for detecting OXP.

12.8.4 ELECTROSPUN CARBON NANOFIBRES/GOLD NANOPARTICLES–MODIFIED GRAPHITE ELECTRODE FOR SALMONELLA ENTERICA SEROVAR DETECTION

Salmonella enterica, a primary cause of gastrointestinal infections, is among the most prevalent foodborne pathogens. The Centers for Disease Control and Prevention (CDC) report that roughly 2% of foodborne illnesses in the United States result from *Salmonella enterica* contamination, causing about 19,000 hospitalizations and 380 deaths each year. Antibiotic resistance is a major concern in *Salmonella enterica* serovars [55–57].

As a novel platform for the electrochemical detection of *Salmonella enterica* serovar, Fathi et al. [58] presented a new, straightforward, and

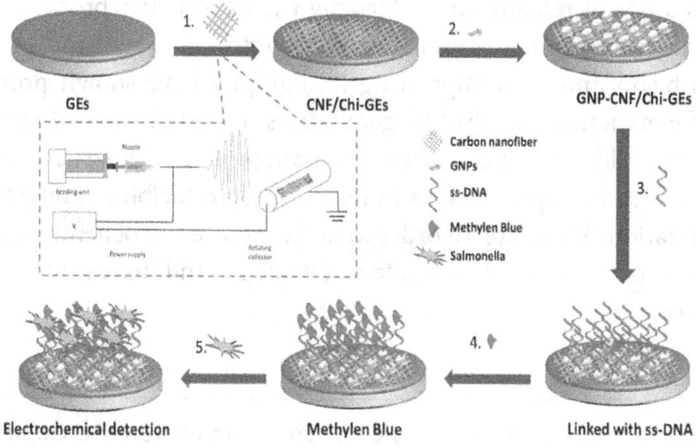

FIGURE 12.18 The fabrication steps of the aptasensor using glassy electrode (GE) was modified with a composite of aptamer-gold nanoparticles, carbon nanofiber, and chitosan (Apt-GNPs/CNF-Chi//GE . (From Fathi, S. et al., *Biointerface Research in Applied Chemistry*, 11, 2020.)

sensitive aptasensor based on a chitosan (Chi)-electrospun carbon nanofibres (CNF)/gold nanoparticles (GNPs)–studded pencil graphite electrode (GE). This voltammetric sensor was developed using a *Salmonella*-specific recognition aptamer DNA sequence (Figure 12.18). The increased surface area at the electrode's active site, which improves electron transport to the electrode surface, is probably what caused the increase in current.

12.9 CONCLUSION

By utilizing less material required for stabilization, nano-electrochemical sensors are intended to improve sensitivity, process catalysis, the ability to react at low potentials, and the speed at which electrons move from the active reaction centre to the electrode. By removing chemical intermediates in electron transmission, the incorporation of nanomaterials into the structure helps simplify electronics. Miniaturized electrochemical probes have distinct benefits in this respect, including the ability to achieve very low detection limits and excellent resolution, as well as ease of integration with other parts and complementing methodologies. In particular, there is an increase in the use of machine learning and deep learning methods like recurrent neural networks (RNNs) and convolutional neural networks (CNNs) in the processing of sensory data. However, because of (i) the extremely high resolution needed for nano-patterning and (ii) the difficulties of reliably characterizing the nanometric probe, it is still very difficult to obtain an exact geometry control at the nanoscale. While electron-beam and nanoimprinting lithography have shown promising results for nanometric chip-like geometries, recent reports of pyrolysed carbon nanoelectrode fabrication for nanopipette-like structures have shown consistent improvements in terms of reproducibility and potential automatization. Therefore, in order to make nano-electrochemical sensors better than traditional sensors, a few advantages and disadvantages must be addressed.

REFERENCES

1. Singh, R., et al., A Review on Recent Trends and Future Developments in Electrochemical Sensing. *ACS Omega*, 2024. **9**(7): p. 7336–7356.
2. Adimule, V., et al., Highly Photoluminescence and Wide Band Gap Insulating Metal Hybrid Nanoparticles Array of Samarium-Doped SrO:CoO: Synthesis, Characterizations and Sensor Characteristics. *Journal of Materials Science: Materials in Electronics*, 2023. **34**(5): p. 442.
3. Baranwal, J., et al. Electrochemical Sensors and Their Applications: A Review. *Chemosensors*, 2022. **10**, DOI: 10.3390/chemosensors10090363.
4. Saputra, H.A., Electrochemical Sensors: Basic Principles, Engineering, and State of the Art. *Monatshefte für Chemie - Chemical Monthly*, 2023. **154**(10): p. 1083–1100.
5. Zhang, Z., E. Boselli, and I. Papautsky Potentiometric Sensor System with Self-Calibration for Long-Term, In Situ Measurements. *Chemosensors*, 2023. **11**, DOI: 10.3390/chemosensors11010048.
6. Shu, J. and D. Tang, Recent Advances in Photoelectrochemical Sensing: From Engineered Photoactive Materials to Sensing Devices and Detection Modes. *Analytical Chemistry*, 2020. **92**(1): p. 363–377.
7. Adimule, V., et al., Materials and Production Technologies in Machinery. 2022, Trans Tech Publications Ltd: Baech.
8. Wang, Y., J. Liu, and F. Lin A Photoelectrochemical Sensor for the Sensitive Detection of Cysteine Based on Cadmium Sulfide/Tungsten Disulfide Nanocomposites. *Nanomaterials*, 2024. **14**, DOI: 10.3390/nano14050427.
9. Qi, H. and C. Zhang, Electrogenerated Chemiluminescence Biosensing. *Analytical Chemistry*, 2020. **92**(1): p. 524–534.
10. Gao, W., et al., Electrogenerated Chemiluminescence for Chronopotentiometric Sensors. *Analytical Chemistry*, 2019. **91**(7): p. 4889–4895.
11. Qian, M., et al. Electrogenerated Chemiluminescence Biosensor for Quantization of Matrix Metalloproteinase-3 in Serum via Target-Induced Cleavage of Oligopeptide. *Biosensors*, 2024. **14**, DOI: 10.3390/bios14040181.
12. Mahale, R.S., et al., Electrochemical Sensor Applications of Nanoparticle Modified Carbon Paste Electrodes to Detect Various Neurotransmitters: A Review. *Applied Mechanics and Materials*, 2022. **908**: p. 69–88.

13. Lazanas, A.C. and M.I. Prodromidis, Electrochemical Impedance Spectroscopy—A Tutorial. *ACS Measurement Science Au*, 2023. **3**(3): p. 162–193.
14. Kanoun, O., et al. A Review of Nanocomposite-Modified Electrochemical Sensors for Water Quality Monitoring. *Sensors*, 2021. **21**, DOI: 10.3390/s21124131.
15. Magar, H.S., R.Y.A. Hassan, and A. Mulchandani Electrochemical Impedance Spectroscopy (EIS): Principles, Construction, and Biosensing Applications. *Sensors*, 2021. **21**, DOI: 10.3390/s21196578.
16. Ozdalgic, B., et al. Emerging Applications of Electrochemical Impedance Spectroscopy in Tear Film Analysis. *Biosensors*, 2022. **12**, DOI: 10.3390/bios12100827.
17. Rajendrachari, S. and D. Ramakrishna, 1 - Functionalized Nanomaterial-Based Electrochemical Sensors: A Sensitive Sensor Platform, in *Functionalized Nanomaterial-Based Electrochemical Sensors*, C.M. Hussain and J.G. Manjunatha, Editors. 2022, Woodhead Publishing. p. 3–25.
18. Wang, Y.-R., et al., High-Sensitivity and Trace-Amount Specimen Electrochemical Sensors for Exploring the Levels of β-Amyloid in Human Blood and Tears. *Analytical Chemistry*, 2021. **93**(22): p. 8099–8106.
19. Somashekharappa, K.K. and S. Rajendrachari, Chapter 1 - Sustainable Development Information Management of Carbon Nanomaterial-Based Sensors, in *Carbon Nanomaterials-Based Sensors*, J.G. Manjunatha and C.M. Hussain, Editors. 2022, Elsevier. p. 3–12.
20. Lin, Q., et al. Electrochemical Impedance Spectroscopy (EIS) Explanation of Single Crystal Cu(100)/Cu(111) in Different Corrosion Stages. *Materials*, 2023. **16**, DOI: 10.3390/ma16041740.
21. Sharath, P.C., et al., Chapter 7 - Metal Oxide Composite-Based 1D Nanomaterials and Its Gas-Sensing Characteristics, in *Complex and Composite Metal Oxides for Gas, VOC and Humidity Sensors*, B.C. Yadav and P. Kumar, Editors. 2024, Elsevier. p. 189–210.
22. O'Rourke, M., et al. Electrochemical Impedance Spectroscopy—A Simple Method for the Characterization of Polymer Inclusion Membranes Containing Aliquat 336. *Membranes*, 2011. **1**: p. 132–148, DOI: 10.3390/membranes1020132.
23. Rajendrachari, S., et al. Assessing the Food Quality Using Carbon Nanomaterial Based Electrodes by Voltammetric Techniques. *Biosensors*, 2022. **12**, DOI: 10.3390/bios12121173.
24. Kim, T., et al., Applications of Voltammetry in Lithium Ion Battery Research. *Journal of Electrochemical Science and Technology*, 2020. **11**(1): p. 14–25.
25. Somashekharappa, K.K., R.B. Halappa, and S. Rajendrachari, Green Chemistry Applications in Electrochemical Sensors, in *Recent Developments in Green Electrochemical Sensors: Design, Performance, and Applications*. 2023, American Chemical Society. p. 23–37.
26. Wang, H.-W., et al. Cyclic Voltammetry in Biological Samples: A Systematic Review of Methods and Techniques Applicable to Clinical Settings. *Signals*, 2021. **2**: p. 138–158 DOI: 10.3390/signals2010012.

27. Molinara, M., et al. A Deep Learning Approach to Organic Pollutants Classification Using Voltammetry. *Sensors*, 2022. **22**, DOI: 10.3390/s22208032.
28. Kshama, S.S., K. Swamynathan, and R. Shashanka, Electrochemical Devices for Soil Analysis, in *Recent Developments in Green Electrochemical Sensors: Design, Performance, and Applications*. 2023, American Chemical Society. p. 121–139.
29. Dăscălescu, D. and C. Apetrei Voltammetric Determination of Levodopa Using Mesoporous Carbon—Modified Screen-Printed Carbon Sensors. *Sensors*, 2021. **21**, DOI: 10.3390/s21186301.
30. Shashanka, R., Carbon Composite Voltammetric Sensors for Food Quality Assessment, in *Electrochemical Sensors Based on Carbon Composite Materials*. 2022, IOP Publishing. p. 8-1–8-13.
31. Bounegru, A.V. and C. Apetrei Voltammetric Sensors Based on Nanomaterials for Detection of Caffeic Acid in Food Supplements. *Chemosensors*, 2020. **8**, DOI: 10.3390/chemosensors8020041.
32. Adimule, V., M.G. Revaigh, and H.J. Adarsha, Synthesis and Fabrication of Y-Doped ZnO Nanoparticles and Their Application as a Gas Sensor for the Detection of Ammonia. *Journal of Materials Engineering and Performance*, 2020. **29**(7): p. 4586–4596.
33. Adimule, V., B.C. Yallur, and A.H.J. Gowda, Chapter 14 - Advanced Sensors Based on Carbon Nanomaterials, in *Carbon Nanomaterials-Based Sensors*, J.G. Manjunatha and C.M. Hussain, Editors. 2022, Elsevier. p. 259–268.
34. Adimule, V., D. Bhowmik, and A.H.J. Gowda, Morphology, Characterization, and Gas Sensor Properties of Sr Doped WO3 Thin Film Nanostructures. *Macromolecular Symposia*, 2021. **400**(1): p. 2100065.
35. Chen, A. and S. Chatterjee, Nanomaterials Based Electrochemical Sensors for Biomedical Applications. *Chemical Society Reviews*, 2013. **42**(12): p. 5425–5438.
36. Pai, M., et al., Optical Graphene for Biosensor Application: A Review. *Applied Mechanics and Materials*, 2022. **908**: p. 51–68.
37. Wang, J.L. and L.J. Meng, Influence of Carbon Nano-Fiber on Mechanical Property of PALC. *Applied Mechanics and Materials*, 2014. **535**: p. 785–787.
38. Chang, C.I., et al., Microstructure and Mechanical Properties of Nano-ZrO2 and Nano-SiO2 Particulate Reinforced AZ31-Mg Based Composites Fabricated by Friction Stir Processing. *Key Engineering Materials*, 2007. **351**: p. 114–119.
39. Zou, B., et al., Effect of Nano-Scale TiN on the Mechanical Properties and Microstructure of Si3N4 Based Ceramic Tool Materials. *Key Engineering Materials*, 2006. **315–316**: p. 154–158.
40. Wu, Q., et al., Mechanical Properties of Nanomaterials: A Review. *Nanotechnology Reviews*, 2020. **9**(1): p. 259–273.
41. Adimule, V., S.S. Nandi, and B.C. Yallur, Devices and Sensors Based on Additively Manufactured Shape-Memory of Hybrid Nanocomposites, in Shape Memory Composites Based on Polymers and Metals for 4D Printing: Processes, Applications and Challenges, M.R. Maurya, et al., Editors. 2022, Springer International Publishing. p. 341–359.

42. Li, X., et al., Fabrication and Mechanical Properties of Nano-Al2O3 Based Composite Ceramic Tool Material. *Advanced Materials Research*, 2011. **335–336**: p. 736–739.
43. Cernat, A., et al., Electrochemical Sensors Based on Carbon Nanomaterials for Acetaminophen Detection: A Review. *Analytica Chimica Acta*, 2015. **886**: p. 16–28.
44. Xie, F., et al., Development of Nano-TiN Modified Cermet Cutter. *Materials Science Forum*, 2004. **471–472**: p. 711–715.
45. Falola, T.O., Nanoparticles Modified Electrodes: Synthesis,Modification, and Characterization—A Review. *World Journal of Nano Science and Engineering*, 2022. **12**: p. 29–62.
46. Adimule, V., et al., Insight into Electro Chemical and Dielectric Properties of Flower Shaped Samarium Embedded Y2O3@SeO2 Nanocomposites for H2O2 Sensor Applications. *Materials Today Communications*, 2024. **38**: p. 108163.
47. Rajendrachari, S., et al., Fabrication of Mechanically Alloyed High Entropy Alloys (20-Fe-20Cr-20Ni-20Mn-20Ti) Modified Carbon Paste Electrode Sensor for the Cyclic Voltammetric Determination of Methyl Orange. *Inorganic Chemistry Communications*, 2024. **164**: p. 112428.
48. Herbei, E.E., P. Alexandru, and M. Busila Cyclic Voltammetry of Screen-Printed Carbon Electrode Coated with Ag-ZnO Nanoparticles in Chitosan Matrix. *Materials*, 2023. **16**, DOI: 10.3390/ma16083266.
49. Lahcen, A.A. and A. Amine, Recent Advances in Electrochemical Sensors Based on Molecularly Imprinted Polymers and Nanomaterials. *Electroanalysis*, 2019. **31**(2): p. 188–201.
50. Saylan, Y., S. Kılıç, and A. Denizli Biosensing Applications of Molecularly Imprinted-Polymer-Based Nanomaterials. *Processes*, 2024. **12**, DOI: 10.3390/pr12010177.
51. Carrera, V., et al., A Simple and Rapid HPLC-MS Method for the Simultaneous Determination of Epinephrine, Norepinephrine, Dopamine and 5-Hydroxytryptamine: Application to the Secretion of Bovine Chromaffin Cell Cultures. J Chromatogr B Analyt Technol Biomed *Life Science*, 2007. **847**(2): p. 88–94.
52. Kong, D., et al., Simultaneous Voltammetry Detection of Dopamine and Uric Acid in Human Serum and Urine with a Poly(Procaterol Hydrochloride) Modified Glassy Carbon Electrode. *Talanta*, 2018. **185**: p. 203–212.
53. Gaya, E., et al. Fe3O4-Nanoparticle-Modified Sensor for the Detection of Dopamine, Uric Acid and Ascorbic Acid. *Chemosensors*, 2023. **11**, DOI: 10.3390/chemosensors11020079.
54. Zahid, A., A. Shah, and S. Munir, Gold Nanoparticles and Surfactant-Based Electrochemical Sensing Platform for the Detection and Electroanalysis of the Anticancer Drug Oxoplatin. *ECS Advances*, 2023. **2**(4): p. 040506.
55. Chai, Y., et al., Design of a Surface-Scanning Coil Detector for Direct Bacteria Detection on Food Surfaces Using a Magnetoelastic Biosensor. *Journal of Applied Physics*, 2013. **114**(10): p. 104504.

56. Klemm, E.J., et al., Emergence of an Extensively Drug-Resistant Salmonella enterica Serovar Typhi Clone Harboring a Promiscuous Plasmid Encoding Resistance to Fluoroquinolones and Third-Generation Cephalosporins. *mBio*, 2018. **9**(1): p. 1–10.
57. D, V.T.N., K. Venkitanarayanan, and A. Kollanoor Johny, Antibiotic-Resistant Salmonella in the Food Supply and the Potential Role of Antibiotic Alternatives for Control. *Foods*, 2018. **7**(10): p. 1–11.
58. Fathi, S., et al., Novel Competitive Voltammetric Aptasensor Based on Electrospun Carbon Nanofibers-Gold Nanoparticles Modified Graphite Electrode for Salmonella enterica serovar Detection. *Biointerface Research in Applied Chemistry*, 2020. **11**: 1–10.

CHAPTER 13

Carbon-Based Nanomaterials

Synthesis, Characterization, and Applications of Electrochemical Sensors and Biosensors

Sanjay Ballur Prasanna, Hari Prasad Badubanahalli Ramesh, Santhosh Arehalli Shivamurthy, Karthik Chimatahalli Shanthakumar, Hareesha Nagarajappa, Varun Donnakatte Neelalochana, Sandeep Shadakshari, and Ren-Jei Chung

13.1 INTRODUCTION

13.1.1 Nanomaterials

"Nano" is a scientific word derived from the Greek word *nanos*, meaning dwarf. The word "nano" was meant to be a very distinct approach in the past; the materialistic property of tiny particles was a fresh topic to the world, and there was a lot of life in it. During a well-known lecture by Nobel laureate Richard P Feynman, he quoted, "There's Plenty of Space at the Bottom" [1]. Nanoscale materials gained ground at the time due to

DOI: 10.1201/9781003512639-13

their reported characteristic features like tuneability, durability, and high aspect of surface area. These materials can be studied in all three dimensions along with the zero-dimensional (0D) geometries, depending on their total shape [2]. The materials gained attention and appreciation from researchers based on how size can influence the physicochemical properties of the material. Their optical properties were also considered eccentric, depicting solitary colours due to varying interaction of nanoparticles (NPs) with light. For example, NPs of gold (Au), platinum (Pt), silver (Ag), and palladium (Pd) with a uniform size of 20 nm show different characteristic colorations of wine-red, yellowish grey, black, and dark black, respectively.

For this study, the synthesis of Au nanoparticles particularly with variable sizes was accomplished. The characteristic colorations and properties of the NPs were also related to differences in size and structure, making them efficient materials for bioimaging applications [3]. As Figure 13.1 indicates, due to variations in aspect ratio, nanoshell diameter, and percentage of gold concentration in nanocages, the colour of the solution varies. Alteration of any of the above-listed factors affects the NPs' absorption properties, and thus, different colours of absorption are observed. NPs are not defined as simple molecules themselves; they are made up of stacked layers, i.e. (a) a surface layer with a series of tiny particles, molecules, metal ions, surfactants, and polymers that can be functionalized; (b) the surface layer, different from the core material due to the difference in functionality; and (c) the central core of the NPs [3]. NPs with the discussed properties have now become the subject of open-door research for many applications, like drug delivery [4], chemical and biological sensing [5], gas sensing [6], CO_2 capture [7], and

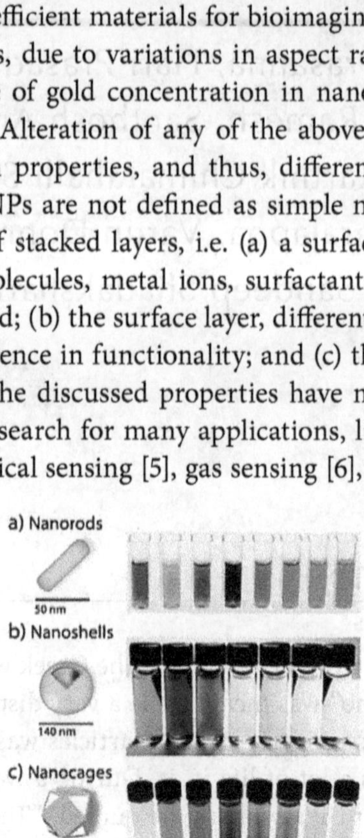

FIGURE 13.1 AuNPs' colour dependency on size and shape. (From Khan, K.S. and Khan, I., *Arab. J. Chem.* **12,** 908–931, 2019.)

other related applications [8]. The design of metal-based nanomaterials is also a line of interest and has gained importance in current research. This modern nanoscience is the preference of many researchers striving to achieve impeccable and novel standards. In the early year of 1857, Au nanoparticles were successfully synthesized and reported by Faraday [9]. With data interpretation and design analysis using the electron microscope, the observation of nanostructures and comprehensive studies on the controlled synthesis of colloidal Au nanomaterials have emerged [10]. Significant efforts were made to examine the controlled synthesis of other kinds of metal nanostructures and their properties. NPs are classified into different groups, like carbon-based NPs, metal NPs, ceramic NPs, semiconductor NPs, and polymeric NPs, based on their morphology, size, and physicochemical properties [11–18].

The properties of the NPs are imparted by their synthesis techniques. Two important techniques, namely, top-down and bottom-up synthesis, are incorporated in nanomaterial synthesis. These two techniques work on simple fusion and fission mechanisms involved in the nanoscale materials.

13.1.2 Top-Down Synthesis

Top-down synthesis is a destructive approach focusing on the disintegration of larger molecules to form smaller particles. Many techniques are known in this process; a simple one is grinding. The other techniques, like chemical and physical vapour decomposition, help in the reduction of larger molecules to smaller particles, as shown in Figure 13.2, which further helps in the Important property detailing [19]. These small particles are further modified into suitable desired NPs. As determined using the Scherer equation, the crystalline size decreases with more milling time. The decrease in particle size or diameter is observed by the results of the scanning electron microscopy (SEM) and X-ray pattern, which follow the effect as stated above in relation to time [20]. Examples of references to the development of nanomaterials using the top-down approach have been demonstrated. One such synthesis of spherical magnetic nanoparticles was from natural iron oxide ore in the presence of oleic acid to produce nanoparticles of varying sizes between 20 and 50 nm by a destructive method [21]. The simple method of the incorporation of the inorganic metal oxide clusters with the organic carbon-based interfacial moiety yields a colloidal carbon spherical particle with easily controlled size. Particles with a narrow size distribution and high dispersion ability were produced by this method of synthesis, as shown in Figure 13.3 [22]. In the

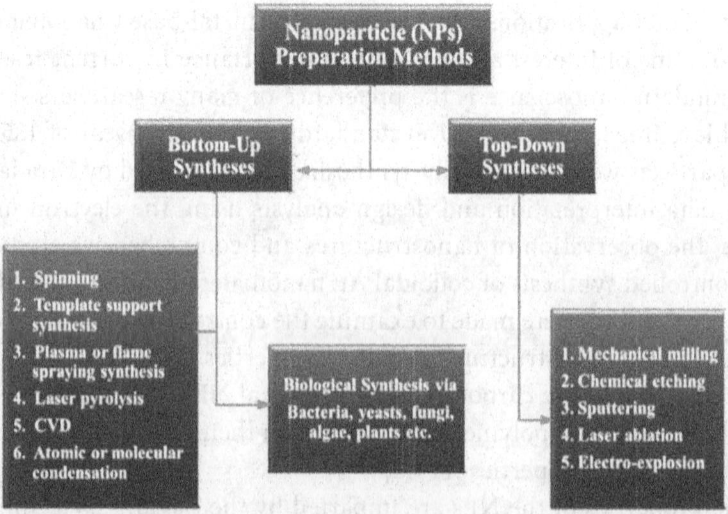

FIGURE 13.2 Synthetic pathways for the preparation of nanoparticles. (From Madkour, L.H., *Adv. Struct. Mater.*, **116**, 393–456, 2019.)

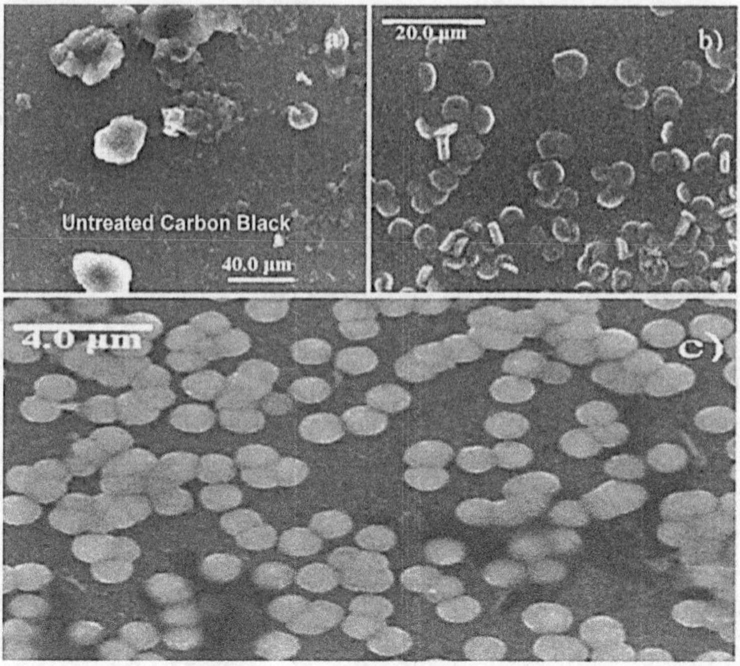

FIGURE 13.3 SEM images of (a) untreated carbon black, (b) and (c) ultrasound in Polyoxometalate solution for 10 min and 1 h, respectively. (From Zahid, M.U. et al., *Mater. Res. Express.*, **5**, 052002, 2018.)

process of synthesis of carbon nanoparticles, the sonication technique is used, and this technique adversely affects the size of the nanoparticles over time by reducing the size of the particles. In the synthesis of transition-metal dichalcogenide nanodots (TMD-NDs), the combination of grinding and sonic top-down methods was also used. They presented an excellent, narrow size distribution on a scale of >10 nm [23]. Recent advances in nanoparticle synthesis have included a top-down laser fragmentation technique that produced nanoparticles in the 5.8 nm $\pm$ 1.1 nm range.

13.1.3 Bottom-Up Synthesis

This synthesis is the inverse of the top-down approach. There is a fusion of the basic raw materials used in this process, which is also called the method of building up. Sol–gel, spinning, biochemical synthesis, and green synthesis are some examples of these types of methods (Iravani, 2011) [24]. One of the examples of the above technique was used by Mogilevsky et al. [25]. Through his technique, he synthesized nanoparticles from metastable minerals composed of TiO_2 along with a graphene matrix. Alizarin's hydroxyl terminals have a good affinity for TiO_2. The functional matrix formed was characterized and confirmed using X-ray diffraction (XRD) data. SEM images showed a change in the size pattern with a rise in temperature, as shown in Figure 13.4 [25].

By both techniques, we can conclude that the obtained NPs are in the range of 100–500 nm. The toxicity of the NPs has a high influence on the efficacy of their applications; thus, focusing on the green and biogenic approach in bottom-up synthesis, which is also considerably inexpensive and cost-effective, is appreciated by many researchers. There is a keen focus on how biological systems can be incorporated into the synthesis of the NPs; there are many examples of how leaf extracts, plant extracts, and household bacteria can be incorporated into the greener approach of NP preparation. This type of practice involves lower pollution, and the particles can be fully monitored for high-potential applications [26]; the plant microorganisms and biogenic extracts can also be used as reducing agents. The typical scheme given in Figure 13.5 shows a simple representation of the nanomaterial synthesis routes [26].

13.1.4 Carbon-Based Nanomaterials

In biosensors and biomedical applications, the importance of electrochemical sensors has given rise to a desire to find ways of integrating the idea of conditioning the electrodes for maximum sensitivity and

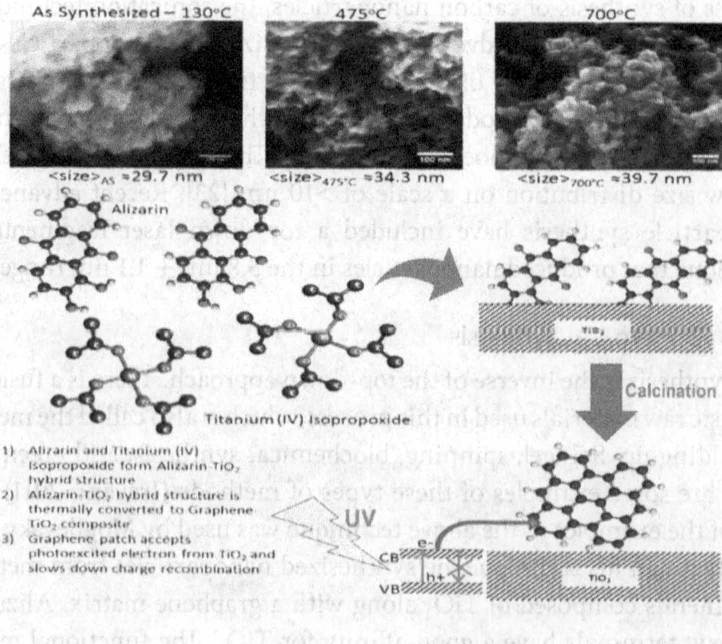

FIGURE 13.4 TiO_2 synthesis via bottom-up technology. SEM images demonstrating the TiO_2 NPs. (From Mogilevsky, G. et al., *ACS Appl. Mater. Interfaces*, **6,** 10638–10648, 2014.)

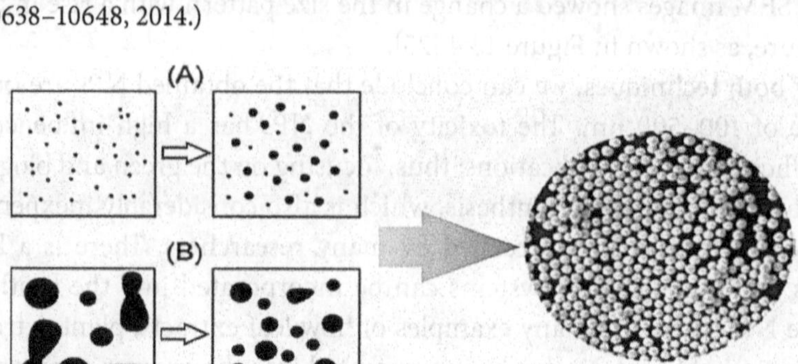

FIGURE 13.5 (A) Bottom-up approach: A molecular precursor is shattered down into simpler atoms of metal that develop into colloids. (B) Top-down approach: Large drops of metal that have been broken into smaller drops. (From Wang, Y. and Xia, Y., *Nano Lett.*, **4,** 2047–2050, 2004.)

selectivity. The materialistic approach to sensors generates strong interest in developing sensors that are more efficient and effective. The sensors are used in various applications in housing, industry, and the environment [27]. If the sensors are mainly used for biomedical applications, then the toxicity of the sensory components and their conditioners is a serious consideration. The major element that allows the monitoring unit to sense and analyse the data is the electrodes. Recent publications are gathering data on the various characteristics of electrode materials. The choice of existing materials is based on their electrical, mechanical, and thermal properties. Silver, gold, aluminium, and carbon are the common materials used in electrode preparation [28, 29]. Graphene probes are the most predictable material for future efficacy in electrochemical sensors. Graphene can simply be defined as a single layer of tightly packed carbon atoms that form the structure of a two-dimensional (2D) honeycomb crystal lattice [30]. Here, we focus on the observations of the previously discussed variable carbon-based nanomaterials and their properties, advantages, and disadvantages regarding toxicity, reusability, efficiency, and selectivity. For their numerous properties and uses, the types of carbon materials based on the dimensional structures are discussed. Carbon nanotubes (CNTs) are graphene sheets that are rolled up. They were first published in the journal *Nature* by the Japanese scientist Sumio Iijima in 1991. Multi-walled CNTs were observed from a carbon arc discharge, and Prof. Sumio Ijima and Donald Bethune individually observed single-walled CNTs a few years later [31]. Based on the wall structure, nanotubes can be classified into single-walled nanotubes (SWNTs) and multi-walled nanotubes (MWNTs). MWNTs are divided into two types: Russian doll and parchment. The inner tube separation of MWNT is 3.4 Å, which is the same as the interlayer separation of graphite [32]. Based on chirality, nanotubes are classified as chiral and achiral. Achiral nanotubes have a plane of the mirror (pom), and they have two parts, a zigzag and an armchair, whereas chiral nanotubes lack a pom. The chiral vector is taken as the circumference of the nanotube, and the vector perpendicular to the chiral vector becomes the axis of the tube.

Since the discovery of CNTs, the study of their extensive properties and the nature of their applications has been a marked focus for many researchers. Recent advances in CNTs are in the field of electroanalysis [33]. Studies have shown that CNTs can drastically increase their importance in biomolecular applications, and their host nature enhances the specificity and selectivity of any drug action. They advance direct electron–transfer mechanisms of proteins. These properties make them a better option in biomedical applications and sensors. CNTs have cylindrical nanostructures with various graphene sheet stacking frameworks, such as stacked platelets, ribbons, or herringbones, as shown in Figure 13.6 [34]. They have micrometre dimensions, while their diameter ranges from a

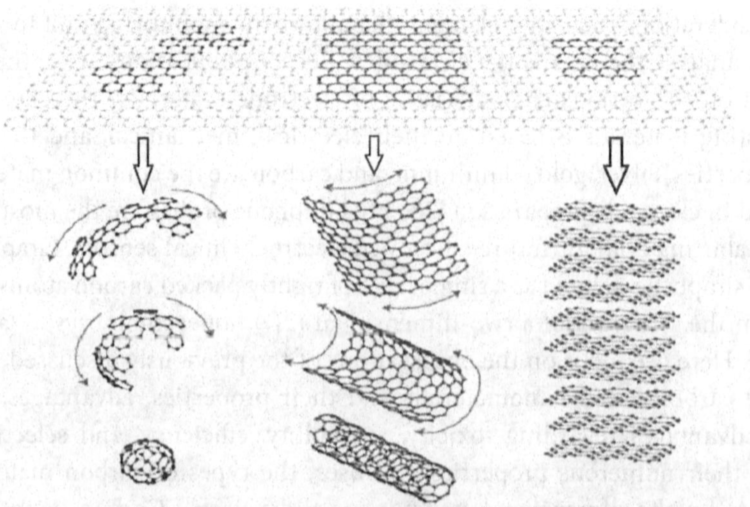

FIGURE 13.6 Different forms of modified graphene. (From Geim, A.K. and Novoselov, K.S., in *Nanoscience and Technology: A Collection of Reviews from Nature Journals*, World Scientific, Singapore 2009.)

few tens to many hundreds of nanometres. Carbon nanofibres (CNFs) have a stacked graphene layer, which forms an outer layer. This sheet gives more strength and more active binding sites when compared with CNTs [35]. CNFs promote more electron transfer from electroactive analytes [36]. In addition, CNFs are unique in the way that it is possible to activate their entire surface region. Activation of CNFs by using nitric acid or electrochemical oxidation may produce a variety of oxygen-bearing functionalities without affecting the structural cohesion of their backbone. These characteristics mean that for biomolecules, CNFs can be ideal immobilization matrixes, thus transmitting electrochemical signals that act as transducers. The application of CNFs, however, has centred primarily on catalyst support and gas storage systems [37]. In previous decades, polymer reinforcements [38], probe tips [39], and many other things were extensively used until the discovery of CNFs [40]. Because of the superior properties of the CNFs, they have attracted all interest since then [41].

13.1.5 Graphene

Graphene is a type of carbon material that has fascinated researchers with its innumerable properties, such as electrical, crystalline, and high-tensile materials. They are made from layers of carbon stacked layer by layer to give a 2D honeycomb crystal lattice structure. Graphene and its interesting properties have been known for the past 60 years as part of the study of allotropes of carbon materials. However, graphene materials and their applications have gained concentrated attention since the formation of the 2D graphene structure [42, 43]. Graphene has applications in the synthesis of anodes and supercapacitors that have low charging time, high efficiency of charge-to-mass ratio, and large surface area. The versatile and flexible characteristics of graphene materials make them unique nanopore materials with high electrical and mechanical strength and excellent thermal conductivity. In the formation of the 2D structure of graphene by stacking of individual layers, there is a cohesive force between graphene molecules, which affects the physiological properties. To reduce this effect, graphene is functionalized accordingly. This tuneable property of graphene makes it ready for high-end applications [44–46]. The functionalization is done through covalent and non-covalent means to improve the surface area of graphene. A few techniques that are achieved by chemical treatment are spin coating, filtration, and layer-by-layer assembly for surface modification, retaining the intrinsic properties [47]. Graphene is a material in high demand in the production of sensors because of its specific properties, like a large surface-area-to-volume ratio, peculiar optical properties, excellent stability of the carrier, and superior electrical and thermal properties when compared with the other carbon allotropes. It is also considered to be environmentally friendly when compared with other metals [48, 49]. Comparatively, graphene-based sensors show higher sensitivity and detection limits than CNTs. However, one disadvantage of this sensor is the lack of linearity of response in detection. The applications of these materials depend on their ability to be functionalized based on the required applications. By incorporating some active materials, including epoxy, carboxyl, and hydroxyl groups, these materials are made efficiently functional, with catalytic active centres along with covalent and non-covalent functionalization anchoring sites [50]. Functionalization of molecular graphene with various functional materials enables the physicochemical properties of the material to be perfectly tuned, resulting in the combination of individual materials. Implementing water-soluble additives can contribute to enhanced aqueous dispersibility, which is beneficial for film-modified electrode preparation [51]; polyelectrolyte decoration may significantly increase the adsorption potential of oppositely charged species [52]; combination of nanoparticles (NPs) with metal can produce special small molecules with selective and catalytically active graphene nanocomposite [53]; combination with elements of biological recognition can help to capture objectives and help transduce signals [54]; integration with transparent plastics can proceed to photosensitive, wearable electrochemical sensors [55], etc. The graphene material is typically designed based on the desired applications by enhancing its characteristic stability, durability, best host nature [56], and excellent electrocatalytic activity; all these properties show the potential for electrode performance in the applications of electrochemical sensors and biomedical applications [56–59]. Recent significant developments applicable to graphene-based nanocomposites in the field of biosensors and electrochemical sensors are reviewed in this chapter. In the synthesis and molecular functionalization of graphene nanocomposites, a prime factor to be considered is the choice of materials that are practical for the operating procedure. Then, some important technological discoveries focusing on different electrochemical techniques are discussed regarding their analytical practice. Finally, we conclude by looking at the existing problems and prospects of electrochemical sensors and biosensors based on graphene nanocomposite.

13.1.6 Graphene Oxides

The synthesis of graphene nanocomposite is achieved by the standard basic principle of the aforementioned bonding or interaction. Among many methods of synthesis, the simplest one is mechanical mixing. Functional materials could be combined by hand grinding with graphene [60], electromagnetic stirring in solutions, or ultrasonic mixing [61], and refluxing or heating is also a common practice [62]. Synthesizing and developing graphene-based functional materials is a more controllable technique; the active graphene surfaces are created in situ by the incorporation of new design and functionalization before being used in synthesis. Electrodeposition is one such special synthetic tool to produce nanocomposite graphene. It is important to electrochemically deposit or co-deposit functional materials on graphene from a solution containing functional precursors [63, 64] because in the case of electrodes in electrochemical sensors, direct deposition of the graphene nanocomposites is important. This reduces further electrode preparation by mechanical modification and its complications. Considering the chemical nature of GO, we have to focus on the impairments that the oxygen group can cause to the mechanical properties when present along with the graphene, and accordingly, considering the presence of the oxygen group will improve the study of GO for potential applications. Assuming that GO is oxygen-functionalized graphene, how the oxygen-bearing groups will impact the mechanical and physical properties of the perfect graphene sheet is an interesting fundamental study, and this information has key importance in its application areas.

13.1.7 Carbon Nanotubes

In the mid-1980s, fullerene chemistry was developed and reported by Smalley and colleagues at Rice University [65]. Fullerenes consist of hexagonal and pentagonal faces with typical geometrical cage structures with carbon nodes. The C_{60} (buckyball) molecule was the first convex-shaped structure with a closed cage of 60 carbon atoms where the adjacent side of a soccer ball-like hexagon is on either side of a pentagon; this structure was named after R. Buckminster Fuller, well known for his geodesic domes architectural design [66]. Their discovery paved the way for new characteristic materials named CNTs, which were synthesized and studied a couple of years later. Solid, slender fullerenes are nanotubes, where the walls of the tubes are hexagonal carbon (graphite structure) and are often capped at each end. Mechanical characteristics of CNTs that surpass those of any previously known materials have been documented by several researchers. While there are various reports on the exact properties of CNTs in the literature, theoretical and experimental findings have shown that the elastic modulus is extremely strong, greater than 1 TPa (the elastic modulus of a diamond is 1.2 TPa); the strength of just a fraction of CNTs was 10–100 times greater than the strength of the strongest steel registered and published at the time. Indeed, CNTs can result in a new class of advanced materials if the stated mechanical properties are validated. To unlock the efficiency of CNTs for use in polymer nanocomposites, the elastic and fracture properties of CNTs and the nanotube/matrix interface interactions must be completely comprehended [67]. Along with their efficient mechanical properties, they also possess superior thermoelectric properties, they show high thermal stability up to 2800 °C in a vacuum, their thermal conductivity measurement is twice as high as that of diamond, and their electrical current–carrying capacity is 1000 times larger when compared with copper wires [68]. The latest evidence has proved and reported a remarkably large elastic modulus [69, 70] in the durability of large elastic strains and fracture strains [71, 72]. Some theoretical studies have also drawn similar conclusions [73–75], but very few connections have been made between theoretical predictions and experimental studies. The formation of carbon nanotube is achieved by the schematic rolling of hexagonal plane graphite sheets [76].

13.1.8 Carbon Nanofibres

Thomas Edison used the first carbon fibre (CF), prepared by carbonizing cotton and bamboo, as the filament of a light bulb in 1879, and since then, CF has gained enormous advancement in basic science research and its innovative applications [77–80]. CNFs are low-density nanofibres with small pore sizes and good breathability. They have shown their potential applications in biosensors, electrochemical storage, and conversions. They have good mechanical properties and act as better hosts for external materials, and this can be used as a potential modification for a better desired application [81–83]. CNFs and traditional carbon fibres have several variants. The primary noticeable characteristic is their length [84, 85]. There are multi-micrometre diameters in the standard CF, while CNFs have diameters of 50–200 nm. A possible difference between CNFs and CFs is the measure of internal diameter (Figure 13.7).

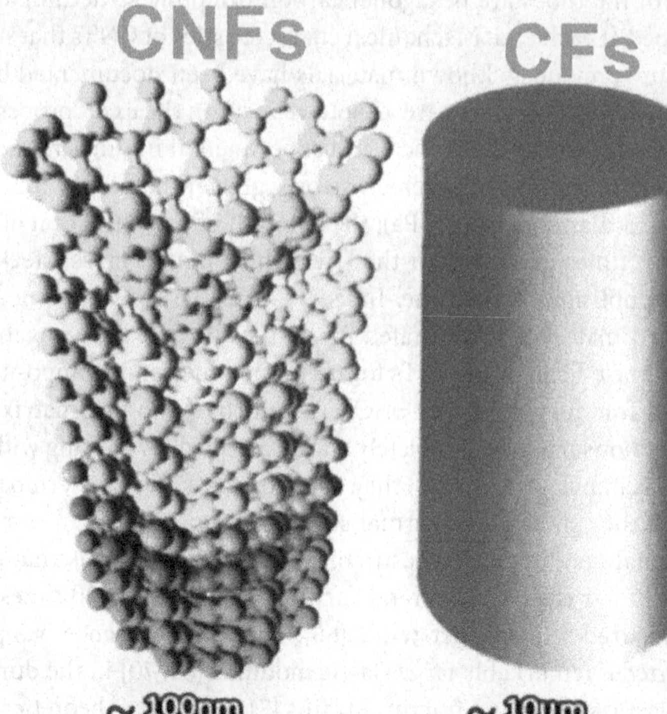

FIGURE 13.7 Schematic example of the differentiation between CNF and CF. (From Zeng, J. et al., *Compos. B Eng.*, **35,** 245–249, 2004.)

In the 1890s, the interaction of carbon-containing gases with hot metal surfaces to produce carbon-based materials was reported [86]; however, recent developments in electron microscopy techniques have resulted in structural details and mechanistic features. The characteristics of these newly grown CNFs have attracted much interest in the future of electrochemical storage and sensor applications. It is important to study the mechanism of the materials when we are focusing on the desired application. This growth of the individual nanofibres in early practice was by the direct individual growth of structures, reported by Baker and colleagues and characterized using a high-resolution electron microscope [87]. The carbon species are produced by the dissolution and dispersion of the adsorbed hydrocarbon on the metal surface. The dispersed hydrocarbons likely precipitate at the rear of the particle to produce CNFs. High-resolution electron microscopy was used in the detailed structural studies of this type of carbon [88–91]. In the reported structure of CNT, the graphene layers are arranged parallel to catalyst particles in an orderly manner. The electrons in CNT are more transparent because there is a thin crystalline layer formed on the walls of the cylindrical structures [92].

13.2 SYNTHESIS

The properties of carbon have shown us its capability, and humans have fully utilized it in the field of sensors, surface area applications, etc. Carbon's existence in different allotropic forms and its numerous oxidation states have made it a suitable material for future research. There are many studies, ongoing or already reported, on the exceptional qualities of carbon in the field of electrochemical sensors. The ability to functionalize carbon-based materials has shown great promise for affording a good hydrophilic surface, and with functional groups like carbonyl, carboxyl, and hydroxyl, these modifications promise to make materials more efficient in functionalizing and increasing the surface area [93]. The synthesis subsection for each material gives us more detail on its properties and helps us to categorize variable carbon materials in a brief discussion.

13.2.1 Graphene

Graphene materials are synthesized by two well-known methods, namely, the top-down approach and bottom-up synthesis, which are discussed in Sections 13.1.1 and 13.1.2. Both methods are considered standard approaches using a carbonaceous gas source. The methods include some specific processes, like chemical conversion, CO reduction, electric arc, and CNT decompression to synthesize graphene [94]. As discussed, top-down synthesis involves the disintegration of the existing bulk source to create the final nanomaterials. The top-down methods are considered to be cost-effective methods. However, this process is limited to the laboratory scale and has limited quality control [95]. Even though manufacture by this method is economically tough to achieve, with a low yield, industries are focusing on mass scale-ups to fulfil their requirements and achieve a robust material. Also, exfoliation of the graphite material for better exposure of the surface (by methods such as sonification, super acid dissolution of graphite, functionalization, electrochemical exfoliation, alkylation of graphene derivatives, reduction of aqueous/organically treated graphene oxide, thermal exfoliation, and chemical reduction of GO) has advanced the efficacy and reusability of the material [96, 97]. On the other hand, the methods of organic synthesis and crystal growth by metallic substrates are used in the production of graphene materials. The morphology, crystallinity, and structure of graphene depend on the small molecules and catalysts involved in the reaction [98]. Although the functionalization of bottom-up-synthesized graphene is considered to be much harder to perform in bulk and with the desired qualities (industrial production departments don't encourage the use of this approach for synthesis if they don't find it economical [99, 100]), bottom-up synthesis is the preferred path for graphene synthesis because it offers remarkable opportunities to achieve the atomic size, durability, stability, good surface area, and strength [101]. Researchers around the world are striving to achieve a synthetic standard fabrication protocol for the large-scale production of bulk graphene at a low cost and with efficient functionality [102, 103].

13.2.2 Carbon Nanotubes

CNTs are exceptional materials that have shown us the reach of nanoscience beyond our expectations. Nanotubes are one of the novel approaches derived by using bottom-up synthesis. CNTs are derived from graphene materials, and their one-dimensional property makes them exceptionally good for electronic, mechanical, and sensitivity applications. They are well known for their diverse structure–property relations, though they have the simplest bonding property and atomic structure [104]. There are two types of nanotubes, depending upon the condition in which the graphene sheets are rolled. A single sheet of graphene is rolled on itself to form a cylindrical tube-shaped single-walled carbon nanotube (SWCNT), and concentric nanotubes held together by van der Waals forces are called multi-walled carbon nanotubes (MWCNTs) [105].

A single-walled nanotube (SWNT) is obtained by rolling a graphene sheet along the lattice vector of the graphene plane, and SWNTs are formed as shown in Figure 13.8 [106]. SWNTs could be metals or semiconductors with relatively large (0.5 eV for a typical 1.5 nm diameter) or small (10 meV) band gaps, even if they have nearly identical diameters [107]. The synthesis of CNTs is studied using three methods: Arc discharge, chemical vapour deposition (CVD), and laser ablation methods [108]. The production of nanotubes has proved to be superior despite considering the by-products that are produced at a lower temperature [109]. For CVD, hydrocarbon gases are used as the source of carbon and metal catalyst particles as "seeds." Comparing the three methods of synthesis, the arc discharge and laser ablation methods have produced SWNTs with a narrow diameter distribution averaging ~1.4 nm. All three methods failed in the bulk production of SWNTs, and the homogeneous diameters

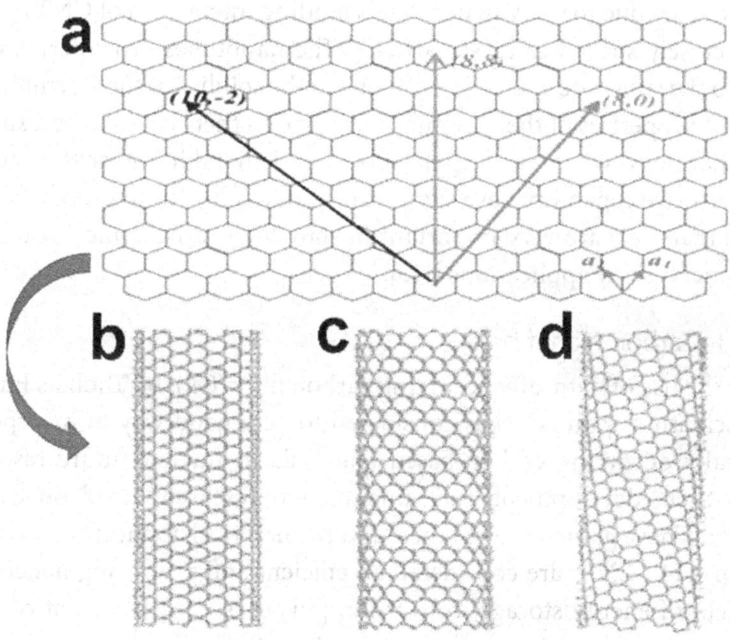

FIGURE 13.8 (a) Schematic honeycomb structure of a graphene sheet, (b) armchair, (c) zigzag, and (d) chiral tubes. (From Jeynes, J., 2007. Single-walled carbon nanotubes functionalised with RNA or DNA (PhD Thesis). University of Surrey, England.)

and chirality of the nanotubes were not maintained. Carbon fibres, filaments, and MWCNTs were synthesized by CVD methods [110, 111]. The synthesized nanotubes were evaluated for their electrical, mechanical, and physical parameters as facilitated by the controlled material synthesis. The implications of nanotube functionalization, molecular sensors, and interfacing with biological molecules were discussed. The pattern development and modification of SWCNTs and MWCNTs have paved the way for a promising orientation of the nanotubes with self-assembly characteristics and active electric field control, and pursued several generations of catalysts, ranging from powdery-supported catalysts to discrete catalytic nanoparticles. All these parameters have paved the way for the development of ordered nanotube arrays or networks at the stage of nanotube synthesis.

Our earlier work has shown that silicon substrates patterned with iron particles in an ordered square region are incorporated in the synthesis of regularly ordered towers consisting of MWCNTs by a CVD method at 700 °C with an ethane carbon source and alumina-supported iron catalyst [112]. The self-assembling of the nanotubes into square rigid bundle structures occurs due to the Van der Waals binding interaction of CNTs grown from closely spaced catalyst particles. The nanotubes grow perpendicularly to the substrate and self-orient due to the solidity of the assembly. The arrayed properties of the nanotube have proved themselves to be a surface area manifesto where their application and demand in the field of storage and sensitive capability have significantly on the high interest of research. These materials are very important in showing good chemical routes for flat panel display applications [113].

13.2.3 Carbon Nanofibres

Since the use of naturally occurring carbon fibre (CF), by Thomas Edison, this scientific approach was considered to be a good way to incorporate naturally occurring carbon-based materials in further future research. Since then, development and fundamental research based on carbon materials have gained importance, and the practical applications are scaling up [114]. CNFs are considered an efficient and promising material in the field of energy storage, conversion purposes, reinforcement of composites, and self-sensing applications [115–118]. Catalytical thermal CVD and electrospinning-influenced heat treatment are the most favourable methods reported in the synthesis of CNFs.

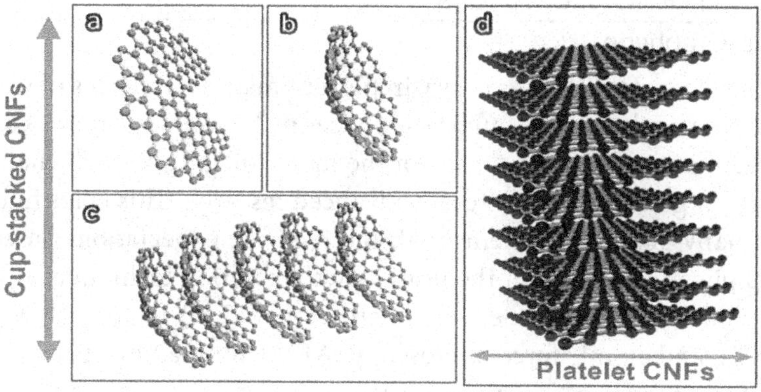

FIGURE 13.9 Schematic demonstration of (a–c) formation of cup-stacked CNF structure and (d) platelet CNF structure. (From V. García-Martínez, V. et al., *Mater. Today. Commun.*, **24,** 100990, 2020.)

13.2.3.1 Synthesis of CNF by Catalytically Active Thermal Chemical Vapor Deposition Growth

Ge and Sattler (1994) provided two types of morphological structure derived from this method [119]. These are cup-stacked and platelet CNF, as shown in Figure 13.9 (a–d) [120]. The carbon sources are important, and at the temperature range of 700 to 1200 K, methane, carbon monoxide, molybdenum, ethyne, or ethene is preferred as the carbon source [121]. The deposition of the graphene layer is achieved by dissolving the metal particles in the hydrocarbon solution and then precipitating that solution on the required metal surface [122].

13.2.3.2 Electrospinning Method in Carbon Nanofibre Preparation

A highly recommended as well as preferred method for the synthesis of CNFs is the electrospinning method, This method generates CNFs that are susceptible to forming web or mat structures in the majority of cases. This structure can be used as a battery or supercapacitor electrode material [123]. Currently reviewed CNF was also under a systematic approach of the electrospinning process. Electrospinning was followed by carbonization via heat treatment, and the structural properties of the CNF were studied [124]. The heat treatment was combined with parameters that controlled crystallinity, morphology, purity, and porosity. These controlled pore structures, which were challenging to achieve, can be used in battery or supercapacitor electrode material. As the overall performance of batteries is highly dependent on the ion transport capabilities of the electrolyte, the battery efficiency is related to the pore structures.

13.3 CHARACTERIZATION

13.3.1 Graphene

Graphene is established as a very important material in terms of surface area and morphology. It can be simply defined it as monoatomic carbon layered material. But the details of the morphology, although challenging, are of great interest in recent advanced research. This helps initially with many possible problems with the required expectations based on our applications. Some of the prior characterization techniques involve SEM and transmission electron microscopy (TEM) for morphological analysis, and atomic force microscopy (AFM) and Raman spectroscopy (RS) for analysis of chemical structure and intra-molecular orientations. Optical microscopy through the colour distinction instigated by the light-interfering conclusion on the SiO_2 substratum has been utilized for the determination of graphene sheets down to one layer deep in the thickness modulated by the graphene layer [125, 126]. Atomic Force Microscopy (AFM) involves transferring graphene onto silicon wafers using a carefully controlled layer of silicon oxide (SiO) of specific thickness. Geim's assembly showed the comparison between the graphene images of optical microscopy and SEM. The novelty of the structure is purely dependent on the packing scale, which is visible in detail only in SEM imaging [127]. SEM of graphene film shows a clearer disparity [128]. AFM is an effective mode for the depiction of graphene. By using AFM, one can easily understand the morphological features of graphene due to its thickness, and also, the number of graphene layers on the substrate can be estimated. It is difficult to achieve the hypothetical width of 0.34 nm due to the variation in tip revulsion among the protecting substratum, the semi-metallic graphene, and the protective layer formed by the efficient adsorption of the thin layer of liquid on the surface of the graphene under atmospheric conditions. Hence, characterization by the AFM method has a very low output, and the value for the size of the graphene film is inaccurate [129].

Raman spectroscopy is a well-known immediate and impressive method for the analysis of graphene with structural and quality parameters. The spectrum analysis of a graphene samples shows the stretching frequency at 1350 cm^{-1} (noted as the D band), 1580 cm^{-1} (G band), and 2700 cm^{-1} (G′ band, also called a 2D band). The D band of graphene is correlated with defects: The lower the D stretching amplitude, the lower the graphene layer defects. It is also correlated with graphene doping [130]. The G band is the graphite $E_{2}g$ mode, arising due to the two-dimensional hexagonal graphite layer of sp^2-bonded carbon atoms. Raman spectroscopy is widely used for characterizing graphene due to the advantage of high-throughput and non-destructive analysis of graphene sheets.

13.3.2 Carbon Nanotubes

The characterization of the CNTs and their surface interpretation and imaging are carried out by non-destructive techniques like SEM and TEM. The imaging of nanoscale structures is achieved by surface probe microscopy (SPM). The molecular- and atomic-level imaging is characterized using AFM. The depth of the atom is measured using a non-optical microscopic technique called scanning tunnelling microscopy (STM). Some electron microscopy, such as neutron diffraction (ND), shows the lattice constants of the metals. X-ray diffraction (XRD) depicts the crystallinity of the surface. The characterization of CNTs is maximal under the base-level properties of the surface; hence, the detailing of the matrix with the consideration of all the above parameters helps in analysing the structure achieved and further, helps in the course of application.

13.3.2.1 Scanning and Transmission Electron Microscopy

The study of the surface morphology of organic and inorganic species helps in creating a better idea of how the structure is arranged at the molecular level. The electron beam produced through SEM on ejection to the surfaces and their interaction through atoms of CNTs reflections are recorded as images of the surfaces [131], estimate the eminence of as-equipped CNTs (Figure 13.10), to validate the existence of functionalization reactions. TEM analysis overcomes the resolution limit of SEM (1–20 nm) [132]. In Transmission Electron Microscopy (TEM), high-energy electrons pass through the sample, and carbon nanotubes (CNTs) transmit these electrons to form an image. High-resolution TEM (HR-TEM) enables detailed analysis of parameters such as interlayer spacing, number of graphene walls, and the distance between sheets in multi-walled carbon nanotubes (MWCNTs) [133]. TEM shows the morphology of the organic and inorganic functional groups that modify the surface of the nanotubes based on their characteristic functionality [134, 135].

13.3.2.2 X-Ray Diffraction

XRD shows the percentage crystallinity of the synthesized sample. The crystallinity is calculated by the Scherrer formula. The X-ray incident on the sample is diffracted. The formation of the defect sites and the new organic bonds influence the diffraction pattern; the concentration of

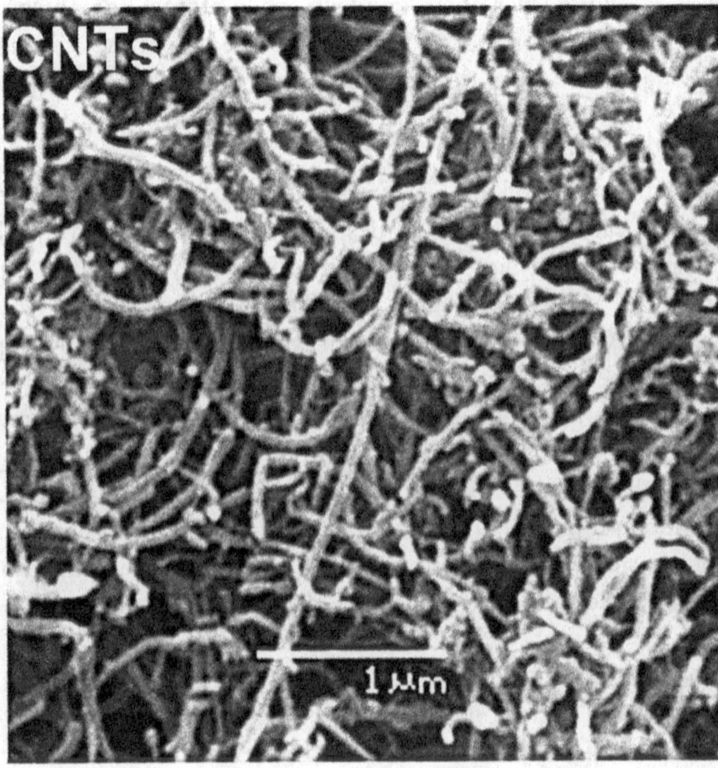

FIGURE 13.10 SEM image of a bulk sample of MWCNTs. (From S. Zomorodbakhsh, Z. et al., *Int. J. Nanomed.*, **15**, 5901–5909, 2020.)

the diffraction pattern also varies according to the new sample topology achieved. Detail regarding the nature of inter-layer stacking and associations based on parameters like chirality of layers and spatial correlations between atoms is an important criterion to focus on in the case of CNTs [136]. In a recent publication, García-Gutiérrez et al. reported the application of XRD in characterizing the structures of both SWCNTs and MWCNTs and associated CNT–polymer nanocomposites at various length scales (single CNTs, bundles of CNTs, and aggregates of CNT bundles) [137].

13.3.2.3 Spectroscopic Techniques

Spectroscopic techniques are important in the meticulous study of synthesized compounds. There are many spectroscopic techniques, and we assign the analysis of the product based on the required validations. In the analysis of nanomaterials, we incorporate techniques such as near-infrared spectroscopy (NIR), Raman spectroscopy, energy dispersive spectroscopy (EDS), X-ray photoelectron spectroscopy (XPS), and fluorescence spectroscopy (FS), which are the most valid techniques for nanotube characterization. Metal impurities are detected at the ppm level by inductively coupled plasma atomic emission spectroscopy (ICP-AES), inductively coupled plasma mass spectroscopy (ICP-MS), and graphite furnace atomic absorption spectrometry (GFAAS).

13.3.2.4 Raman Spectroscopy

RS is non-destructive, non-invasive, and an important spectroscopic technique. It is also the most influential characterization method in the case of nanomaterials and the purity of CNTs [138]. Recent advances over 10 years in RS history (1997–2007) show RS as an important tool for CNT characterization [139]. Previously reported articles on the application and performance of CNTs are considered important metrology in industrial standardization and excellent management of CNTs for their highly efficient applications. The characteristics of CNTs are derived from dispersed photons emitted by the sample on subjected to high-intensity laser light, which is associated with the unique vibrational modes of the Raman spectrum. CNTs show two primary first-order bands (Figure 13.11) [140].

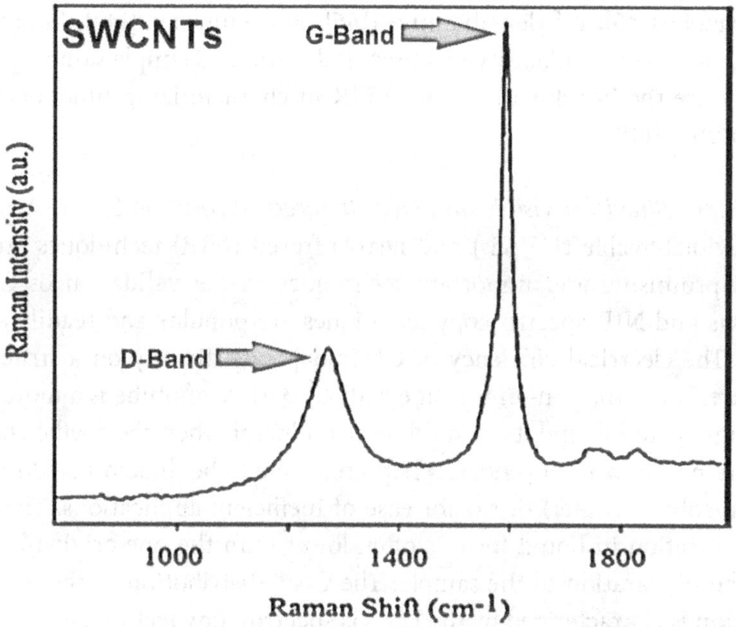

FIGURE 13.11 D and G bands of SWCNTs are illustrated by Raman spectra. (From Singh, B.P. et al., *Surf. Coat. Technol.*, **206**, 1319–1326, 2011.)

At approximately 1500–1600 cm^{-1}, the G band (graphic band) is detected, related to the vibrations in the graphene sheet planes. Depending on the cause of excitation, the D band is detected in the spectrum around 1300–1350 cm^{-1} and is connected to defects in the sidewalls of the CNTs. RS helps in the measurement of the physical strength, size, and structure of the CNTs formed. It is possible to obtain structural and size determination of CNTs by RS. The configuration of the CNT changes with respect to the graphene sheets stacked and rolled to form the CNT [141]. With variation in the diameter (n) and the chiral angle (m), their values give raise to different configurations, such as, zig-zag, chiral, and armchair [142].

13.3.2.5 Infrared and Fourier Transform Infrared Spectroscopy

Here, this type of spectroscopic technique is mainly used to analyse the impurities arising from amalgamation and spot organic molecules on the surface of the carbon nanotube [143]. Functionalization may be a common next step to boost solubility and build up efficiency in CNTs. This can be achieved by binding a range of useful additions to the CNT sidewalls via chemical reactions or physical interactions [144]. Some functionalization can be verified by using Fourier transform infrared (FTIR) analysis, for example to assess the functionalization of MWCNTs by amino benzothiazole [145], ethylenediamine [146], and chitosan [147]. Simplicity, speed, low cost, availability in a prepared form, and simple sample preparation are the benefits of IR and FTIR in characterizing functionalized nanotubes [148].

13.3.2.6 Ultraviolet-Visible and Near-Infrared Absorption Spectroscopy

Ultraviolet-visible (UV-vis) and near-infrared (NIR) techniques are the most promising and important techniques in the validation of CNTs. UV-vis and NIR spectroscopy techniques are popular and readily available. The electrical efficiency of CNTs depends directly on a structural feature. If n = m, or n–m is split evenly by 3, the nanotube is a more conducting material, and it is a mild semiconductor when the coefficients (n, m) are n = m or n–m [149]. CNTs prerequisite to be disseminated in universal solvent (water) or say for ease of inefficient applications. The final concentration in liquid form is often lower than the one originally used for the preparation of the sample. The CNT distribution in the ultimate solution is characterized by the UV-vis spectroscopy technique.

To accomplish this goal, the optical spectroscopy results have been accordingly used in three ways: (i) With certain criteria, under analysis at a single wavelength, the calculations were coordinated with the dispersion of the nanotubes; (ii) in other cases, the absorbance/resonance ratios were used in wavelength range testing; (iii) in the remaining cases, the whole spectrum was considered [150]. The absorption at 253 nm had a direct correlation with the bulk of nanotubes distributed in the solution. Attal et al. reported a similar method to achieve the same objective, but at a different wavelength (273 nm) [151]. For SWCNT solubilization, CNT dispersions from peak absorption at 972 and 1710 nm were reported in the comparison of a wide variety of surfactants, including bile salts [152].

13.3.2.7 X-Ray Photoelectron Spectroscopy

X-ray photoelectron spectroscopy (XPS) is used in elemental analysis of the surface of the nanotubes. It is a quantitative approach that shows the properties of the surface of the nanotubes. An X-ray laser is used to illuminate the sample and calculate the amount and kinetic energies of evolving electrons. XPS has commonly been used to study the structural alteration of the nanotube films by chemical reactions that are carried out to functionalize these structures. By XPS, we can find any impurities or functionalization on the surface of the carbon nanotube; one such example is that fluorination observed on the outer layer might migrate to the inner walls due to the deterioration of graphene [153]. In this process, we can characterize the depth of amino-functionalized SWCNTs by two separate methods [154]. Nitrogen doping of CNTs on the CNT sidewall by immobilization of different biological molecules has also been investigated by XPS [155].

13.4 CARBON NANOFIBRES

Here we discuss important techniques for the validation of carbon nanomaterials. The surface of the materials is characterized, and this is very important in the case of CNFs. The lateral arrangement and stacking of the layers provide insights into the structural strengths and weaknesses, with particular attention given to surface resolution. The most suitable and novel techniques in this case study involve SPM, infrared spectroscopy (IR), STM, and XPS.

13.4.1 Scanning Probe Microscopy

In comparison, in the case of large surface areas, IR and XPS are used for validation. For small surfaces, research proves that SPM is the perfect required analytical technique with high lateral resolution. SPM is one of the techniques for analysis at the atomic level of resolution, like STM and AFM. Using these methods, more efficient and sensitive nanomaterials are generated comparatively with more efficient and wide surface areas and yield methodical values in surface characteristics. Interpreting the results can be challenging due to the high-resolution nature of the scans, often requiring multiple images to ensure accurate analysis and stabilization of the observed matrix.

13.4.1.1 Scanning Tunnelling Microscopy

The STM technique helps in characterizing the conducting surface material, and the conducting property of the surfaces can also be traced using vertical sub-Angstrom resolution. The reorganization of the CNF surface at the atomic level after prior oxidation was addressed by STM observations [156]. STM images of the nanofibre surfaces were recorded before and after 5 minutes of treatment with oxygen plasma, resulting in two key assumptions regarding the composition of the surface. First, on the untreated surface of CNF, on the local scale of the graphite, atomic parameters like atomic length were analysed, but the long-range atomic order was unavailable for analysis. Second, oxygen plasma treatment led to the extinction of the short-range baseline graphitic order, resulting in a large and uniform rise in the surface disorder of CNF at the atomic scale. The results align with previous studies of oxygen plasma etching of Highly Oriented Pyrolytic Graphite (HOPG) surfaces that have contributed to atomic defects [157]. Ultimately, the incorporation of functional groups like oxygen moieties into the surface of the CNF has influenced the atomic defects introduced in the plasma, and these defects are rectified by AFM, XPS, IR, and other techniques.

Post-growth heat treatment can enhance surface graphitization. Paredes et al. have reported the CNF surface and its graphitization mechanism at various annealing temperatures using STM characterization [158]. The STM images represented strong, atomic-level support for CNF surface development, consistent with spatially averaged characterization findings from XRD and Raman spectroscopy. For more detail, the STM showed a nano-level transformation of the surface and its platelet morphology to a wider area, and also, the wider-sided nanofibres were transformed into a more convenient flat-surfaced atomic structure at 28,000 °C. This imaging helps in precisely viewing thin defective crystalline mass as a large defective crystallite by the use of true graphic realms. Endo et al. have reported graphitization showing similar results to that of STM imaging [157]. Also,

Yoon et al. showed their STM imaging of graphitized CNF platelets, which are constructed of carbon nanorods, where the graphene layer has edges with closed caps [159].

13.4.1.2 Infrared Spectroscopy

IR spectroscopy is a highly sensitive and easy technique for functional surface characterization with the addition of an interferometer. Depths from a few centimetres to the 100 nm shallow range can be measured below the surface by surface-sensitive IR techniques. In the case of CNFs, the spectra express "bulk" chemical details, as the particles and surface functionalities are in the order of 20–200 nm in diameter. However, this technique is useful for distinguishing the functional groups on the surface area. In a typical example reported by Ros et al., the untreated bamboo-type surface structure was compared with herringbone-type CNFs using IR transmission [160]. The results suggested that the surface of herringbone-type CNFs is more vulnerable to oxidation than that of bamboo-type CNFs, reporting a higher number of efficient reactive edges on the terminating sidewalls of the graphitic surface. The herringbone fibres showed evidence of carboxyl groups with the stretching vibrations at 1717–1712 peaks, and when annealed there was a loss of CO_2 by mass calculation. In Figure 13.12, the two types of fibre are compared by keeping each sample's transmission levels the same. Overall, the report validates CNFs that have a deficient-rich structure, the bamboo, and herringbone fibres have reported the carbon-hydrogen stretching frequencies at (3012, 2947, 2917, and 2846 cm^{-1}) [161]. In addition, the difference in the surface structure is calculated by comparison of the structure before and after processing treatments. In several examples, the surface fibre is oxidized to create sites of reactive carboxylic acid that are valuable for further functionalization. Considering several methods tested for surface oxidation, the most efficient way was by treating the surface with combined nitric and sulphuric acid [162]. In addition, the IR spectra showed that the defective sites are the first susceptible places where we can observe oxidation or functionalization, and that carboxy and carboxylic anhydride groups on the surfaces are well oxidized from prior carbonyl groups and other oxides. Also studied the immobilization resulting on the surface of the CNFs due to the covalent bonding of metal complexes [163]. This covalent cluster was achieved by anthranilic acid with oxidized fibres. FTIR was used by Han et al. to verify covalent clusters formed the oxidized MWCNTs with alkane thiolate-coated gold nanoparticles; the results demonstrated the significance of linker molecules in the development of composite materials [164].

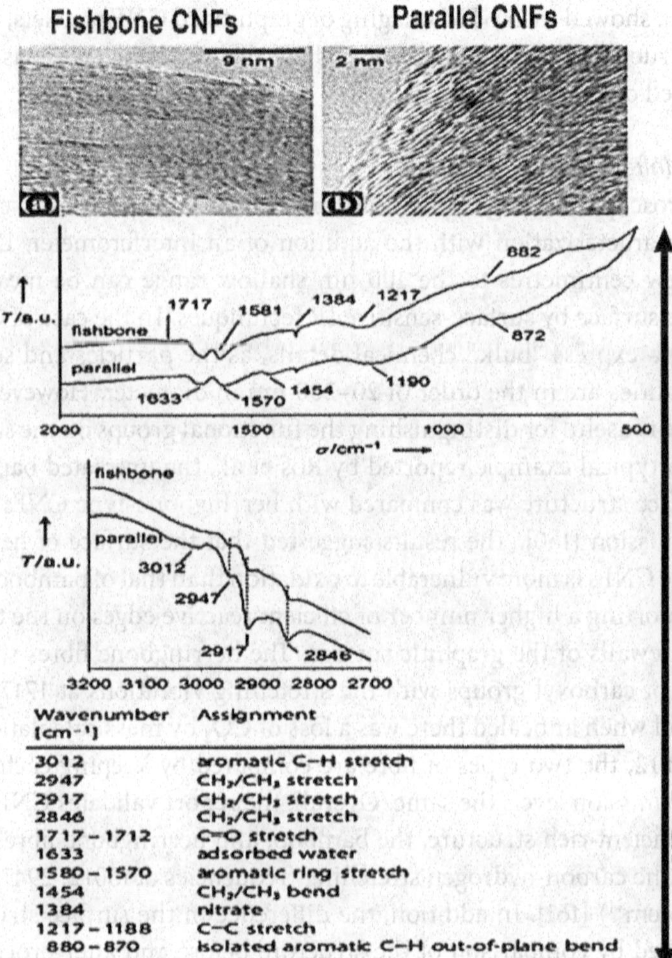

FIGURE 13.12 TEM images of (a) fishbone and (b) parallel CNFs with the equivalent IR spectroscopy outcomes below. (From Klein, K.L., *J. Appl. Phys.*, **103**, 061301, 2008.)

Furthermore, FTIR has shown the results of thermal annealing of the surface of the MWCNTs at 3000 °C to remove the organic shells from Au-nanoparticles. The covalent grafting of polyether ketones on the phosphoric acid–treated surfaces of both MWCNTs and CNFs has been demonstrated by the FTIR technique [165].

13.4.1.3 X-Ray Photoelectron Spectroscopy

This technique is considered the most versatile, durable, sensitive, "workhorse" technique for surface analysis. It has a 30 μm/mm resolution of the lateral surface. The core-level photoelectrons may tend to escape through a few atomic layers using an X-ray energy source within the keV range. With the exception of hydrogen and helium, the data produced help all elements to be classified with a sensitivity better than 1 at.%, in comparison to AES, which, with a cantered scanning electron beam, will test areas down to 10 nm. This method, when considered for a wide sample range, only produces average data for the spatial analysis. XPS characterization is the greatly preferred technique for the study of carbon nanostructures. Therefore, we have focused and chosen papers merely on the CNFs and their surface functionalization with oxidation treatment, metal adsorption, and impurity analysis. Hoshi et al. reported the purity of untreated CNFs with a peak of C 1s of 284.7 eV [166]. There were fewer than 5% oxygen-bonded sites on the pure surface with C-C XPS graphite peaks (284.5 eV), and C-O bonding showed a peak at 288 eV. Toebes et al. showed that there is an increase in the surface area of the inner tubular channel, which is longitudinal to the fibre, due to the high influence of acid oxidation [167]. The amount of oxygen below the 2–3-nm sub-surface level of the CNF was also recorded. Considering the above observations, the overall content of oxygen depends on the type of oxidizing agent and the time of treatment, as well as the number of classes of acids.

Xia et al. used XPS analysis to examine additional secondary fibre growth on CNFs under the influence of iron oxide particle deposition [168]. This secondary nanofibre growth enhanced the specific surface area of nanocomposites, adjusting the morphology to the required applications. CNFs were functionalized with oxygen groups by exposing the surface to oxygen plasma. Next, for the preparation of nanofibre-supported iron oxide nanoparticles, ferrocene was deposited by the CVD method under good oxidizing conditions. The sintered metallic iron nanoparticles were subsequently catalysed with diameters in the 10 to 20 nm range by secondary CNFs. XPS actively indicated the specific variations in the surface of the CNFs due to the chemical variations in the coalescence and sintering of metallic iron particles [169].

13.4.1.4 Electron Microscopy

Probably the Scanning electron microscopy (SEM) technique is the most precise approach for the topographical and morphological analysis of surface. TEM and scanning transmission electron microscopy (STEM) likely depict the surface layers through their high-resolution imaging (HRTEM) and demonstrate the atomic structure of the surface interface.

Both TEM and STEM relay the entire nanostructure surface analysis data.

13.4.1.4.1 Scanning Electron Microscopy This technique is considered the most suitable for surface imaging. The visibility of the secondary electrons depends on the material and the material bulkiness. Normally, the typical imaging depth varies from 5 to 50 nm. Images including both the surface and the bulk may interfere with the proper interpretation of the surface morphology. The image quality depends on the surface, contrast, and effective brightness. The edges of the surface may also be highlighted by surface charging, resulting in scan raster distortion; this may lead to uncertainty in the surface analysis. However, for the study of surface morphology and CNF-related process control research, there are several cases of the use of SEM. Just two experiments are listed for brevity.

An example reported by Klein et al. showed how SEM analyses are used to record shifts or changes in surface morphology in vertically aligned forms of carbon nanofibers (VACNF) for preparing catalysts [170]. In another example, a Cu–Ni catalyst was studied, and there was a change in the surface morphology with the change in the concentrations of the alloy mixtures. A minor change in Cu concentration in the Ni-rich alloy of the catalyst resulted in a change in morphology [171].

13.4.1.4.2 Energy Dispersive X-Ray Microanalysis This technique helps in measuring the energy of photons by the analysis of the X-rays emitted after the process of ionization inside the specimen under analysis. These X-rays are recorded by the use of a semiconductor, which provides a histogram displaying the emitted X-ray diffractions from the surface sample under analysis. The obtained spectrum is used in the qualitative and quantitative analysis of the elements along with the chemical nature of the surface structure. Therefore, along with electron microscopy imaging, EDX provides supportive chemical particulars. The lateral spatial resolution depends on the energy source and the nature of the specimen. The measure of both the spatial resolution range and depth resolution is considered, with a range of a few millimetres to tens of nanometres, and for the depth, the resolution is considered as the fraction of the incident beam range with a scale of less than hundreds of nanometres.

This technique has also demonstrated the change in the morphology of the surface structures and the constitution and site of the catalyst, as reported and explained by Klein et al. [172]. From the EDX information, it was shown how Cu–Ni catalyst particles were drawn to the base of the nanofibre from its alloy catalyst phase [173].

13.4.1.4.3 Transmission Electron Microscopy TEM analysis provides us with detailed data on the surface and inner structure of the nanoscale sample of interest. Unlike the real space (image) and reciprocal space obtained by the chemical and electronic data provided by EDX and Electron Energy Loss Spectroscopy (EELS), TEM provides wide and deep detail for both the surface and the inner structure of CNFs. For example, Ominami et al. explained the structural analysis of the action of the different catalysts on the nanofibre (Ni versus Pd) [174]. Variations in the catalyst significantly affect the surface graphitic structure and, as reported by Ominami et al., result in minimal interfacial resistance with the substrate. In this analysis, quality improvements in the surface structure and electrical conductivity of Pd-catalysed fibres were reported when compared with Ni-catalysed fibres. The two different morphologies coordinated with the individual catalyst [175]. The Ni-catalysed fibres showed graphitic planes, and they ended on the sidewalls of the fibre in a herringbone pattern, while the Pd-catalysed fibres showed multiwalled nanotube structures on the outer fibre walls [176].

In previous work, Han et al. have shown how IR spectroscopy [177] is a good example of surface characterization, like the TEM technique. The study of the availability of the functional sites was depicted in one such example, where gold nanoparticles on the surface of the fibre indicated the functional sites available. In addition, the high-resolution TEM (HRTEM) image depicts a good resolution of how gold nanoparticles are terminated by the graphite planes [178].

13.5 APPLICATIONS IN AN ELECTROCHEMICAL SENSOR

13.5.1 Graphene

Graphene materials are fascinating materials that have gained a lot of attention in the past decade due to their innumerable properties, like conductivity, thermal stability, and durability. Many applications of graphene materials are reported with low detection limits and good linear range. A novel approach was reported by Tirawattanakoson R et al., who synthesized nanocomposites for the detection of total antioxidant capacity (TAC) based on a 2,2-diphenyl-1-picrylhydrazyl (DPPH) assay. He used a Gr/PEDOT:PSS/SPCE (graphene/poly-3,4ethylenedioxythiophene/poly styrene sulphonate/modified screen-printed carbon electrode) matrix for the study. The low detection limit (LOD) and the calibration curve of TAC were recorded to be 0.59 μM (S/N = 3) and 5–30 μM range [179]. Graphene (Gr) and polyaniline (PANI) nanocomposites were fabricated onto the electrode. This matrix application influenced the surface area enhancement on the electrode, which was confirmed under SEM analysis. When electrodes were modified with Nafion/G/PANI nanocomposites, the linear range (LR) was recorded to be 1–300 mg L^{-1}; the LOD measured for Zn(II), Cd(II) was 1.0 mg L^{-1}, and for Pb(II), it was as low as 0.1 mg L^{-1}. This was reported by Nipapan Ruecha et al. [180]. Sohee Lee et al. described a new, efficient graphene-modified electrode with Fe_2O_3 nanoparticles. This modification enhanced the electrocatalytic activity and sensitivity towards trace metals like Zn^{2+}, Cd^{2+}, and Pb^{2+}, with a LR of 1–100 μg L^{-1} and LOD for each metal of 0.11, 0.08, and 0.07 μg L^{-1}, respectively. In a real-sample analysis, electrodes were coated with a Bi/Fe_2O_3/G/GCE (bismuth, iron oxide, graphene, glassy carbon electrode) matrix to improve their sensitivity and efficacy towards metal ions [181]. In another study, Bin He et al. claimed advantages for a nanocomposite matrix (Fe_3O_4/GN/GE/GCE) along with natural extracts in the detection of metal ions. Natural extracts such as garlic extract (GE) were prepared and fabricated on the GCE and used in the detection of Pb^{2+} in wastewater. Fe_3O_4/Graphene Nanosheets in the matrix helped enhance the electrical properties, and the use of natural extracts showed good complexation with Pb^{2+}. The electrode showed two dynamic linear ranges in optimum conditions, one from 0.001 to 0.5 nM and the other from 0.5 to 1000 nM, along with a favourable LOD of 0.0123 pM (S/N = 3) [182]. A simple electrochemical sensor was prepared using graphene nanocomposites that were sensitive to the concentrations of Ca^{2+} in meat with good stability and reproducibility. Reported by Xiaoquan Fan et al., this electrode showed a linear range between 5×10^{-8} and 3×10^{-4} M, and the LOD value was 3.9×10^{-8} M [183]. Kunc hen et al. reported a GCE probe that was fabricated with a graphene-based nanocomposite, i.e. graphene-PANI/Tyr/Nafion/GCE (Tyr = tyrosine); this nanocomposite showed an efficient adsorption capacity for hydroquinone and also good electrocatalytic property, sensitivity, and good adsorption capacity towards hydroquinone. Under optimal conditions, this electrode showed a linear reduction peak current varying with phenol concentration, which further depicts a linear regression equation ranging from I = -4.887×10^{-4} C to 5.331×10^{-6} C (mol L^{-1}) with a correlation coefficient of 0.9963 and LOD of 2.00×10^{-4} mol L^{-1} (S/N = 3) [184]. A novel approach for the synthesis of an efficient electrode with PSS (poly(sodium styrene sulphonate)) and graphene nanocomposites was reported by Cong Wang et al. for its high sensitivity to hydrazine. This electrode showed a linear range from 3.0 to 300 μmol L^{-1} with an LOD of 1 μmol L^{-1} [185]. Xiaoqiang Wang et al. reported the synthesis of a composite film named poly(3,4-ethylene dioxythiophene)-sulfonated graphene (SG-PEDOT) by a one-step electrochemical method. Further incorporation of gold nanoparticles (AuNPs) as an electrode material into the above nanocomposite enabled the efficient detection of L-cysteine (Cys). The composite SG-PEDOT film was shown to provide a rough surface makeup for AuNP particle deposition, and hence, there was an incompetent interaction of the substrate with Cys, whereas the composite adorned with AuNPs exhibited better electrocatalytic efficiency than the composite decorated with SG-PEDOT only, with the peak current attributed to the Cys concentration varying from 0.1 to 382 mM and a detection limit of 0.02 mm (S/N = 3) [186]. A simple TiO_2-graphene electrochemical-based sensor was reported by Fan Y et al. in 2011. This sensor showed good sensitivity to adenine and guanine detection, with a wide LR of 0.5–200 μM and an LOD of 0.10 and 0.15 μM, respectively [187]. Han Y et al. synthesized an efficient, rapid, electrochemical-based sensor that was cost-effective and showed high sensitivity in nitrite detection. Under optimized conditions, this sensor showed a dynamic LR of 5.0 μM to 5.0 mM and an LOD of 1.0 μM [188]. A cyclic voltammogram of the ferri/ferrocyanide redox couple at a graphene-modified electrode showed a good enhancement in the current intensity when compared with graphite and GCE electrodes. This achievement was reported by Yang-RaeKim et al. The sensitivity of the electrode for dopamine was confirmed by using ascorbic acid (1 mM). The LR was recorded between 4 and 100 μM with an LOD of 2.64 μM [189].

13.5.2 Graphene Oxide

The functionalization of graphene has created more interest in propounding its properties and applications. There are a lot of reports showcasing the efficiency of functionalized graphene materials. Shabi Abbas Zaidi reported an efficient electrochemical-based sensor for the detection of adrenaline in urine. He fabricated a GCE with Na-rGO/GCE (nicotinamide reduced graphene oxide) matrix, which showed a wide LR between 0.015 and 40 μM and an LOD of 3 nM [190]. A novel approach was taken by Eun Jung Lee et al, who achieved a nanocomposite matrix formed between graphene oxides (GS) and gold nanoparticles, which was fabricated on an electrode for efficient detection of norepinephrine. The authors also explained how the sensitivity of detection increases with the use of ascorbic acid (AA). Even with a high concentration of AA, this electrode reported an extensive detection range of 0.2–10 μM with an LOD of 200 nM [191]. Another example of a novel approach involved using graphene oxide and porphyrin–based nanocomposites fabricated on an electrode, which was incorporated into the environmental monitoring of 17 β-oestradiol. This approach was reported by FC Moraes et al, This electrode showed a linear range varying from 0.1 to 1.0 μmol L^{-1} with the highly anticipated LOD of 5.3 nmol L^{-1} (1.4 μg L^{-1}) [192]. Shabi Abbas Zaidi reported an efficient, highly specific electrochemical-based sensor for the detection of adrenaline in urine. He fabricated a GCE with Na-rGO/GCE (nicotinamide reduced graphene oxide) matrix, which showed a wide LR between 0.015 and 40 μM and an LOD of 3 nM. Li, J Kuan, et al. reported an efficient GCE that was fabricated with graphene oxide material and showed good sensitivity for 4-nitrophenol (4-NP). The linear peak current range was between 0.1 and 120 μM, and the LOD was reported as 0.02 μM (S/N = 3) [193]. A GO-based biosensor was developed for the detection of dopamine (DA). It was a label-free, bio photoinduced charge transfer (PCT) and near-IR fluorescent biosensor reported by Chen, JL et al., with an LOD of 94 nM. Also, a relative standard deviation of 2.0% was reported for 11 replicates of 2.0 μM DA concentration [194]. The combination of bimetallic nanoparticles like Au–Pt with rGO was reported by Zhonggang Liu et al. The conductivity was enhanced by the rGO, and the electrocatalytic activity was boosted by the use of bimetallic nanoparticles; hence, the combination matrix showed a high selectivity of 7.35 μA μM^{-1} and sensitivity, with a convenient LOD of 2.88 nM [195]. Gao F et al. reported an efficient GCE fabricated with GO by a covalent coupling method. This electrode was used in the detection of DA and showed a peak current for concentration range from 1.0 to 15.0 μM, with an LOD of 0.27 μM [196].

13.5.3 Carbon Nanotubes

The electrochemical-based applications of the composite-derived CNTs are a hot topic that has gained a lot of interest due to their high sensitivity and selectivity, reported by many scientists. Many such examples have shown their unique properties and applications. Alam et al. reported an efficient electrochemical-based probe with graphene oxide and CNTs functionalized with β-cyclodextrin. The reported sensor showed efficient sensitivity of detection of bisphenol (BPA) in drinking water, with an LOD of 6 nM, and a two-step linear detection range was reported, from 0.05 to 5 μM and from 5 to 30 μM [197]. Single-walled CNT–functionalized glassy carbon electrodes are incorporated with inefficient oxidation of phenolic compounds by electrochemical method and with the absence of influence by any catalysts and enzyme in the probe. This was reported by Govindhan et al. reported that single-walled carbon nanotube (SWCNT)-functionalized glassy carbon electrodes enabled efficient electrochemical oxidation of phenolic compounds without the need for catalysts or enzymes. The system demonstrated high sensitivity—135.08, 94.90, and 110.38 μA μM^{-1}—for the simultaneous detection of p-cresol, catechol, and p-nitrophenol, respectively. The method also achieved remarkably low detection limits of 2.3, 3.7, and 7.7 nM for these analytes [198]. Yang et al. reported a novel approach to the determination of electrochemical sensors for the analysis of chloramphenicol (CAP). The multi-walled carbon electrode was functionalized with molecularly imprinted polymer (MIP), three-dimensional porous graphene (P-r-GO), and mesoporous carbon (CKM-3). The sensor showed an excellent sensitivity for CAP, with a detection limit of 1.0×10^{-10} mol L^{-1}, and the linear range was reported to be 5.0×10^{-9}–5×10^{-7} mol L^{-1} and 5.0×10^{-7}–4.0×10^{-6} mol L^{-1} [199]. The functionalized electrode (AuNP/MIP/f-MWCNT/GCE) reported by Razaei et al., used in warfarin detection, showed a linear range of 0.031–0.616 ng mL^{-1} and an LOD of 0.024 ng mL^{-1} [200]. Ciprofloxacin was efficiently identified using functionalized magnetic MWCNTs and MIP. This novel approach was achieved by Bagheri et al, The linear detection range was 0.005–0.85 μmol L^{-1}, and the detection limit was recorded to be 0.0017 μmol L^{-1} [201]. Yuan et al. explained that the functionalization of MWCNTs with MIP amplified the electric signal responses in metronidazole (MNZ) measurement. This probe showed an exceptional LOD of 4.92×10^{-5} mg L^{-1} (S/N = 3) [202]. A hybrid double-walled CNT and graphene prepared by Duoc et al. showed an exceptional LOD of 0.287 ppb and an LR from 1 to 10 ppb [203]. MWCNTs/GNPs (graphene nanoparticles) were mixed to form a hybrid composite material, which was used in efficiently detecting Pb^{2+}. This matrix showed a linear range from 5.0×10^{-11} to 1.0×10^{-14} M, with a detection limit of 4.3×10^{-15} M recorded [204]. Arumugasamy et al. reported that MWCNTs were modified with graphene quantum dots (GQDs) to form a matrix and then coated on a GCE (GQDs@MWCNTs/GCE). This sensor probe was used in the oxidation of dopamine with good electrocatalytic activity, with a lower detection limit of 95 and 110 nM (S/N = 3) and a dynamic linear range of 0.25–250 μM [205]. An MIP-based quartz crystal microbalance (QCM) electrode was reported by

Yang et al. It showed a wide linear range from 2.0×10^{-6} to 1.0×10^{-3} mol L^{-1} with an exceptionally low detection limit of 4.4×10^{-7} mol L^{-1} (S/N = 3) [206]. An efficient dopamine detection sensor was synthesized and reported by Huang et al. The sensor was coated with an electrode matrix prepared using GQDs-MWCNTs. This composite material–based electrode showed high electrochemical sensitivity for the detection of DA, with a linear range of 0.005 to 100.0 μM and a lower detection limit of 0.87 nM (S/N = 3) [207]. MWCNTs blended with graphdiyne (GPY) formed an efficient matrix for coating a GCE, as reported by Xiao et al. This sensor was used in the determination of low concentrations of DA. The sensor showed promising results, with an estimated detection limit of 8.6 nM (S/N = 3) and linear response in the concentration range of 0.02 to 20.0 μM [208]. Bagheri et al. reported a novel electrochemical sensor–based nanosheet probe with a (Cu/MWCNT/RGO) matrix. The fabrication of the matrix showed a wide linear range from 0.1 to 75 μM with an LOD of 30 nM and 20 nM for nitrite and nitrate ions, respectively [209]. Au-PtNP nanocomposites were fabricated on the surface of an f-MWCNT to obtain an efficient electrode for sensing diclofenac, as reported by Eteya et al. The prepared probe showed a linear range concentration from 0.5 to 1000 μM, and the LOD was found to be 0.3 μM [210]. A new approach to the preparation of the nanocomposite matrix on the GCE was achieved by Atta et al. The matrix was formed by coating with successive layers of CNTs, poly(hydroquinone) (PHQ) and benzo-12-crown-4(CE) and showed a linear range of 0.005–20 μM with a detection limit of 0.221 nM [211]. Mao et al. reported a new MWCNTs and chitosan-nickel (CTS–Ni)–based nanocomposite–fabricated electrode, which was used in the detection of metronidazole, with a LR concentration of 0.1 to 150 μmol L^{-1} and an exceptional LOD of 0.025 μmol L^{-1} [212].

5.4 Carbon Nanofibres

Huang et al. reported an efficient immunosensor electrode fabricated with Au–PEI–CNFs (PEI = polyethyleneimine) nanocomposites, which showed a wide LR from 0.05 to 25 ng mL^{-1} and an LOD of 0.027 ng mL^{-1} (S/N = 3) [213]. Le et al. reported gold-coated worm-like nanowires along with CNFs-Gr (Au WNWs/CNFs–Gr), which showed a wide LR of 1.98 μM to 3.77 mM and a low LOD of 1.24 μM along with high sensitivity of 836 μA cm^{-2} mM^{-1} [214]. Zhu et al. reported composites of silver nanoparticles (AgNPs) fabricated onto CNFs showing a broad dynamic range of 0.01–50 mM and an LOD of 3 μM (S/N = 3) [215]. A Ni/CNF electrode, prepared and reported by Adabi et al., displayed a linear range of 0.002–5 mM and an LOD of 0.57 μM [216]. An electrospun carbon nanofibre (ECNF) sensor was reported by Rahmani et al. and was used to analyse milk samples to detect aflatoxin M1(AFM1). This reported biosensor showed excellent stability, high sensitivity, and reproducibility, with a linear range of 1–100 pg mL^{-1} and an LOD of 0.6 pg mL^{-1} [217]. Jahromi et al. reported a novel electrochemical sensor based on a CNFs/SPE (screen-printed electrode) used in sensing tramadol. This sensor showed a wide LR of 0.05–100 nM maintained under optimal conditions, with an LOD of 0.016 nM [218]. An electrochemical sensor derived from Prussian blue analogues for sensing phenol was reported by Yin et al., who incorporated cobalt–iron selenides doped onto porous carbon nanofibre ($CoFe_2Se_4$/PCF). This electrode showed a linear range between 0.5 and 200, 0.5 and 190, and 5 and 350 μM and an LOD of 0.13, 0.15 and 1.36 μM for Hydroquinone, Catechol and Resorcinol, respectively [219]. A new approach involved incorporating β-cyclodextrin-based Cu–MOF (metal organic frameworks) with the chelate complexation of ETB isomers (SS-Ethambutol/RR-ETB). A GCE was fabricated with composites of (CD-CuMOF) and CNF to prepare an efficient electrochemical sensor (CD-CuMOF-CNF-GCE). This was reported by Upadhyay et al., and the sensor showed a wide LR in the range of 1.0×10^{-7} to 1×10^{-4} M and 5.0×10^{-7} to 2.5×10^{-4} M with an LOD of 3.10×10^{-8} M and 8.52×10^{-8} M for RR-ETB and SS-ETB isomers, respectively [220]. Sasal et al. achieved an electrochemical-based sensor to detect the concentration of paracetamol (PA) in environmental water samples. The electrode produced was a screen-printed carbon electrode fabricated with CNFs (SPCE/CNFs). This electrode showed two linear PA concentration ranges, from 2.0×10^{-9} to 5.0×10^{-8} mol L^{-1} ($r = 0.9991$) and from 1.0×10^{-7} to 2.0×10^{-6} mol L^{-1} ($r = 0.9994$) with an LOD of 5.4×10^{-10} mol L^{-1} [221]. Duan et al. achieved an electrochemical sensor wherein CNFs were fabricated with FeCo alloys. This sensor showed an LR of 0.08–400 μm and an LOD of 2.7 nm [222]. Xiaojing et al. achieved and reported a sensor made by fabrication of multilayer ultrathin films (UTFs) containing Ni–Al layered double hydroxide (LDH) and with carbon fibres synthesized by electrospinning technology. This sensor showed a good LR between 0.2 and 120 μM and an LOD of 0.067 μM [223]. An enzyme-free glucose sensor was reported by Ding et al. This sensor used CNFs decorated with $CuCo_2O_4$. The electrode showed two efficient, wide linear ranges of 0.01–0.5 mM and 0.5–1.5 mM, with high selectivity of 2932 and 708 μA mM^{-1} cm^{-2} [224]. Fakude et al. developed an aptasensor with a CNF host; the fabrication was done by immobilizing a biotinylated aptamer selected for Cd (II) and streptavidin on a screen-printed electrode. This sensor showed a good LR between 2 and 100 ppb and an LOD of 0.11 ppb [225]. An electrochemical sensor was prepared by using CNF and functionalizing it with flake-like structured erbium molybdate (Er_2MoO_6)/functionalized carbon black (f-CNF) composite was synthesized via a facile sonochemical synthesis method. An f-CNF/Er_2MoO_6-modified SPE was fabricated for the electrochemical detection of phenothiazine (PTZ). This was reported by Liu et al., and this sensor showed a broad dynamic LR from 0.025 to 80 μM with the LOD recorded to be 0.008 μM [226]. Villan et al. reported efficient electro-etched CNFs that were hydrothermally processed with bottlebrush nanowires like zinc oxide to form a CNF–ZnO matrix; then, the matrix was coated onto a GCE to make a CNF–ZnO–GCE probe. This sensor showed an appreciable linear response range of 2–123 nM and an ultra-LOD of 1 nM [227]. Afzali et al. achieved an electrochemical sensor with the electrode made from CNF fabricated CuO composites (CuO/CNF); it showed the concentration in LR from 1.0 to 100.0 nM, with an LOD of 0.25 nM [228].

13.6 FUTURE PERSPECTIVE

A vast literature has been published on carbon-based materials and their engineering applications. We must make sure the research we are hoping for in order to better our future takes an environmental approach. Hence, more attention is being focused on carbon materials and their allotropic forms as raw materials in nanomaterial probes. There is much focus on exploring the physical and chemical aspects and the integrity of graphene materials. Graphene and its two-dimensional materials have been reported to have novel properties, like the highest charge transport, thermal conductivity, and quantum hall effect at room temperature. These properties are far more advanced, increase the functionality of the graphene materials, and are now taking their place in advanced research when compared with other conventional three-dimensional materials research previously reported in the field of nanomaterials. Considering the research in the last decade, the development of the research focus in the field of CNTs, CNF, and graphene materials has vastly increased, and they have the potential to surpass interest in the field of silicon-based microelectronics. Now, graphene materials are focused on the microscale and nanoscale of electronics, and these materials have the potential to build scope in industries and biomedical applications. Synthesis of defect-free pristine graphene was reported by improvising a method using Scotch tape, but the bulk synthesis of the graphene and its characterization were a challenge, and alternative methods like CVD have produced defect-free single-layer and bilayer graphene. These properties of graphene have helped in the application to semiconductor devices. The semi-metallic properties and bilayer nanoribbons of graphene show their applications as transistors and field effect transistors (FETs). Controlled oxidation and reduction of the graphite material give us GO; further, on functionalization, they help in increasing the band gap, conductivity, and thermal stability of the material. The mass production of GO and graphene, the chemical exfoliation of graphite to form GO, and their thermal and chemical reduction have provided a method for large-scale production of graphene oxide in an efficient, cost-effective way, but this bulk production is toxic to human health. Hence, evaluation of the toxicity of graphene-based nanomaterials is also a strong concern. CNTs and CNFs have shown a high elastic property and good modulus strength, and their thermal stability has also been reported. Simple characterization techniques are involved in the detection of graphene materials; the fabrication of the materials must be simple and quantified. The future aim of the review is to study the toxicity of the reported materials, and advanced techniques still have to be incorporated for the smooth production, characterization, and application of the carbon-based nanomaterials purely based on the type of functionalization, energy gap, and thermal stability, so we have to concentrate more on these issues and come to a bright conclusion.

13.7 CONCLUSION

Carbon-based nanomaterials are promising electro-sensory probes that have shown high efficacy. The materials discussed, like CNFs, CNTs, graphene, and graphene oxide materials, are alternatives to the carbon allotropes, where these materials show their importance based on their desired applications. The functionalization of the graphene materials has led to the study of change in surface graphene layers and change in the morphological cavities and active sites. The high durability and extended surface area structures of the CNFs and CNTs have been studied and reported for their valuable applications. The advanced practice in the synthesis of carbon-based probes has been discussed, and the applications of the probes have been reported. The LR and the LOD of various matrices are reported, and when compared with other sensors, these advanced carbon-based probes show the lowest LOD. These materials are being investigated for toxicity, and based on this, their application to biomedical applications will be standardized. This review was undertaken to gather information on the recent advances in carbon-based materials, their synthesis, and their implementation in sensor applications. Advancement in the research must continue to be kept on track, which will influence aspiring young researchers.

REFERENCES

1. L.H. Madkour, Introduction to nanotechnology (NT) and nanomaterials (NMs), *Adv Struct Mater* **116** (2019) 1–47. https://doi.org/10.1007/978-3-030-21621-4_1
2. C.E. Hay, J. Lee, D.S. Silvester, Formation of 3-dimensional gold, copper and palladium microelectrode arrays for enhanced electrochemical sensing applications, *Nanomaterials* **9** (2019) 1170. https://doi.org/10.3390/nano9081170
3. Khan, K. Saeed, I. Khan, Nanoparticles: Properties, applications and toxicities, *Arab J Chem* **12** (2019) 908–931. https://doi.org/10.1016/j.arabjc.2017.05.011
4. J.A. Hubbell, A. Chilkoti, Nanomaterials for drug delivery, *Science* **337** (2012) 303–305. https://doi.org/10.1126/science.1219657
5. B. Zhang, J.Y. Sun, M.Y. Ruan, P.X. Gao, Tailoring two-dimensional nanomaterials by structural engineering for chemical and biological sensing, *Sens Actuators Rep* **2** (2020) 100024. https://doi.org/10.1016/j.snr.2020.100024
6. Y. Zeng, S. Lin, D. Gu, X. Li, Two-dimensional nanomaterials for gas sensing applications: The role of theoretical calculations, *Nanomaterials* **8** (2018) 851. https://doi.org/10.3390/nano8100851
7. X. Wang, T. He, J. Hu, M. Liu, The progress of nanomaterials for carbon dioxide capture via the adsorption process, *Environ Sci Nano* **8** (2021) 890–912. https://doi.org/10.1039/d0en01140a
8. Hamid Nurfaizey, N. Abdul Munajat, F. Kejuruteraan Mekanikal, Effect of electrospinning distance and applied voltage on the production of polyacrylonitrile electrospun fibres, (2020) 94–96.
9. G. Li, M. Lu, S. Li, M. Yang, Y. Zhang, Y. Zhang, H. Wang, W. Yang, A novel fluorescent "off–on" sensing strategy for Hg (II) in water based on functionalized gold nanoparticles, *Chemosphere* **303** (2022) 135174. https://doi.org/10.1016/j.chemosphere.2022.135174
10. P. Suchomel, L. Kvitek, R. Prucek, A. Panacek, A. Halder, S. Vajda, R. Zboril, Simple size-controlled synthesis of Au nanoparticles and their size-dependent catalytic activity, *Sci Rep* **8** (2018) 1–11. https://doi.org/10.1038/s41598-018-22976-5
11. R. Shashanka, D. Chaira, B.E.K. Swamy, Electrocatalytic response of duplex and yittria dispersed duplex stainless steel modified carbon paste electrode in detecting folic acid using cyclic voltammetry, *Int J Electrochem Sci* **10** (2015) 5586–5598. https://doi.org/10.1016/s1452-3981(23)17279-8

12. R. Shashanka, B.E. Kumara Swamy, Biosynthesis of silver nanoparticles using leaves of *Acacia melanoxylon* and their application as dopamine and hydrogen peroxide sensors, *Phys Chem Res* **8** (2020) 1–18. https://doi.org/10.22036/pcr.2019.205211.1688
13. S. Rajendrachari, N. Basavegowda, R. Vinaykumar, D. Narsimhachary, P. Somu, M.-J. Lee, Electrocatalytic determination of methyl orange dye using mechanically alloyed novel metallic glass modified carbon paste electrode by cyclic voltammetry, *Inorg Chem Commun* **155** (2023) 111010. https://doi.org/10.1016/j.inoche.2023.111010
14. R. Shashanka, Investigation of optical and thermal properties of CuO and ZnO nanoparticles prepared by *Crocus sativus* (saffron) flower extract, *J Iran Chem Soc* **18** (2021) 415–427. https://doi.org/10.1007/s13738-020-02037-3
15. V. Adimule, K. Sharma, P. Lakshminarayana, S. Rajendrachari, R. Keri, F.A. Alharthi, Fabrication of carbon nanospheres functionalized Sm^2O^3/Co^3O^4 novel hybrid nanocomposites for selective and sensitive amperometric detection of hydrogen peroxide, *J Appl Electrochem* **54** (2024) 323–340. https://doi.org/10.1007/s10800-023-01972-2
16. S. Rajendrachari, P. Taslimi, A.C. Karaoglanli, O. Uzun, E. Alp, G.K. Jayaprakash, Photocatalytic degradation of rhodamine B (RhB) dye in wastewater and enzymatic inhibition study using cauliflower-shaped ZnO nanoparticles synthesized by a novel one-pot green synthesis method, *Arab J Chem* **14** (2021) 103180. https://doi.org/10.1016/j.arabjc.2021.103180
17. Halfadji, M. Naous, S. Rajendrachari, Y. Ceylan, K.B. Ceylan, P.V.R. Shekar, Effective investigation of electro-catalytic, photocatalytic, and antimicrobial properties of porous CuO nanoparticles green synthesized using leaves of *Cupressocyparis leylandii*, *J Mol Struct* **1301** (2024) 137318. https://doi.org/10.1016/j.molstruc.2023.137318
18. S. Rajendrachari, R. Shrinivas Mahale, V. Adimule, M. Gulen, V. Mahesh, N. Basavegowda, Fabrication of mechanically alloyed high entropy alloys (20-Fe-20Cr-20Ni-20Mn-20Ti) modified carbon paste electrode sensor for the cyclic voltammetric determination of methyl orange, *Inorg Chem Commun* **164** (2024) 112428. https://doi.org/10.1016/j.inoche.2024.112428
19. L.H. Madkour, Synthesis methods for 2D nanostructured materials, nanoparticles (NPs), nanotubes (NTs) and nanowires (NWs), *Adv Struct Mater* **116** (2019) 393–456. https://doi.org/10.1007/978-3-030-21621-4_12
20. V. Uvarov, I. Popov, Metrological characterization of X-ray diffraction methods at different acquisition geometries for determination of crystallite size in nano-scale materials, *Mater Charact* **85** (2013) 111–123. https://doi.org/10.1016/j.matchar.2013.09.002
21. Shah, A.T. Khalil, K. Ahmad, J. Iqbal, H. Shah, Z.K. Shinwari, M. Maaza, Biogenic nanoparticles: Synthesis, mechanism, characterization and applications, in: *Biogenic Nanoparticles for Cancer Theranostics*, Elsevier, 2021: pp. 27–42. https://doi.org/10.1016/B978-0-12-821467-1.00010-0
22. Ijaz, E. Gilani, A. Nazir, A. Baloch, Detail review on chemical, physical and green synthesis, classification, characterizations and applications of nanoparticles, *Green Chem Lett Rev* **13** (2020) 59–81. https://doi.org/10.1080/17518253.2020.1802517

23. M.U. Zahid, E. Pervaiz, A. Hussain, M.I. Shahzad, M.B.K. Niazi, Synthesis of carbon nanomaterials from different pyrolysis techniques: A review, *Mater Res Express* **5** (2018) 052002. https://doi.org/10.1088/2053-1591/aac05b
24. S. Iravani, Green synthesis of metal nanoparticles using plants, *Green Chem* **13** (2011) 2638. https://doi.org/10.1039/c1gc15386b
25. G. Mogilevsky, O. Hartman, E.D. Emmons, A. Balboa, J.B. Decoste, B.J. Schindler, I. Iordanov, C.J. Karwacki, Bottom-up synthesis of anatase nanoparticles with graphene domains, *ACS Appl Mater Interfaces* **6** (2014) 10638–10648. https://doi.org/10.1021/am502322y
26. Y. Wang, Y. Xia, Bottom-up and top-down approaches to the synthesis of monodispersed spherical colloids of low melting-point metals, *Nano Lett* **4** (2004) 2047–2050. https://doi.org/10.1021/nl048689j
27. Haroun, X. Le, S. Gao, B. Dong, T. He, Z. Zhang, F. Wen, S. Xu, C. Lee, Progress in micro/nano sensors and nanoenergy for future AIoT-based smart home applications, *Nano Express* **2** (2021) 022005. https://doi.org/10.1088/2632-959x/abf3d4
28. S. Sharma, S. Shekhar, S. Gautam, B. Sharma, A. Kumar, P. Jain, Carbon-based nanomaterials as novel nanosensors, in: *Nanofabrication for Smart Nanosensor Applications*, Elsevier, 2020: pp. 323–347. https://doi.org/10.1016/B978-0-12-820702-4.00014-3
29. P.M. Anjana, M.R. Bindhu, M. Umadevi, R.B. Rakhi, Antibacterial and electrochemical activities of silver, gold, and palladium nanoparticles dispersed amorphous carbon composites, *Appl Surf Sci* **479** (2019) 96–104. https://doi.org/10.1016/j.apsusc.2019.02.057
30. Z. Zhen, H. Zhu, Structure and properties of graphene, in: *Graphene: Fabrication, Characterizations, Properties and Applications*, Elsevier, 2018: pp. 1–12. https://doi.org/10.1016/B978-0-12-812651-6.00001-X
31. S. Basu-Dutt, M.L. Minus, R. Jain, D. Nepal, S. Kumar, Chemistry of carbon nanotubes for everyone, *J Chem Educ* **89** (2012) 221–229. https://doi.org/10.1021/ed1005163
32. P.K. Ojha, D. Mandal, K. Roy, QSPR modeling of adsorption of pollutants by carbon nanotubes (CNTs), in: *Methods in Pharmacology and Toxicology*. Springer, 2020: pp. 477–511. https://doi.org/10.1007/978-1-0716-0150-1_20
33. A.K. Geim, K.S. Novoselov, The rise of graphene, in: *Nanoscience and Technology: A Collection of Reviews from Nature Journals*, World Scientific, 2009: pp. 11–19. https://doi.org/10.1142/9789814287005_0002
34. N. Jaiswal, I. Tiwari, Recent build outs in electroanalytical biosensors based on carbon-nanomaterial modified screen printed electrode platforms, *Anal Methods* **9** (2017) 3505–3520. https://doi.org/10.1039/c7ay01276d
35. J. Wang, Y. Lin, Functionalized carbon nanotubes and nanofibers for biosensing applications, *TrAC Trends Anal Chem* **27** (2008) 619–626. https://doi.org/10.1016/j.trac.2008.05.009
36. J. Dusza, G. Blugan, J. Morgiel, J. Kuebler, F. Inam, T. Peijs, M.J. Reece, V. Puchy, Hot pressed and spark plasma sintered zirconia/carbon nanofiber composites, *J Eur Ceram Soc* **29** (2009) 3177–3184. https://doi.org/10.1016/j.jeurceramsoc.2009.05.030

37. G. Mittal, K.Y. Rhee, V. Mišković-Stanković, D. Hui, Reinforcements in multi-scale polymer composites: Processing, properties, and applications, *Compos B Eng* **138** (2018) 122–139. https://doi.org/10.1016/j.compositesb.2017.11.028
38. J.H. Hafner, C.L. Cheung, C.M. Lieber, Growth of nanotubes for probe microscopy tips, *Nature* **398** (1999) 761–762. https://doi.org/10.1038/19658
39. M. Marafi, A. Stanislaus, Spent hydroprocessing catalyst management: A review, *Resour Conserv Recycl* **53** (2008) 1–26. https://doi.org/10.1016/j.resconrec.2008.08.005
40. Q. Zheng, A. Javadi, R. Sabo, Z. Cai, S. Gong, Polyvinyl alcohol (PVA)–cellulose nanofibril (CNF)–multiwalled carbon nanotube (MWCNT) hybrid organic aerogels with superior mechanical properties, *RSC Adv* **3** (2013) 20816. https://doi.org/10.1039/c3ra42321b
41. S.K. Tiwari, V. Kumar, A. Huczko, R. Oraon, A. De Adhikari, G.C. Nayak, Magical allotropes of carbon: Prospects and applications, *Taylor & Francis* **41** (2016) 257–317. https://doi.org/10.1080/10408436.2015.1127206
42. S. Deng, V. Berry, Wrinkled, rippled and crumpled graphene: An overview of formation mechanism, electronic properties, and applications, *Mater Today* **19** (2016) 197–212. https://doi.org/10.1016/j.mattod.2015.10.002
43. H. Wang, C.M.B. Holt, Z. Li, X. Tan, B. Shalchi Amirkhiz, Z. Xu, B.C. Olsen, T. Stephenson, D. Mitlin, Graphene-nickel cobaltite nanocomposite asymmetrical supercapacitor with commercial level mass loading, *Springer* **2012** (2012) 605–617. https://doi.org/10.1007/s12274-012-0246-x
44. Ambrosi, C.K. Chua, A. Bonanni, M. Pumera, Electrochemistry of graphene and related materials, *Chem Rev* **114** (2014) 7150–7188. https://doi.org/10.1021/cr500023c
45. F. Akbar, M. Kolahdouz, S. Larimian, B. Radfar, H.H. Radamson, Graphene synthesis, characterization and its applications in nanophotonics, nanoelectronics, and nanosensing, *J Mater Sci Mater Electron* **26** (2015) 4347–4379. https://doi.org/10.1007/s10854-015-2725-9
46. S. Sharif, K.S. Ahmad, F. Rehman, Z. Bhatti, K.H. Thebo, Two-dimensional graphene oxide based membranes for ionic and molecular separation: Current status and challenges, *J Environ Chem Eng* **9** (2021) 105605. https://doi.org/10.1016/j.jece.2021.105605
47. G. Reina, J.M. González-Domínguez, A. Criado, E. Vázquez, A. Bianco, M. Prato, Promises, facts and challenges for graphene in biomedical applications, *Chem Soc Rev* **46** (2017) 4400–4416. https://doi.org/10.1039/c7cs00363c
48. L. Ma, M. Zhou, C. He, S. Li, X. Fan, C. Nie, H. Luo, L. Qiu, C. Cheng, Graphene-based advanced nanoplatforms and biocomposites from environmentally friendly and biomimetic approaches, *Green Chem* **21** (2019) 4887–4918. https://doi.org/10.1039/c9gc02266j
49. J. Liu, J. Tang, J.J. Gooding, Strategies for chemical modification of graphene and applications of chemically modified graphene, *J Mater Chem* **22** (2012) 12435. https://doi.org/10.1039/c2jm31218b

50. F. Wang, L.T. Drzal, Y. Qin, Z. Huang, Enhancement of fracture toughness, mechanical and thermal properties of rubber/epoxy composites by incorporation of graphene nanoplatelets, *Compos Part A Appl Sci Manuf* **87** (2016) 10–22. https://doi.org/10.1016/j.compositesa.2016.04.009
51. H. Liu, J. Wu, C. Liu, B. Pan, N.H. Kim, J.H. Lee, Differently-charged graphene-based multilayer films by a layer-by-layer approach for oxygen gas barrier application, *Compos B Eng* **155** (2018) 391–396. https://doi.org/10.1016/j.compositesb.2018.08.137
52. S. Liu, B. Yu, S. Wang, Y. Shen, H. Cong, Preparation, surface functionalization and application of Fe^3O^4 magnetic nanoparticles, *Adv Colloid Interface Sci* **281** (2020) 102165. https://doi.org/10.1016/J.CIS.2020.102165
53. M. Hasanzadeh, N. Shadjou, M. de la Guardia, Iron and iron-oxide magnetic nanoparticles as signal-amplification elements in electrochemical biosensing, *TrAC or Trends Analyt Chem* **72** (2015) 1–9. https://doi.org/10.1016/j.trac.2015.03.016
54. C. Zhu, S. Dong, Recent progress in graphene-based nanomaterials as advanced electrocatalysts towards oxygen reduction reaction, *Nanoscale* **5** (2013) 1753. https://doi.org/10.1039/c2nr33839d
55. Y. Guo, X. Liu, S. Peng, X. Jiang, K. Xu, C. Chen, Z. Wang, C. Dai, W. Chen, A review of wearable and unobtrusive sensing technologies for chronic disease management, *Comput Biol Med* **129** (2021) 104163. https://doi.org/10.1016/J.COMPBIOMED.2020.104163
56. N.M. Julkapli, S. Bagheri, Graphene supported heterogeneous catalysts: An overview, *Int J Hydrogen Energy* **40** (2015) 948–979. https://doi.org/10.1016/j.ijhydene.2014.10.129
57. S. Liu, J. Tian, L. Wang, X. Sun, Microwave-assisted rapid synthesis of Ag nanoparticles/graphene nanosheet composites and their application for hydrogen peroxide detection, *J Nanopart Res* **13** (2011) 4539–4548. https://doi.org/10.1007/S11051-011-0410-3
58. S. Guo, D. Wen, Y. Zhai, S. Dong, E. Wang, Platinum nanoparticle ensemble-on-graphene hybrid nanosheet: One-pot, rapid synthesis, and used as new electrode material for electrochemical sensing, *ACS Nano* **4** (2010) 3959–3968. https://doi.org/10.1021/NN100852H
59. Q. Zhang, L. Jiang, H. Wang, J. Liu, J. Zhang, Y. Zheng, F. Li, C. Yao, S. Hou, Hollow Graphitized Carbon Nanocage Supported Pd Catalyst with Excellent Electrocatalytic Activity for Ethanol Oxidation, *ACS Sustain Chem Eng* **6** (2018) 7507–7514. https://doi.org/10.1021/ACSSUSCHEMENG.8B00208
60. B. Alinejad, K. Mahmoodi, Synthesis of graphene nanoflakes by grinding natural graphite together with NaCl in a planetary ball mill, *Functional Mater Lett* **10** (2017) 1750047. https://doi.org/10.1142/S1793604717500473
61. M. Khaleghi, K. Didehban, M. Shabanian, Simple and fast preparation of graphene oxide@ melamine terephthaldehyde and its PVC nanocomposite via ultrasonic irradiation: Chemical and thermal resistance study, *Ultrason Sonochem* **43** (2018) 275–284. https://doi.org/10.1016/j.ultsonch.2017.12.049

62. K. Chen, S. Song, D. Xue, Beyond graphene: Materials chemistry toward high performance inorganic functional materials, *J Mater Chem A* **3** (2015) 2441–2453. https://doi.org/10.1039/C4TA06989G
63. C. Liu, K. Wang, S. Luo, Y. Tang, L. Chen, Direct electrodeposition of graphene enabling the one-step synthesis of graphene–metal nanocomposite films, *Wiley Online Library* **7** (2011) 1203–1206. https://doi.org/10.1002/smll.201002340
64. M. Hilder, O. Winther-Jensen, B. Winther-Jensen, D.R. MacFarlane, Graphene/zinc nano-composites by electrochemical co-deposition, *Phys Chem Chem Phys* **14** (2012) 14034. https://doi.org/10.1039/c2cp42385e
65. Amato A first sighting of buckyballs in the wild, *Science* **257** (1992) 167. https://doi.org/10.1126/SCIENCE.257.5067.167/ASSET/47D85995-463D-4D00-82B4-DF8053E2D271/ASSETS/SCIENCE.257.5067.167.FP.PNG
66. T.C. Dinadayalane, J. Leszczynski, Remarkable diversity of carbon-carbon bonds: Structures and properties of fullerenes, carbon nanotubes, and graphene, *Struct Chem* **21** (2010) 1155–1169. https://doi.org/10.1007/S11224-010-9670-2
67. M. Paradise, T. Goswami, Carbon nanotubes – production and industrial applications, *Mater Des* **28** (2007) 1477–1489. https://doi.org/10.1016/J.MATDES.2006.03.008
68. L. Dai, D.W. Chang, J.B. Baek, W. Lu, Carbon nanomaterials for advanced energy conversion and storage, *Small* **8** (2012) 1130–1166. https://doi.org/10.1002/SMLL.201101594
69. M.R. Falvo, G.J. Clary, R.M. Taylor, V. Chi, F.P. Brooks, S. Washburn, R. Superfine, Bending and buckling of carbon nanotubes under large strain, *Nature* **389** (1997) 582–584. https://doi.org/10.1038/39282
70. B.I. Yakobson, M.P. Campbell, C.J. Brabec, J. Bernholc, High strain rate fracture and C-chain unraveling in carbon nanotubes, *Comput Mater Sci* **8** (1997) 341–348. https://doi.org/10.1016/S0927-0256(97)00047-5
71. S. Ruan, P. Gao, X. Yang, T.X. Yu, Toughening high performance ultrahigh molecular weight polyethylene using multiwalled carbon nanotubes, *Polymer (Guildf)* **44** (2003) 5643–5654. https://doi.org/10.1016/S0032-3861(03)00628-1
72. W. Ding, L. Calabri, K.M. Kohlhaas, X. Chen, D.A. Dikin, R.S. Ruoff, Modulus, fracture strength, and brittle vs. plastic response of the outer shell of arc-grown multi-walled carbon nanotubes, *Springer* **47** (2007) 25–36. https://doi.org/10.1007/s11340-006-9344-6
73. C. Li, T.W. Chou, A structural mechanics approach for the analysis of carbon nanotubes, *Int J Solids Struct* **40** (2003) 2487–2499. https://doi.org/10.1016/S0020-7683(03)00056-8
74. I. Suarez-Martinez, A. Felten, J.J. Pireaux, C. Bittencourt, C.P. Ewels, Transition metal deposition on graphene and carbon nanotubes, *J Nanosci Nanotechnol* **9** (2009) 6171–6175. https://doi.org/10.1166/jnn.2009.1557
75. R.S. Alencar, W. Cui, A.C. Torres-Dias, T.F.T. Cerqueira, S. Botti, M.A.L. Marques, O.P. Ferreira, C. Laurent, A. Weibel, D. Machon, D.J. Dunstan, A.G. Souza Filho, A. San-Miguel, Pressure-induced radial collapse in

few-wall carbon nanotubes: A combined theoretical and experimental study, *Carbon* **125** (2017) 429–436. https://doi.org/10.1016/J.CARBON.2017.09.044

76. M. Paradise, T. Goswami, Carbon nanotubes – production and industrial applications, *Mater Des* **28** (2007) 1477–1489. https://doi.org/10.1016/j.matdes.2006.03.008
77. B. Zhang, F. Kang, J.M. Tarascon, J.K. Kim, Recent advances in electrospun carbon nanofibers and their application in electrochemical energy storage, *Prog Mater Sci* **76** (2016) 319–380. https://doi.org/10.1016/J.PMATSCI.2015.08.002
78. E. Hammel, X. Tang, M. Trampert, T. Schmitt, K. Mauthner, A. Eder, P. Pötschke, Carbon nanofibers for composite applications, *Carbon* **42** (2004) 1153–1158. https://doi.org/10.1016/j.carbon.2003.12.043
79. K. De Jong, J.G.-C. Reviews, undefined 2000, Carbon nanofibers: Catalytic synthesis and applications, *Taylor & Francis* **42** (2000) 481–510. https://doi.org/10.1081/CR-100101954
80. C. Ma, Y. Song, J. Shi, D. Zhang, X. Zhai, M. Zhong, Q. Guo, L. Liu, Preparation and one-step activation of microporous carbon nanofibers for use as supercapacitor electrodes, *Carbon* **51** (2013) 290–300. https://doi.org/10.1016/J.CARBON.2012.08.056
81. P.A. Tran, L. Zhang, T.J. Webster, Carbon nanofibers and carbon nanotubes in regenerative medicine, *Adv Drug Deliv Rev* **61** (2009) 1097–1114. https://doi.org/10.1016/j.addr.2009.07.010
82. S.A. Chowdhury, M.C. Saha, S. Patterson, T. Robison, Y. Liu, S.A. Chowdhury, M.C. Saha, Y. Liu, S. Patterson, T. Robison, Highly conductive polydimethylsiloxane/carbon nanofiber composites for flexible sensor applications, *Wiley Online Library* **4** (2018). https://doi.org/10.1002/admt.201800398
83. E. Hammel, X. Tang, M. Trampert, T. Schmitt, K. Mauthner, A. Eder, P. Pötschke, Carbon nanofibers for composite applications, *Carbon* **42** (2004) 1153–1158. https://doi.org/10.1016/J.CARBON.2003.12.043
84. S. Kumar, T. Rath, R. Mahaling, C.S. Reddy, C.K. Das, K.N. Pandey, R.B. Srivastava, S.B. Yadaw, Study on mechanical, morphological and electrical properties of carbon nanofiber/polyetherimide composites, *Mater Sci Eng B* **141** (2007) 61–70. https://doi.org/10.1016/j.mseb.2007.06.002
85. J. Zeng, B. Saltysiak, W.S. Johnson, D.A. Schiraldi, S. Kumar, Processing and properties of poly(methyl methacrylate)/carbon nanofiber composites, *Compos B Eng* **35** (2004) 245–249. https://doi.org/10.1016/J.COMPOSITESB.2003.08.009
86. L. Feng, N. Xie, J. Zhong, Carbon nanofibers and their composites: A review of synthesizing, properties and applications, *Materials* 7 (2014) 3919–3945. https://doi.org/10.3390/ma7053919
87. P.E. Anderson, N.M. Rodríguez, Influence of the support on the structural characteristics of carbon nanofibers produced from the metal-catalyzed decomposition of ethylene, *Chem Mater* **12** (2000) 823–830. https://doi.org/10.1021/cm990582n

88. T.-J. Zhao, D. Chen, Y.-C. Dai, W.-K. Yuan, A. Holmen, Synthesis of dimethyl oxalate from CO and CH^3ONO on carbon nanofiber supported palladium catalysts, *Ind Eng Chem Res* **43** (2004) 4595–4601. https://doi.org/10.1021/ie030728z
89. D.W. Kim, H.S. Kil, J. Kim, I. Mochida, K. Nakabayashi, C.K. Rhee, J. Miyawaki, S.H. Yoon, Highly graphitized carbon from non-graphitizable raw material and its formation mechanism based on domain theory, *Carbon* **121** (2017) 301–308. https://doi.org/10.1016/J.CARBON.2017.05.086
90. S. Helveg, P.L. Hansen, Atomic-scale studies of metallic nanocluster catalysts by in situ high-resolution transmission electron microscopy, *Catal Today* **111** (2006) 68–73. https://doi.org/10.1016/j.cattod.2005.10.019
91. P.L. Gai, E.D. Boyes, S. Helveg, P.L. Hansen, S. Giorgio, C.R. Henry, Atomic-resolution environmental transmission electron microscopy for probing gas-solid reactions in heterogeneous catalysis, *MRS Bull* **32** (2007) 1044–1050. https://doi.org/10.1557/MRS2007.214
92. Fonseca, K. Hernadi, P. Piedigrosso, J.F. Colomer, K. Mukhopadhyay, R. Doome, S. Lazarescu, L.P. Biro, P. Lambin, P.A. Thiry, D. Bernaerts, J.B. Nagy, Synthesis of single- and multi-wall carbon nanotubes over supported catalysts, *Appl Phys A Mater Sci Process* **67** (1998) 11–22. https://doi.org/10.1007/S003390050732
93. K. Shen, X. Chen, J. Chen, Y. Li, Development of MOF-derived carbon-based nanomaterials for efficient catalysis, *ACS Catal* **6** (2016) 5887–5903. https://doi.org/10.1021/ACSCATAL.6B01222
94. Z. Li, L. Wang, Y. Li, Y. Feng, W. Feng, Carbon-based functional nanomaterials: Preparation, properties and applications, *Compos Sci Technol* **179** (2019) 10–40. https://doi.org/10.1016/J.COMPSCITECH.2019.04.028
95. J.R. McFarland, J.M. Reilly, H.J. Herzog, Representing energy technologies in top-down economic models using bottom-up information, *Energy Econ* **26** (2004) 685–707. https://doi.org/10.1016/J.ENECO.2004.04.026
96. R.G. Hobbs, N. Petkov, J.D. Holmes, Semiconductor nanowire fabrication by bottom-up and top-down paradigms, *Chem Mater* **24** (2012) 1975–1991. https://doi.org/10.1021/CM300570N
97. S. Liu, T.H. Zeng, M. Hofmann, E. Burcombe, J. Wei, R. Jiang, J. Kong, Y. Chen, Antibacterial activity of graphite, graphite oxide, graphene oxide, and reduced graphene oxide: Membrane and oxidative stress, *ACS Publications* **5** (2011) 6971–6980. https://doi.org/10.1021/nn202451x
98. D.R. Dreyer, A.D. Todd, C.W. Bielawski, Harnessing the chemistry of graphene oxide, *Chem Soc Rev* **5288** (2014) 5288. https://doi.org/10.1039/c4cs00060a
99. R. Muñoz, C. Gómez-Aleixandre, Review of CVD synthesis of graphene, *Chem Vapor Deposition* **19** (2013) 297–322. https://doi.org/10.1002/cvde.201300051
100. X.-Y. Wang, A. Narita, K. Müllen, Precision synthesis versus bulk-scale fabrication of graphenes, *Nat Rev Chem* **2** (2018) 0100. https://doi.org/10.1038/s41570-017-0100

101. J. Lu, J.W. Elam, P.C. Stair, Atomic layer deposition—Sequential self-limiting surface reactions for advanced catalyst "bottom-up" synthesis, *Surf Sci Rep* **71** (2016) 410–472. https://doi.org/10.1016/J.SURFREP.2016.03.003
102. N. Kumar, R. Salehiyan, V. Chauke, O. Joseph Botlhoko, K. Setshedi, M. Scriba, M. Masukume, S. Sinha Ray, Top-down synthesis of graphene: A comprehensive review, *FlatChem* **27** (2021) 100224. https://doi.org/10.1016/j.flatc.2021.100224
103. L. Lin, B. Deng, J. Sun, H. Peng, Z. Liu, Bridging the gap between reality and ideal in chemical vapor deposition growth of graphene, *Chem Rev* **118** (2018) 9281–9343. https://doi.org/10.1021/acs.chemrev.8b00325
104. S. Kumar, M. Nehra, D. Kedia, N. Dilbaghi, K. Tankeshwar, K.-H. Kim, Carbon nanotubes: A potential material for energy conversion and storage, *Prog Energy Combust Sci* **64** (2018) 219–253. https://doi.org/10.1016/j.pecs.2017.10.005
105. K.M. Liew, M.F. Kai, L.W. Zhang, Mechanical and damping properties of CNT-reinforced cementitious composites, *Compos Struct* **160** (2017) 81–88. https://doi.org/10.1016/J.COMPSTRUCT.2016.10.043
106. J. Jeynes, Single walled carbon nanotubes functionalised with RNA or DNA, 2007
107. J. Hicks, A. Tejeda, A. Taleb-Ibrahimi, M.S. Nevius, F. Wang, K. Shepperd, J. Palmer, F. Bertran, P. Le Fèvre, J. Kunc, W.A. de Heer, C. Berger, E.H. Conrad, A wide-bandgap metal–semiconductor–metal nanostructure made entirely from graphene, *Nat Phys* **9** (2013) 49–54. https://doi.org/10.1038/nphys2487
108. Kumar, K. Singh, O.P. Pandey, One step synthesis and growth mechanism of carbon nanotubes, *J Mater Sci Technol* **30** (2014) 112–116. https://doi.org/10.1016/J.JMST.2013.09.005
109. L. Su, J. Wang, F. Yu, Z. Sheng, H. Chang, C. Pak, Continuous production of single-wall carbon nanotubes by spray pyrolysis of alcohol with dissolved ferrocene, *Chem Phys Lett* **420** (2006) 421–425. https://doi.org/10.1016/j.cplett.2005.12.046
110. C. Journet, W.K. Maser, P. Bernier, A. Loiseau, M. Lamy de la Chapelle, S. Lefrant, P. Deniard, R. Lee, J.E. Fischer, Large-scale production of single-walled carbon nanotubes by the electric-arc technique, *Nature* **388** (1997) 756–758. https://doi.org/10.1038/41972
111. S. Sharma, S.C. Lakkad, Morphology study of carbon nanospecies grown on carbon fibers by thermal CVD technique, *Surf Coat Technol* **203** (2009) 1329–1335. https://doi.org/10.1016/j.surfcoat.2008.10.043
112. S. Saxena, A.K. Srivastava, Carbon nanotube-based sensors and their application, in: *Nano-Optics*, Elsevier, 2020: pp. 265–291. https://doi.org/10.1016/B978-0-12-818392-2.00010-X
113. H. Dai, Carbon nanotubes: Synthesis, integration, and properties, *Acc Chem Res* **35** (2002) 1035–1044. https://doi.org/10.1021/ar0101640
114. Kenry, C.T. Lim, Nanofiber technology: Current status and emerging developments, *Prog Polym Sci* **70** (2017) 1–17. https://doi.org/10.1016/J.PROGPOLYMSCI.2017.03.002

115. Z. Niu, W. Yuan, Highly efficient thermo- and sunlight-driven energy storage for thermo-electric energy harvesting using sustainable nanocellulose-derived carbon aerogels embedded phase change materials, *ACS Sustain Chem Eng* 7 (2019) 17523–17534. https://doi.org/10.1021/acssuschemeng.9b05015
116. M. Raimondo, L. Guadagno, L. Vertuccio, C. Naddeo, G. Barra, G. Spinelli, P. Lamberti, V. Tucci, K. Lafdi, Electrical conductivity of carbon nanofiber reinforced resins: Potentiality of tunneling atomic force microscopy (TUNA) technique, *Compos B Eng* **143** (2018) 148–160. https://doi.org/10.1016/J.COMPOSITESB.2018.02.005
117. C. Marolla, Information and communication technology for sustainable development, *CRC Press*, 2018. https://doi.org/10.1201/9781351045230
118. M.S. Konsta-Gdoutos, C.A. Aza, Self sensing carbon nanotube (CNT) and nanofiber (CNF) cementitious composites for real time damage assessment in smart structures, *Cem Concr Compos* **53** (2014) 162–169. https://doi.org/10.1016/J.CEMCONCOMP.2014.07.003
119. Q. Zhang, X.-M. Xie, S.-Y. Wei, Z.-Z. Zhu, L.-S. Zheng, S.-Y. Xie, Q. Zhang, X. Xie, S. Wei, Z. Zhu, L. Zheng, S. Xie, The synthesis of conical carbon, *Small Methods* **5** (2021) 2001086. https://doi.org/10.1002/SMTD.202001086
120. V. García-Martínez, M. R. Gude, S. Calvo, M.R. Martínez-Miranda, A. Ureña, Influence of graphene nanoplatelets on curing kinetics and rheological properties of a benzoxazine resin, *Mater Today Commun* **24** (2020) 100990. https://doi.org/10.1016/j.mtcomm.2020.100990
121. L. Feng, N. Xie, J. Zhong, Carbon nanofibers and their composites: A review of synthesizing, properties and applications, *Materials* **7** (2014) 3919–3945. https://doi.org/10.3390/ma7053919
122. S. Tripathi, Role of nanocatalysts in synthesis of carbon nanofiber, *Carbon Nanofibers: Fundamentals and Applications* (2021) 49–74. https://doi.org/10.1002/9781119769149.CH3
123. Y. Chen, A. Amiri, J.G. Boyd, M. Naraghi, Y. Chen, A. Amiri, J.G. Boyd, M. Naraghi, Promising trade-offs between energy storage and load bearing in carbon nanofibers as structural energy storage devices, *Wiley Online Library* **29** (2019). https://doi.org/10.1002/adfm.201901425
124. H. Niu, J. Zhang, Z. Xie, X. Wang, T. Lin, Preparation, structure and supercapacitance of bonded carbon nanofiber electrode materials, *Carbon* **49** (2011) 2380–2388. https://doi.org/10.1016/J.CARBON.2011.02.005
125. B. De, S. Banerjee, K.D. Verma, T. Pal, P.K. Manna, K.K. Kar, Carbon nanotube as electrode materials for supercapacitors, *Springer Series in Materials Science* **302** (2020) 229–243. https://doi.org/10.1007/978-3-030-52359-6_9
126. B. Murugesan, N. Pandiyan, M. Arumugam, J. Sonamuthu, S. Samayanan, C. Yurong, Y. Juming, S. Mahalingam, Fabrication of palladium nanoparticles anchored polypyrrole functionalized reduced graphene oxide nanocomposite for antibiofilm associated orthopedic tissue engineering, *Appl Surf Sci* **510** (2020) 145403. https://doi.org/10.1016/J.APSUSC.2020.145403

127. S.-J. Lee, E. Kawakami, A. Karn, M. Ahmad, H. An, Y. Seung Kim, A.J. Marsden, M. Phillips, N.R. Wilson, Friction force microscopy: A simple technique for identifying graphene on rough substrates and mapping the orientation of graphene grains on copper, *IopscienceIop.Org* **24** (2013) 255704–255713. https://doi.org/10.1088/0957-4484/24/25/255704
128. A. Resta, T. Leoni, C. Barth, A. Ranguis, C. Becker, T. Bruhn, P. Vogt, G. Le Lay, Atomic structures of silicene layers grown on Ag(111): Scanning tunneling microscopy and noncontact atomic force microscopy observations, *Scientific Rep* **3** (2013) 1–7. https://doi.org/10.1038/srep02399
129. B. Alimkhanuly, J. Sohn, I.J. Chang, S. Lee, Graphene-based 3D XNOR-VRRAM with ternary precision for neuromorphic computing, *NPJ 2D Mat Appl* **5** (2021) 1–10. https://doi.org/10.1038/s41699-021-00236-x
130. A.Y. Lee, K. Yang, N.D. Anh, C. Park, S.M. Lee, T.G. Lee, M.S. Jeong, Raman study of D* band in graphene oxide and its correlation with reduction, *Appl Surf Sci* **536** (2021) 147990. https://doi.org/10.1016/j.apsusc.2020.147990
131. B.J. Inkson, Scanning electron microscopy (SEM) and transmission electron microscopy (TEM) for materials characterization, in: *Materials Characterization Using Nondestructive Evaluation (NDE) Methods* (2016) 17–43. https://doi.org/10.1016/B978-0-08-100040-3.00002-X
132. S. Zomorodbakhsh, Y. Abbasian, M. Naghinejad, M. Sheikhpour, The effects study of isoniazid conjugated multi-wall carbon nanotubes nanofluid on *Mycobacterium tuberculosis*, *Int J Nanomed* **15** (2020) 5901–5909. https://doi.org/10.2147/IJN.S251524
133. T. Zhao, C. Hou, H. Zhang, R. Zhu, S. She, J. Wang, T. Li, Z. Liu, B. Wei, Electromagnetic wave absorbing properties of amorphous carbon nanotubes, *Scientific Rep* **4** (2014) 1–7. https://doi.org/10.1038/srep05619
134. S.M. Xue, Z.L. Xu, Y.J. Tang, C.H. Ji, Polypiperazine-amide nanofiltration membrane modified by different functionalized multiwalled carbon nanotubes (MWCNTs), *ACS Appl Mater Interfaces* **8** (2016) 19135–19144. https://doi.org/10.1021/ACSAMI.6B05545
135. Q. Li, L. Dong, J. Fang, C.X.-A. nano, Property–structure relationship of nanoscale ionic materials based on multiwalled carbon nanotubes, *ACS Publications* **4** (2010) 5797–5806. https://doi.org/10.1021/nn101542v
136. Z. Zhang, B. Liu, Y.W. Zhang, K.C. Hwang, H. Gao, Ultra-strong collagen-mimic carbon nanotube bundles, *Carbon* **77** (2014) 1040–1053. https://doi.org/10.1016/J.CARBON.2014.06.020
137. M. García-Gutiérrez, A. Nogales, J. Hernández, X-ray scattering applied to the analysis of carbon nanotubes, polymers and nanocomposites, (2007)
138. R.A. DiLeo, B.J. Landi, R.P. Raffaelle, Purity assessment of multiwalled carbon nanotubes by Raman spectroscopy, *J Appl Phys* **101** (2007). https://doi.org/10.1063/1.2712152
139. M. Zdrojek, W. Gebicki, C. Jastrzebski, T. Melin, A. Huczko, Studies of multiwall carbon nanotubes using Raman spectroscopy and atomic force microscopy, *Solid State Phenomena* **99–100** (2004) 265–268. https://doi.org/10.4028/WWW.SCIENTIFIC.NET/SSP.99-100.265

140. B.P. Singh, S. Samal, S. Nayak, S.M. Majhi, L. Besra, S. Bhattacharjee, The production of a multi-walled carbon nanotube/hexamethylene diisocyanate nanocomposite coating on copper by electrophoretic deposition, *Surf Coat Technol* **206** (2011) 1319–1326. https://doi.org/10.1016/J.SURFCOAT.2011.08.054
141. D. Janas, Towards monochiral carbon nanotubes: A review of progress in the sorting of single-walled carbon nanotubes, *Mater Chem Front* **2** (2018) 36–63. https://doi.org/10.1039/C7QM00427C
142. Jorio, C. Fantini, M.S.S. Dantas, M.A. Pimenta, A.G. Souza Filho, A.G. Souza Filho, V.W. Brar, G. Dresselhaus, M.S. Dresselhaus, A.K. Swan, M.S. Ünlü, B.B. Goldberg, R. Saito, Linewidth of the Raman features of individual single-wall carbon nanotubes, *Phys Rev B Condens Matter Mater Phys* **66** (2002) 1154111–1154118. https://doi.org/10.1103/PHYSREVB.66.115411
143. A.S. Skwarecki, M.G. Nowak, M.J. Milewska, Synthetic strategies in construction of organic low molecular-weight carrier-drug conjugates, *Bioorg Chem* **104** (2020) 104311. https://doi.org/10.1016/J.BIOORG.2020.104311
144. I.O. MacIel, N. Anderson, M.A. Pimenta, A. Hartschuh, H. Qian, M. Terrones, H. Terrones, J. Campos-Delgado, A.M. Rao, L. Novotny, A. Jorio, Electron and phonon renormalization near charged defects in carbon nanotubes, *Nat Mat* **7** (2008) 878–883. https://doi.org/10.1038/nmat2296
145. Hariharasubramanian, Y. Dominic Ravichandran, R. Rajesh, R. Rajkumari, G. Kalai Selvan, S. Arumugam, Functionalization of multi-walled carbon nanotubes with 6-aminobenzothiazole and their temperature-dependent magnetic studies, *Taylor & Francis* **22** (2014) 874–886. https://doi.org/10.1080/1536383X.2013.794339
146. Z. Zhao, Z. Yang, Y. Hu, J. Li, X. Fan, Multiple functionalization of multi-walled carbon nanotubes with carboxyl and amino groups, *Appl Surf Sci* **276** (2013) 476–481. https://doi.org/10.1016/J.APSUSC.2013.03.119
147. J. Venkatesan, Z.-J. Qian, B. Ryu, N. Ashok Kumar, S.-K. Kim, Preparation and characterization of carbon nanotube-grafted-chitosan – Natural hydroxyapatite composite for bone tissue engineering, *Carbohydr Polym* **83** (2011) 569–577. https://doi.org/10.1016/j.carbpol.2010.08.019
148. P.R. Chang, P. Zheng, B. Liu, D.P. Anderson, J. Yu, X. Ma, Characterization of magnetic soluble starch-functionalized carbon nanotubes and its application for the adsorption of the dyes, *J Hazard Mater* **186** (2011) 2144–2150. https://doi.org/10.1016/J.JHAZMAT.2010.12.119
149. S. Attal, R. Thiruvengadathan, O.R.-A. chemistry, Determination of the concentration of single-walled carbon nanotubes in aqueous dispersions using UV–Visible absorption spectroscopy, *ACS Publications* **78** (2006) 8098–8104. https://doi.org/10.1021/ac060990s
150. S. Meng, W.L. Wang, P. Maragakis, E. Kaxiras, Determination of DNA-base orientation on carbon nanotubes through directional optical absorbance, *Nano Lett* **7** (2007) 2312–2316. https://doi.org/10.1021/NL070953W
151. L. Silvestro, A. Ruviaro, G. Lima, P. De Matos, A.R.G. De Azevedo, S.N. Monteiro, P. Gleize, Influence of ultrasonication of functionalized carbon nanotubes on the rheology, hydration, and compressive strength of Portland cement pastes, (2021). https://doi.org/10.3390/ma

152. M. Erkens, S. Cambré, E. Flahaut, F. Fossard, A. Loiseau, W. Wenseleers, Ultrasonication-induced extraction of inner shells from double-wall carbon nanotubes characterized via in situ spectroscopy after density gradient ultracentrifugation, *Carbon* **185** (2021) 113–125. https://doi.org/10.1016/j.carbon.2021.07.075
153. N.K. Mehra, V. Mishra, N.K. Jain, A review of ligand tethered surface engineered carbon nanotubes, *Biomaterials* **35** (2014) 1267–1283. https://doi.org/10.1016/J.BIOMATERIALS.2013.10.032
154. A.O. Obeid, F. Al-Yusufy, S.A. Al-Aghbari, O.M. Al-Shuja'a, Y. Gaber, Z.A. Al-Washal, Synthesis, characterization and study of DC electrical conductivity of poly composite, (2019). https://doi.org/10.26434/chemrxiv.11441313
155. S. Daniel, T.P. Rao, K.S. Rao, S.U. Rani, G.R.K. Naidu, H.Y. Lee, T. Kawai, A review of DNA functionalized/grafted carbon nanotubes and their characterization, *Sens Actuators B Chem* **122** (2007) 672–682. https://doi.org/10.1016/J.SNB.2006.06.014
156. J. Paredes, A. Martínez-Alonso, J.M.D. Tascón, Application of scanning tunneling and atomic force microscopies to the characterization of microporous and mesoporous materials, *Microporous and Mesoporous Mat* **65** (2003) 93–126. https://doi.org/10.1016/j.micromeso.2003.07.001
157. K.L. Klein, A.V. Melechko, T.E. McKnight, S.T. Retterer, P.D. Rack, J.D. Fowlkes, D.C. Joy, M.L. Simpson, Surface characterization and functionalization of carbon nanofibers, *J Appl Phys* **103** (2008) 061301. https://doi.org/10.1063/1.2840049
158. J.I. Paredes, A. Martínez-Alonso, J.M.D. Tascón, Characterization of microporosity and mesoporosity in carbonaceous materials by scanning tunneling microscopy, *Langmuir* **17** (2001) 474–480. https://doi.org/10.1021/LA0008535
159. S. Lee, J.R. Hahn, B.C. Ku, J. Kim, Effect of carbon nanofiber structure on crystallization kinetics of polypropylene/carbon nanofiber composites, *Bull Korean Chem Soc* **32** (2011) 2369–2376. https://doi.org/10.5012/BKCS.2011.32.7.2369
160. S.-H. Yoon, S. Lim, S. Hong, I. Mochida, B. An, K. Yokogawa, Carbon nanorod as a structural unit of carbon nanofibers, *Carbon* **42** (2004) 3087–3095. https://doi.org/10.1016/j.carbon.2004.07.022
161. K.L. Klein, A.V. Melechko, T.E. McKnight, S.T. Retterer, P.D. Rack, J.D. Fowlkes, D.C. Joy, M.L. Simpson, Surface characterization and functionalization of carbon nanofibers, *J Appl Phys* **103** (2008). https://doi.org/10.1063/1.2840049
162. J.I. Paredes, M. Burghard, A. Martínez-Alonso, J.M.D. Tascón, Graphitization of carbon nanofibers: Visualizing the structural evolution on the nanometer and atomic scales by scanning tunneling microscopy, *Springer* **80** (2005) 675–682. https://doi.org/10.1007/s00339-004-3109-9
163. Z. Yu, Synthesis of carbon nanofibers and carbon nanotubes, (2005).
164. J. Jang, J. Bae, M. Choi, S.-H. Yoon, Fabrication and characterization of polyaniline coated carbon nanofiber for supercapacitor, *Carbon* **43** (2005) 2730–2736. https://doi.org/10.1016/j.carbon.2005.05.039

165. T. Kimura, H. Koizumi, H. Kinoshita, T. Ichikawa, Formation of carbon nanofibers from decacyclene by ion beam irradiation, *Nucl Instrum Methods Phys Res B* **236** (2005) 474–481. https://doi.org/10.1016/J.NIMB.2005.04.022
166. J.I. Paredes, M. Burghard, A. Martínez-Alonso, J.M.D. Tascón, Graphitization of carbon nanofibers: Visualizing the structural evolution on the nanometer and atomic scales by scanning tunneling microscopy, *Appl Phys A* **80** (2005) 675–682. https://doi.org/10.1007/s00339-004-3109-9
167. J. Li, Y. Lin, Nanomaterials for sensing and electrocatalysis, *J Nanosci Nanotechnol* **9** (2009) 2173–2174. https://doi.org/10.1166/JNN.2009.SE1A
168. S.J. Oh, H.J. Lee, D.K. Keum, S.W. Lee, D.H. Wang, S.Y. Park, L.S. Tan, J.B. Baek, Multiwalled carbon nanotubes and nanofibers grafted with polyetherketones in mild and viscous polymeric acid, *Polymer (Guildf)* **47** (2006) 1132–1140. https://doi.org/10.1016/J.POLYMER.2005.12.064
169. L.I. van der Wal, S.J. Turner, J. Zečević, Developments and advances in in situ transmission electron microscopy for catalysis research, *Catal Sci Technol* **11** (2021) 3634–3658. https://doi.org/10.1039/D1CY00258A
170. J. Fowlkes, A. Melechko, K. Klein, P.D. Rack, D.A. Smith, D.K. Hensley, M.J. Doktycz, M.L. Simpson, Control of catalyst particle crystallographic orientation in vertically aligned carbon nanofiber synthesis, *Carbon* **44** (2006) 1503–1510. https://doi.org/10.1016/j.carbon.2005.12.010
171. S.S. Perevoznikov, I.V. Yakovlev, L.S. Tsybulskaya, O.B. Lapina, Synthesis and composition study of electrochemically deposited Ni-P coating with increased surface area, *Coatings* **11** (2021) 1071. https://doi.org/10.3390/COATINGS11091071
172. K.L. Klein, A.V. Melechko, P.D. Rack, J.D. Fowlkes, H.M. Meyer, M.L. Simpson, Cu–Ni composition gradient for the catalytic synthesis of vertically aligned carbon nanofibers, *Carbon* **43** (2005) 1857–1863. https://doi.org/10.1016/J.CARBON.2005.02.027
173. K.L. Klein, A.V. Melechko, T.E. McKnight, S.T. Retterer, P.D. Rack, J.D. Fowlkes, D.C. Joy, M.L. Simpson, Surface characterization and functionalization of carbon nanofibers, *J Appl Phys* **103** (2008). https://doi.org/10.1063/1.2840049
174. Y. Ominami, Q. Ngo, M. Suzuki, A.J. Austin, C.Y. Yang, A.M. Cassell, J. Li, Interface characteristics of vertically aligned carbon nanofibers for interconnect applications, *Appl Phys Lett* **89** (2006). https://doi.org/10.1063/1.2423241
175. Y. Ominami, Q. Ngo, A.J. Austin, H. Yoong, C.Y. Yang, A.M. Cassell, B.A. Cruden, J. Li, M. Meyyappan, Structural characteristics of carbon nanofibers for on-chip interconnect applications, *Appl Phys Lett* **87** (2005) 233105. https://doi.org/10.1063/1.2137873
176. Q. Ngo, T. Yamada, M. Suzuki, Y. Ominami, A.M. Cassell, Li Jun, M. Meyyappan, C.Y. Yang, Structural and electrical characterization of carbon nanofibers for interconnect via applications, *IEEE Trans Nanotechnol* **6** (2007) 688–695. https://doi.org/10.1109/TNANO.2007.907400

177. L. Han, W. Wu, F.L. Kirk, J. Luo, M.M. Maye, N.N. Kariuki, Y. Lin, C. Wang, G.J. Zhong, A direct route toward assembly of nanoparticle-carbon nanotube composite materials, *Langmuir* **20** (2004) 6019–6025. https://doi.org/10.1021/LA0497907
178. A. Romero, M.P. Lavin-Lopez, M.L. Sánchez-Silva, J.L. Valverde, A. Paton-Carrero, Effects of oxidizing procedures on carbon nanofibers surface and dispersability in an epoxy resin, *Mater Chem Phys* **243** (2020) 122571. https://doi.org/10.1016/J.MATCHEMPHYS.2019.122571
179. R. Tirawattanakoson, P. Rattanarat, N. Ngamrojanavanich, N. Rodthongkum, O. Chailapakul, Free radical scavenger screening of total antioxidant capacity in herb and beverage using graphene/PEDOT: PSS-modified electrochemical sensor, *J Electroanal Chem* **767** (2016) 68–75. https://doi.org/10.1016/j.jelechem.2015.11.037
180. N. Ruecha, N. Rodthongkum, D.M. Cate, J. Volckens, O. Chailapakul, C.S. Henry, Sensitive electrochemical sensor using a graphene–polyaniline nanocomposite for simultaneous detection of Zn(II), Cd(II), and Pb(II), *Anal Chim Acta* **874** (2015) 40–48. https://doi.org/10.1016/J.ACA.2015.02.064
181. S. Lee, J. Oh, D. Kim, Y. Piao, A sensitive electrochemical sensor using an iron oxide/graphene composite for the simultaneous detection of heavy metal ions, *Talanta* **160** (2016) 528–536. https://doi.org/10.1016/j.talanta.2016.07.034
182. B. He, X. Shen, J. Nie, X. Wang, F. Liu, W. Yin, C. Hou, D. Huo, H. Fa, Electrochemical sensor using graphene/Fe3O4 nanosheets functionalized with garlic extract for the detection of lead ion, *J Solid State Electro* **22** (2018) 3515–3525. https://doi.org/10.1007/s10008-018-4041-9
183. X. Fan, L. Xing, P. Ge, L. Cong, Q. Hou, Q. Ge, R. Liu, W. Zhang, G. Zhou, Electrochemical sensor using gold nanoparticles and plasma pretreated graphene based on the complexes of calcium and Troponin C to detect Ca2+ in meat, *Food Chem* **307** (2020) 125645. https://doi.org/10.1016/j.foodchem.2019.125645
184. K. Chen, Z.L. Zhang, Y.M. Liang, W. Liu, A graphene-based electrochemical sensor for rapid determination of phenols in water, *Sensors* **13** (2013) 6204–6216. https://doi.org/10.3390/S130506204
185. C. Wang, L. Zhang, Z. Guo, J. Xu, H. Wang, K. Zhai, X. Zhuo, A novel hydrazine electrochemical sensor based on the high specific surface area graphene, *Microchim Acta* **169** (2010) 1–6. https://doi.org/10.1007/S00604-010-0304-6
186. X. Wang, Y. Wen, L. Lu, J. Xu, L. Zhang, Y. Yao, H. He, A novel L-cysteine electrochemical sensor using sulfonated graphene-poly(3,4-ethylenedioxythiophene) composite film decorated with gold nanoparticles, *Wiley Online Library* **26** (2014) 648–655. https://doi.org/10.1002/elan.201300551
187. E.B. Bahadır, M.K. Sezgintürk, Applications of graphene in electrochemical sensing and biosensing, *TrAC or Trends Analyt Chem* **76** (2016) 1–14. https://doi.org/10.1016/j.trac.2015.07.008

188. Y. Han, R. Zhang, C. Dong, F. Cheng, Y. Guo, Sensitive electrochemical sensor for nitrite ions based on rose-like AuNPs/MoS2/graphene composite, *Biosens Bioelectron* **142** (2019) 111529. https://doi.org/10.1016/j.bios.2019.111529
189. Y.-R. Kim, S. Bong, Y.-J. Kang, Y. Yang, R.K. Mahajan, J.S. Kim, H. Kim, Electrochemical detection of dopamine in the presence of ascorbic acid using graphene modified electrodes, *Biosens Bioelectron* **25** (2010) 2366–2369. https://doi.org/10.1016/j.bios.2010.02.031
190. S.A. Zaidi, Utilization of an environmentally-friendly monomer for an efficient and sustainable adrenaline imprinted electrochemical sensor using graphene, *Electrochim Acta* **274** (2018) 370–377. https://doi.org/10.1016/J.ELECTACTA.2018.04.119
191. E. Lee, J. Choi, S. Um, B.O.-K.J. of Chemical, Electrochemical sensor for selective detection of norepinephrine using graphene sheets-gold nanoparticle complex modified electrode, *Springer* **34** (2017) 1129–1132. https://doi.org/10.1007/s11814-016-0363-4
192. F.C. Moraes, B. Rossi, M.C. Donatoni, K.T. de Oliveira, E.C. Pereira, Sensitive determination of 17β-estradiol in river water using a graphene-based electrochemical sensor, *Anal Chim Acta* **881** (2015) 37–43. https://doi.org/10.1016/j.aca.2015.04.043
193. J. Li, D. Kuang, Y. Feng, F. Zhang, Z. Xu, M. Liu, A graphene oxide-based electrochemical sensor for sensitive determination of 4-nitrophenol, *J Hazard Mater* **201–202** (2012) 250–259. https://doi.org/10.1016/J.JHAZMAT.2011.11.076
194. J.L. Chen, X.P. Yan, K. Meng, S.F. Wang, Graphene oxide-based photo-induced charge transfer label-free near-infrared fluorescent biosensor for dopamine, *Anal Chem* **83** (2011) 8787–8793. https://doi.org/10.1021/AC2023537/SUPPL_FILE/AC2023537_SI_001.PDF
195. Z. Liu, H. Forsyth, N. Khaper, A. Chen, Sensitive electrochemical detection of nitric oxide based on AuPt and reduced graphene oxide nanocomposites, *Analyst* **141** (2016) 4074–4083. https://doi.org/10.1039/C6AN00429F
196. H. Jin, C. Zhao, R. Gui, X. Gao, Z. Wang, Reduced graphene oxide/nile blue/gold nanoparticles complex-modified glassy carbon electrode used as a sensitive and label-free aptasensor for ratiometric electrochemical sensing of dopamine, *Anal Chim Acta* **1025** (2018) 154–162. https://doi.org/10.1016/J.ACA.2018.03.036
197. A.U. Alam, M.J. Deen, Bisphenol A electrochemical sensor using graphene oxide and β-cyclodextrin-functionalized multi-walled carbon nanotubes, *Anal Chem* **92** (2020) 5532–5539. https://doi.org/10.1021/ACS.ANALCHEM.0C00402
198. M. Govindhan, T. Lafleur, B. Adhikari, Electrochemical sensor based on carbon nanotubes for the simultaneous detection of phenolic pollutants, *Electroanalysis* **27** (2015) 902–909. https://doi.org/10.1002/elan.201400608
199. G. Yang, F. Zhao, Electrochemical sensor for chloramphenicol based on novel multiwalled carbon nanotubes@molecularly imprinted polymer, *Biosens Bioelectron* **64** (2015) 416–422. https://doi.org/10.1016/J.BIOS.2014.09.041

200. B. Rezaei, O. Rahmanian, A.A. Ensafi, An electrochemical sensor based on multiwall carbon nanotubes and molecular imprinting strategy for warfarin recognition and determination, *Sens Actuators B Chem* **196** (2014) 539–545. https://doi.org/10.1016/j.snb.2014.02.037
201. H. Bagheri, A. Hajian, M. Rezaei, A. Shirzadmehr, Composite of Cu metal nanoparticles-multiwall carbon nanotubes-reduced graphene oxide as a novel and high performance platform of the electrochemical sensor for simultaneous determination of nitrite and nitrate, *J Hazard Mater* **324** (2017) 762–772. https://doi.org/10.1016/J.JHAZMAT.2016.11.055
202. Y. Liu, J. Liu, H. Tang, J. Liu, B. Xu, F. Yu, Y. Li, Fabrication of highly sensitive and selective electrochemical sensor by using optimized molecularly imprinted polymers on multi-walled carbon nanotubes for metronidazole measurement, *Sens Actuators B Chem* **206** (2015) 647–652. https://doi.org/10.1016/J.SNB.2014.10.019
203. P.N.D. Duoc, N.H. Binh, T. Van Hau, C.T. Thanh, P. Van Trinh, N.V. Tuyen, N. Van Quynh, N. Van Tu, V. Duc Chinh, V. Thi Thu, P.D. Thang, P.N. Minh, N. Van Chuc, A novel electrochemical sensor based on double-walled carbon nanotubes and graphene hybrid thin film for arsenic(V) detection, *J Hazard Mater* **400** (2020) 123185. https://doi.org/10.1016/j.jhazmat.2020.123185
204. D. Li, C. Chen, X. Guo, C. Liu, W. Yang, A simple electrochemiluminescence aptasensor using a GCE/NCQDs/aptamers for detection of Pb, *Environ Technol* **43** (2022) 2270–2277. https://doi.org/10.1080/09593330.2021.1871661
205. S.K. Arumugasamy, S. Govindaraju, K. Yun, Electrochemical sensor for detecting dopamine using graphene quantum dots incorporated with multiwall carbon nanotubes, *Appl Surf Sci* **508** (2020) 145294. https://doi.org/10.1016/J.APSUSC.2020.145294
206. Y. Yang, G. Fang, G. Liu, M. Pan, X. Wang, L. Kong, X. He, S. Wang, Electrochemical sensor based on molecularly imprinted polymer film via sol–gel technology and multi-walled carbon nanotubes-chitosan functional layer for sensitive determination of quinoxaline-2-carboxylic acid, *Biosens Bioelectron* **47** (2013) 475–481. https://doi.org/10.1016/j.bios.2013.03.054
207. Q. Huang, X. Lin, L. Tong, Q.X. Tong, Graphene quantum dots/Multiwalled carbon nanotubes composite-based electrochemical sensor for detecting dopamine release from living cells, *ACS Sustain Chem Eng* **8** (2020) 1644–1650. https://doi.org/10.1021/ACSSUSCHEMENG.9B06623
208. J. Xiao, H. Wang, C. Li, K. Deng, X. Li, A simple dopamine sensor using graphdiyne nanotubes and shortened carbon nanotubes for enhanced preconcentration and electron transfer, *Microchem J* **160** (2021) 105755. https://doi.org/10.1016/J.MICROC.2020.105755
209. H. Bagheri, A. Hajian, M. Rezaei, A. Shirzadmehr, Composite of Cu metal nanoparticles-multiwall carbon nanotubes-reduced graphene oxide as a novel and high performance platform of the electrochemical sensor for simultaneous determination of nitrite and nitrate, *J Hazard Mater* **324** (2017) 762–772. https://doi.org/10.1016/j.jhazmat.2016.11.055

210. M.M. Eteya, G.H. Rounaghi, B. Deiminiat, Fabrication of a new electrochemical sensor based on AuPt bimetallic nanoparticles decorated multi-walled carbon nanotubes for determination of diclofenac, *Microchem J* **144** (2019) 254–260. https://doi.org/10.1016/J.MICROC.2018.09.009
211. N.F. Atta, A. Galal, A.R. El-Gohary, Crown ether modified poly(hydroquinone)/carbon nanotubes based electrochemical sensor for simultaneous determination of levodopa, uric acid, tyrosine and ascorbic acid in biological fluids, *J Electroanal Chem* **863** (2020) 114032. https://doi.org/10.1016/j.jelechem.2020.114032
212. Mao, H. Li, L. Yu, X. Hu, Electrochemical sensor based on multi-walled carbon nanotubes and chitosan-nickel complex for sensitive determination of metronidazole, *J Electroanal Chem* **799** (2017) 257–262. https://doi.org/10.1016/j.jelechem.2017.05.049
213. Y. Huang, F. Zhu, J. Guan, W. Wei, L. Zou, Label-free amperometric immunosensor based on versatile carbon nanofibers network coupled with au nanoparticles for aflatoxin B1 detection, *Biosensors* **11** (2020) 5. https://doi.org/10.3390/BIOS11010005
214. H.T. Le, D.T. Tran, N.H. Kim, J.H. Lee, Worm-like gold nanowires assembled carbon nanofibers-CVD graphene hybrid as sensitive and selective sensor for nitrite detection, *J Colloid Interface Sci* **583** (2021) 425–434. https://doi.org/10.1016/j.jcis.2020.09.068
215. Y. Zhu, Q. Miao, B. Han, Electrospinning synthesis of ag nanoparticles-doped carbon nanofibers for voltammetric determination of H2O2, *Int J Electrochem Sci* **16** (2021) 210226. https://doi.org/10.20964/2021.02.38
216. M. Adabi, M.A.-J. of D.S. and Technology, Electrodeposition of nickel on electrospun carbon nanofiber mat electrode for electrochemical sensing of glucose, *Taylor & Francis* **42** (2019) 262–269. https://doi.org/10.1080/01932691.2019.1678483
217. H.R. Rahmani, M. Adabi, K.P. Bagheri, G. Karim, Development of electrochemical aptasensor based on gold nanoparticles and electrospun carbon nanofibers for the detection of aflatoxin M1 in milk, *J Food Meas Charact* **15** (2021) 1826–1833. https://doi.org/10.1007/S11694-020-00780-Y
218. Z. Jahromi, E. Mirzaei, A. Savardashtaki, M. Afzali, Z. Afzali, A rapid and selective electrochemical sensor based on electrospun carbon nanofibers for tramadol detection, *Microchem J* **157** (2020) 104942. https://doi.org/10.1016/j.microc.2020.104942
219. D. Yin, J. Liu, X. Bo, L. Guo, Cobalt-iron selenides embedded in porous carbon nanofibers for simultaneous electrochemical detection of trace of hydroquinone, catechol and resorcinol, *Anal Chim Acta* **1093** (2020) 35–42. https://doi.org/10.1016/j.aca.2019.09.057
220. S.S. Upadhyay, N.S. Gadhari, A.K. Srivastava, Biomimetic sensor for ethambutol employing β-cyclodextrin mediated chiral copper metal organic framework and carbon nanofibers modified glassy carbon electrode, *Biosens Bioelectron* **165** (2020) 112397. https://doi.org/10.1016/j.bios.2020.112397

221. Sasal, K. Tyszczuk-Rotko, M. Wójciak, I. Sowa, First electrochemical sensor (screen-printed carbon electrode modified with carboxyl functionalized multiwalled carbon nanotubes) for ultratrace determination of diclofenac, *Materials* **13** (2020) 781. https://doi.org/10.3390/MA13030781
222. D. Duan, Y. Ding, L. Li, G. Ma, Rapid quantitative detection of melatonin by electrochemical sensor based on carbon nanofibers with embedding FeCo alloy nanoparticles, *ECS Meeting Abstracts* **MA2020-01** (2020) 1939–1939. https://doi.org/10.1149/MA2020-01271939MTGABS/META
223. X. Si, C. Wang, L. Li, D. Ye, Y. Wei, Y. Ding, Preparation of nickel-aluminum hydrotalcite nanosheet-coated carbon nanofibers and their application in the detection of salidroside, *Microchem J* **155** (2020) 104652. https://doi.org/10.1016/j.microc.2020.104652
224. Y. Ding, H. Sun, C. Ren, M. Zhang, K. Sun, A nonenzymatic glucose sensor platform based on specific recognition and conductive polymer-decorated CuCo2O4 carbon nanofibers, *Materials* **13** (2020) 2874. https://doi.org/10.3390/MA13122874
225. C. Fakude, O. Arotiba, F. Arduini, N.M., Flexible polyester screen-printed electrode modified with carbon nanofibers for the electrochemical aptasensing of cadmium (II), *Electroanalysis* **32** (2020) 2650–2658. https://doi.org/10.1002/elan.202060070
226. X. Liu, J.-H. He, R. Sakthivel, R.-J. Chung, Rare earth erbium molybdate nanoflakes decorated functionalized carbon nanofibers: An affordable and potential catalytic platform for the electrooxidation of phenothiazine, *Electrochim Acta* **358** (2020) 136885. https://doi.org/10.1016/j.electacta.2020.136885
227. A.T.E. Vilian, K.S. Ranjith, S.J. Lee, S.-K. Hwang, R. Umapathi, C.W. Oh, Y.S. Huh, Y.-K. Han, Controllable synthesis of bottlebrush-like ZnO nanowires decorated on carbon nanofibers as an efficient electrocatalyst for the highly sensitive detection of silymarin in biological samples, *Sens Actuators B Chem* **321** (2020) 128544. https://doi.org/10.1016/j.snb.2020.128544
228. M. Afzali, A. Mostafavi, T. Shamspur, Sensitive detection of colchicine at a glassy carbon electrode modified with magnetic ionic liquid/CuO nanoparticles/carbon nanofibers in pharmaceutical and plasma samples, *J Iran Chem Soc* **17** (2020) 1753–1764. https://doi.org/10.1007/S13738-020-01894-2

Index

5-Fluorouracil, 171–172
5-hydroxytryptamine, *see* Serotonin

Acetylcholine (ACh), 62
Adeno-associated viruses (AAV), 67
Adsorptive stripping voltammetry
(AdSV), 85
Air quality monitoring, 261–262
Amperometry, 3–5, 117, 158, 231–232
Anabolic drugs, 113–114
Antimony film electrodes (SbFEs), 100
Arc discharge, 341
Ascorbic acid (AA), 61
Atrazine, 162

Bacteriophages, 65
Bioreceptors, 18–20
Biotin, 298–299
Bisphenol A (BPA), 22
Black titanium oxide nanotube (BTNT), 61

Carbon-based nanomaterials, 42, 62–63,
227–228, 291–294, 331–334;
see also Nanomaterials
- applications, 355–362
- carbon nanofibres, 43, 338–339, 349–355
- carbon nanotubes, 43, 226, 337
- characterization, 344–349
- future perspective, 362–363
- graphene, 42–43, 221, 225–226, 334–336
- graphene oxides, 221, 225, 336–337
- synthesis, 339–343

Carbon black–polyethylene (CBPCE), 97
Carbon paste electrodes (CPEs), 87
Carmoisine E-122 dye, 97
Caspase-3 (Cas-3), 66
Cdot, 292
Ceftriaxone sodium (CFX), 295
Chemical vapour deposition (CVD), 341
Chronoamperometry (CA), 4–5
Coating-based techniques, 10–12
Composite nanomaterials, 153
Conductometric sensors, 85–86, 116–117,
158
Convolutional neural networks (CNNs), 321
Coprecipitation approach, 267
Counter electrode, 8
Cyclic voltammetry (CV), 84, 156, 311

Dengue virus, 68
Differential pulse voltammetry (DPV),
84, 156
Digital technologies, 48
Doxorubicin (DOX), 70
Drug delivery systems, 276

Electrochemical impedance
spectrochemical technique, 6–7,
117–118, 157, 229, 298
Electrochemical sensors
- advanced electrode modification, 8–9
  - bioreceptor, 18–20
  - coating-based techniques, 10–12
  - electrochemical deposition, 10
  - metal nanomaterials, 15–17
  - nanomaterials, 12–14
  - polymeric nanomaterials, 17–18
  - self-assembled monolayers, 9–10
- applications
  - MIP for ferritin detection, 25, 318–319
  - nanoparticle-modified sensor, 319–320

electrochemical characterization, 317–318
electrochemical detection techniques, 3
amperometric, 3–5
electrochemical impedance spectrochemical, 6–7
potentiometric, 5–6
voltammetric, 7–8
electrochemical methods
amperometry, 158
conductometry, 158
cyclic voltammetry, 156, 311–313
differential pulse voltammetry, 156
electrochemical impedance spectroscopy, 157, 307–311
electrogenerated chemiluminescence, 306
photoelectrochemistry, 305–306
potentiometry, 158, 305
square wave voltammetry, 157
mechanical characterization, 316–317
miniaturized, 20–23
multisensors, 23–25
*vs.*nanomodified, 315–316
Electrochemiluminescence (ECL), 277
Electrodeposition, 10
Electron microscopy, 353–355
Electrospinning method, 343
Energy Dispersive X-Ray Microanalysis, 354–355
Environmental sustainability, 258–260
challenges, 272–273
essential role of, 273
future research and development, 273–274
importance of, 260–261
Epirubicin (EP), 170
Epithelial cell adhesion molecule (EpCAM), 67

Fabrication techniques of toxic dyes, 79–81
advancements, 103
future prespective, 103
physical and chemical properties, 81–82
principles of EC sensors, 82–84
types of, 84–85
surface modification, 86–87
synthetic dyes, 81–82
types of electrodes, 87–88
film/GCE-based, 100–102
modified carbon paste, 88–96
screen-printed, 96–100
Fenitrothion (FNT), 161, 295, 298
Fluorine doped tin oxide (FTO), 82
Food contaminants, 112–113
developments, 132
electrochemical devices, 116–118
electrode materials, 118
conducting polymers, 124–126
covalent organic frameworks, 128
metal–organic framework, 123–124
molecularly imprinted polymers, 126–128
nanomaterials, 118–123
food additives, 115
heavy metal ions, 115
mycotoxins, 114
pathogens, 115–116
pesticides, 113
veterinary drugs, 113–114

Gentamicin sulfate (GNS), 169
Glassy carbon electrode (GCE), 159
Glucose sensors, 246–247
biosensor, 246, 248
cobalt-based nanomaterial, 250
future perspectives, 251–253
glassy carbon electrode, 247
graphene, 247
metal-organic framework, 249–250
microelectrodes, 251
nanocrystalline zinc oxide, 251
polyoxometalates, 248
Glyphosate, 296
Gold nanoparticles (AuNPs), 15, 41, 58, 222, 270
Graphene, 14, 269, 339–340, 344–345, 355–357

Hepatitis A virus (HAV), 68
human adenovirus type 5 (hAdV5), 67
Hybrid nanoparticles, 16–17, 222–223
Hydroquinone acid (HQ), 61

Impedimetric sensors, 85
Indium doped tin oxide (ITO), 82
Infrared spectroscopy, 351–352
Inorganic nanomaterials, 154
Iron oxide (Fe_3O_4), 44

Laser ablation methods, 341
Laser-based methods, 266
Linear sweep voltammetry (LSV), 84–85
Lithographic process, 266

Medical waste detection, 218–219
  applications, 225
    biosensors, 228
    chemical contaminants, 228
    future directions, 229
    heavy metal monitoring, 227–228
    pathogen identification, 226–227
    pharmaceutical detection, 224
  challenges, 232–233
  future perspectives, 233
  nanomaterial-based electrodes, 220
    carbon-based, 221–222
    in electrochemical sensing, 220–221
    metal and metal oxide, 222
  nano sensors, 229–230
    amperometric, 230
    applications, 231–232
    impedance spectroscopy, 231
    voltammetric techniques, 229, 230
Metronidazole (MTZ), 62–63
Microneedle-associated electrochemical sensor array (MNESA), 24
Miniature Electrophoresis Electrochemical Protein Sensor (MEEPS), 24
Molecularly imprinted kanamycin (KAN), 128
Molecularly imprinted polymers (MIPs), 67, 126–128
Multiplexed sensing, 48
Multiwalled carbon nanotubes (MWCNTs), 13

Nanomaterials
  advancements, 71–72
  advantages, 270–271
  applications of, 44–46
  biomedical applications
    cancer diagnosis, 66–67
    detection of microbes, 64–65
    detection of toxins, 65–66
    drug delivery systems, 70–71
    heavy metal detection, 69–70
    viral infection diagnosis, 67–68
  challenges, 48–49, 274–275, 299
  definition, 327–328
  future perspectives, 47–48, 275–278
  integration of, 177
  limitations, 272, 274–275
  One-Dimensional, 153–153, 195–196
  principle of, 269–270
  properties of, 263–265
    composition, 155
    dimensionality, 153
    medical diagnostics, 155
    surface chemical functionalization, 151–152
    synthesis of, 152–153
  role of, 38–40, 194–197
  synthesis techniques, 266
    bottom-up, 331
    Green, 267–269
    top-down, 329–331
  Three-Dimensional, 153–154, 197
  transformative impact, 49–50
  Two-Dimensional, 153–154, 196–197
  types of, 40–44
    carbon, 62–63, 244–245, 291–294; *see also* Carbon-based nanomaterials
    composite-based, 245
    metal based, 58–61, 119–123, 245–246, 295–299
    organic based, 246
    silica-based, 294–296
  unique properties of, 35–38
  Zero-Dimensional, 153, 195
Nanoparticle, 243
Naphthalene tetracarboxylic dianhydride (NTDA), 162
Near-infrared (NIR) techniques, 348–349
Nicotinamide adenine dinucleotide (NADH), 293

Organic nanomaterials, 154
Organochlorine pesticides, 113
Organophosphorus pesticides, 113
Oxoplatin (OXP), 319–320

Palladium nanoparticles (PdNPs), 59–60
paracetamol (PCM), 166–168
Parathyroid hormone (PTH), 206
Pencil graphite electrodes (PGEs), 167
Pharmaceutical and biomedical diagnosis
  anti-cancer medications, 170–172
  atrazine, 162–163
  biomarkers, 173–174
  gentamicin sulfate, 169
  glassy carbon electrode, 159
  organophosphorus, 161
  paracetamol, 166–169
  phosphate-buffered saline, 159–160
  primary aromatic amines, 163–165
  serotonin, 176–177
Platinum nanoparticles (PtNPs), 15–16, 42, 222
Point-of-care diagnostics, 48, 276
Polydiphenylamine (PDPA), 294
polymeric nanomaterials, 17–18, 224, 270
Potentiometric sensors, 5–6, 117, 158, 305
Primary aromatic amines (PAAs), 163

Quantum dots (QD), 36–37, 195, 228, 243, 270
Quartz crystal microbalance (QCM), 359

Raman spectroscopy, 347–348
Recurrent neural networks (RNNs), 321
Reference electrode, 8
Reversible hydrogen electrode (RHE), 82
Ruthenium phthalocyanine, 296

Salmonella enterica, 320–321
Salmonella typhimurium, 64, 65
Saturated calomel electrode (SCE), 82
Scanning electron microscopy (SEM), 344, 353
Scanning probe microscopy (SPM), 349–350
Scanning tunnelling microscopy (ATM), 350–351
Screen-printed electrodes (SPEs), 88
Self-assembled monolayers (SAMs), 9–10
Sensor technology, 261–262
Serotonin (ST), 176–177
Silver nanoparticles (AgNPs), 16, 41, 58–59, 222, 225, 270
Single-walled carbon nanotubes (SWCNTs), 13
Sol–gel process, 267
Square wave voltammetry (SWV), 84, 157
Staphylococcal enterotoxin B (SEB), 65–66
Supramolecular nanotechnology, 292
Survivin (Sur), 67

Titanium dioxide ($TiO_{2)}$, 43
Titanium oxide nanotubes (TNTs), 61
Topotecan (TP), 170
Transducers, 1
Transmission electron microscopy (TEM), 344, 355
Triphenylmethane dyes, 81–82

Ultrasmall cerium oxide nanoparticles (USCeOxNPs), 61
Ultraviolet-visible (UV-vis) techniques, 348–349

Vitamins and hormones
  carbon nanomaterial–based sensors, 203–206
  metal and metal oxide–based sensors, 200–203
  polymer nanocomposite-based sensors, 206–208
  precision, 199
  sensitivity, 199
  significance of, 197–200
    clinical diagnostics, 198
    environment and food quality, 199–200
    portability and cost-effectiveness, 199
    quick response analysis, 199
    sensitivity and specificity, 198–199
  utilization of nanomaterials, 208

voltammetric techniques, 7–8, 84, 117, 230
Voltammogram, 291

Water quality monitoring, 262
Working electrode, 8

X-ray diffraction (XRD), 331, 345–346
X-ray photoelectron spectroscopy (XPS), 347
X-Ray photoelectron spectroscopy, 352–353

Zinc oxide (ZnO), 43, 270